Lecture Notes in Mathematics

Edited by A. Dold and B. Eckmann

T0213187

899

Analytic Number Theory

Proceedings of a Conference Held at
Temple University, Philadelphia, May 12–15, 1980

Edited by M.I. Knopp

Springer-Verlag
Berlin Heidelberg New York 1981

Editor
Marvin I. Knopp
Department of Mathematics, Temple University
Philadelphia, PA 19122, USA

AMS Subject Classifications (1980): 10-06

ISBN 3-540-11173-5 Springer-Verlag Berlin Heidelberg New York
ISBN 0-387-11173-5 Springer-Verlag New York Heidelberg Berlin

Printing and binding: Beltz Offsetdruck, Hemsbach/Bergstr.
2141/3140-543210

DEDICATION

For what he has accomplished and will yet accomplish, for his unswerving dedication to the best in mathematics, and for the example he has been and will continue to be ...

this volume is dedicated to

EMIL GROSSWALD

with admiration and friendship.

ACKNOWLEDGEMENTS

The research conference in Analytic Number Theory, held at Temple University, May 12-15, 1980 owes its considerable success to the co-operative efforts of a good many people. I am indebted both to George Wheeler, Dean of the College of Liberal Arts at Temple, whose financial support made the conference possible, and to Leon Steinberg, Chairman of the Mathematics Department, who made himself and his office available to me in the planning stages and during the hectic days of the conference itself.

I am aware as well that a good many members of the Mathematics Department performed a number of thoughtful services during the conference, often without being asked, sometimes without my knowledge. My thanks to all of these colleagues.

The responsibility for the typing and final preparation of the manuscript was placed in the most capable hands of Ms. Gerry Sizemore, Administrative Assistant/ Supervisor of the Mathematics Department. In this she was ably assisted by Donna Sowell and Patricia Libbey. Even a casual perusal of the volume reveals the excellence of their work, their patient, painstaking attention to detail in the preparation of the articles, many of which present formidable difficulties to the typist.

Finally, I must thank my wife Josephine for her continued support and encouragement in whatever I undertake. It was she who urged me initially to undertake the editing of this collection as a tribute to our friend Emil Grosswald. The final result is a tribute to her as well.

ACKNOWLEDGEMENTS

The research conference [illegible] Number Theory [illegible] at [illegible] University [illegible] May 12-15, 1980, [illegible] the [illegible] of the [illegible] [illegible] was [illegible] with [illegible] [illegible] of the [illegible] [illegible] whose [illegible] support [illegible] the conference possible [illegible] [illegible] and [illegible] in the [illegible] and its [illegible] [illegible] the [illegible] [illegible] of the conference [illegible].

[illegible] [illegible] Mathematics Department [illegible] [illegible] during the [illegible] [illegible] [illegible] [illegible] thanks to [illegible] these [illegible].

[illegible] [illegible] the [illegible] [illegible] [illegible] of the [illegible] [illegible] [illegible] of the [illegible] [illegible] [illegible] Mathematics Department [illegible] [illegible] the [illegible] [illegible] [illegible] [illegible] [illegible] [illegible] many of which [illegible] difficult [illegible] type.

Finally, [illegible] [illegible] and [illegible] [illegible] [illegible] [illegible] the [illegible] [illegible] [illegible] as well.

FOREWORD

The conference which led to this collection of papers in Analytic Number Theory was organized to mark the retirement of Emil Grosswald from Temple University after a distinguished career as a mathematician and teacher. The reputation of the man among his colleagues accounts for the remarkable response by the American mathematical community to the announcement of the conference, which ultimately attracted far more attention and participation than I had dared to hope in the initial stages of planning. The presence at the conference of many of Grosswald's former students, who travelled to Philadelphia from many distant places, expressed far better than I could the influence he has had upon mathematics.

The diversity of the papers in this volume - ranging over a good part of the vast domain of contemporary number theory - reflects the breadth of Grosswald's own mathematical interests (which are not limited to number theory). Their quality is a fitting tribute to his constant and clear-minded adherence to the highest standards in the publication of research.

For the most part, the papers published here were presented at the conference, but because of the great demand for speaking time, a few of them were not. On the other hand, a number of the papers delivered at the conference could not be included here, for a variety of reasons. These are listed at the end of the Table of Contents. There were four one-hour lectures, presented by Paul Erdös, Patrick Gallagher, Joseph Lehner and Harold Stark. Other talks were thirty minutes in length.

At the dinner closing the conference Emil Grosswald spoke movingly about the many mathematicians he has known and of his own scientific development. Revealing his characteristic humility, he emphasized those things he did not accomplish mathematically rather than his many important contributions. Characteristic as well was his presentation earlier, during the conference itself, of a paper describing not his own work, but previously unpublished work of his teacher, Hans Rademacher.

Emil's remarks at the closing dinner were informed by more than humility; they were motivated as well by that refusal to be satisfied, that drive to learn more and discover more always associated with the most creative scholars. In referring to his "failures" (a term, surely, no-one else would apply to Grosswal's work), Emil was expressing his honest sense that he might have done more with greater effort. Those of us in attendance who have been privileged to know Emil well should not have been surprised at the rigor of the standards implicit in these remarks - standards applied above all to himself. Yet there was something astonishing in his words that evening and one could only react to them by dismissing as implausible the idea of even greater effort on his part.

As all of us could have anticipated, for Grosswald "retirement" has taken place in a formal, institutional sense only. Retirement from mathematics would be as unthinkable for him as the rejection of an old friend. And, indeed, he has continued to exhibit more energy in his work and enthusiasm for mathematics than do a good many of us who are considerably younger. (His case - along with several others I know - highlights the lunacy of mandatory retirement laws, uniformly applied.) We are fortunate to be able to look forward to Emil's continuing influence, internationally, upon the mathematical scene, exerted through his books, his research papers and his astonishingly extensive correspondence with mathematicians - both well established and emerging - all over the world.

Emil Grosswald sets an example - as a mathematician and as a human being - that is difficult to live up to. He has been and remains a most demanding, exhilarating colleague.

September, 1981 Marvin Knopp

CONTENTS

CONFERENCE PAPERS NOT INCLUDED IN THIS VOLUME

Patrick Gallagher,
"Invariants for finite groups."

William Adams,
"Multiples of points on elliptic curves and continued fractions."

C.J. Mozzochi,
"An analytic sufficiency condition for Goldbach's conjecture with minimum redundancy."

Andrew Odlyzko,
"On the nonexistence of Siegel zeros for some number fields."

Carl Pomerance,
"On the distribution of pseudoprimes."

Michael Razar,
"Zeta functions in several complex variables."

Daniel Reich,
"Gaussian Poincaré polynomials."

Daniel Shanks,
"Dedekind zeta functions which have sums of Epstein zeta functions as factors."

Thomas Shemanske,
"The basis problem for modular forms on $\Gamma_0(N)$."

E. Straus,
"Integer valued analytic functions."

ABSTRACT

AN ORTHONORMAL SYSTEM AND ITS LEBESGUE CONSTANTS -
(A lost and found manuscript).

In 1921 H. Rademacher wrote two papers on orthonormal systems. In the first one, the author discusses quite general systems and also defines the functions now commonly known as Rademacher functions; the paper appeared in 1922 (Math. Annalen, v. 87, p. 112-138). The second paper contains the completion of the system of Rademacher functions, theorems on expansions of arbitrary functions in series of the complete system and properties of the Lebesgue constants, both, for ordinary summation and for the first Cesaro means. This paper was never published and is discussed here. At Rademacher's death the manuscript had vanished from sight and only recently it miraculously reappeared.

Emil Grosswald

AN ORTHONORMAL SYSTEM AND ITS LEBESGUE CONSTANTS

(A lost and found manuscript)

E. Grosswald

Temple University

In 1921 Hans Rademacher wrote two papers on orthonormal functions. In the first [7], published in 1922, he discusses properties of fairly general systems; there he also introduces the functions known to-day under the name of Rademacher functions.

This system has much in common with the trigonometric functions, or the Haar functions. We recall that these are defined as follows (see [2]): Let $i_n^{(m)}$ be the interval $\frac{m-1}{2^n} \le t < \frac{m}{2^n}$; then $h_n^{(m)}(t) = 2^{(n-1)/2}$ in $i_n^{(2m-1)}$, $h_n^{(m)}(t) = -2^{(n-1)/2}$ in $i_n^{(2m)}$, $h_n^{(m)}(t) = 0$ otherwise. The new system differs from both, the trigonometric and the Haar system, among others, by not being complete.

In the second paper, Rademacher completed that system and studied the problem of expanding arbitrary functions in series of the complete system, its Lebesgue constants, summability of those series by first Cesaro means and similar topics.

This second paper was never published. Immediate publication may have been postponed, due to a negative, critical appraisal of the manuscript by I. Schur. Then, just a few months later, Walsh's paper [9] on the same topic appeared and Rademacher set aside any idea of ever publishing his own.

Many years later, while he was teaching at the University of Pennsylvania, a graduate student was working on a related problem. With his well known generosity, Rademacher offered the student his own, old manuscript, in case it could be of some help. In this way, the manuscript came to light and some people had the opportunity to read it. In their opinion, the manuscript represented a very valuable work, of high esthetic appeal and they expressed their regret, that it had never been published.

Again, many years passed. Rademacher retired from the University of Pennsylvania in 1962 and went to New York. In 1967 he was stricken by a cruel illness and died in February 1969. Shortly afterwards, Professor Gian-Carlo Rota suggested to the present writer (a former student of Rademacher) to edit the collected papers of Hans Rademacher.

This suggestion was accepted. Also, remembering mentioned unpublished manuscript, every effort was made to locate it, in order to include it in the "Collected Papers" [8], but the manuscript had vanished from sight. It was not to be found among the posthumous papers, nor was it in the possession of the former student. The only thing that the editor of the "Collected Papers" could do was to mention the likely existence of that manuscript in the comments to Paper 13 in [8].

Again the years passed - in fact, more than ten. Then, during the summer of 1979, something surprizing happened. One day, the present author found in the mail a large envelope, without mention of a sender. Also, the envelope contained no letter; it did contain, however, the long lost manuscript (manu-script - not typed, but handwritten!), as well as an earlier draft of the same. The present author could only conjecture the identity of the anonymous sender(s) and offered him/them an easy opportunity to identify himself/themselves. This opportunity having been declined, it appears only fair to respect this wish to remain anonymous. The present writer wants to avail himself of this opportunity to thank the anonymous sender(s) for the service rendered to mathematics, by helping to rescue the manuscript from oblivion.

A rapid perusal of the manuscript seemed to confirm the impressions of those, who had seen it in the past; consequently, an attempt to have it published seemed justified. This would have been easy in the "Collected Papers", as a posthumous manuscript. It appears, however, almost impossible at present. Indeed, the problems discussed had all been solved most satisfactorily already some 60 years ago and later research went much beyond them.

In view of the fact that, if the manuscript was to be published, that may have to be in English, the present author undertook to translate it from the original German. While doing this, it became apparent that the manuscript is far from ready for publication. The meanings to be conveyed are clear enough, but some proofs are missing, others are barely sketched, the pages contain partly irrelevant computations, etc. For these reasons, it may well be the case that, instead of a publication of the manuscript as it is, there is more interest in following up on a certain suggestion by Professors König and Lamprecht, editors of the Archiv der Mathematik. This suggestion is to write up a short exposition of the contents of the manuscript and compare its results with those of Walsh and more recent work. It is the purpose of the present note to do just that.

In the first place, it is necessary to observe that Walsh obtained his results independently, even without knowledge of Rademacher's first paper. Indeed, he refers back to the older work of Haar [2], just like Rademacher himself in his first paper [7]; also, the point of view is completely different.

In fact, Walsh deduces most of his results, by representing his newly defined functions (now generally known as Walsh functions) as finite linear combinations of Haar functions and then invokes known theorems concerning the latter. In this way he obtains immediately the completeness of the new system. Also, if $F(x)$ is continuous on $(0,1)$ and if one sets $a_i^{(j)} = \int_0^1 F(x)\phi_i^{(j)}(x)dx$ ($j = 1,2,\ldots,2^{i-1}$; here $\phi_i^{(j)}(x)$ are the newly introduced functions of Walsh), then $\sum_{i,j} a_i^{(j)}\phi_i^{(j)}(x)$ converges uniformly to $F(x)$, provided that one groups the terms so that all superscripts j that correspond to a given i, are kept in the sum.

Rademacher, on the other hand, defines his functions, by using the binary expansion of the independent variable x. For $0 \leq x < 1$ and $\nu = 1,2,\ldots$, one has $x = \sum_{\nu=1}^{\infty} e_\nu(x)/2^\nu$, $e_\nu(x) = 0$, or 1. If one sets $\psi_\nu(x) = 2e_\nu(x)-1$ for x and ν as before and completes the definitions by setting $\psi_\nu(1) = -1$ and $\psi_o(x) = 1$, then $x = \sum_{\nu=0}^{\infty} \psi_\nu(x)/2^{\nu+1}$. This series converges absolutely and uniformly on $0 \leq x \leq 1$; hence, one can square it, elevate it to the cube, etc. In this way one obtains representations of all successive powers of x by absolutely and uniformly convergent series of products of the functions $\psi_\nu(x)$. Specifically, if for $n = 2^r+s$, $0 < s \leq 2^r$, we set $x_n(x) = \psi_{r+1}(x)x_s(x)$, then all powers of x are represented by absolutely and uniformly convergent series in the $x_n(x)$; this is sufficient to guarantee that the system of the $x_n(x)$ is complete. The orthonormality of the $x_n(x)$ easily follows from their definition.

In order to compare the results of Rademacher and Walsh, one has to observe that Rademacher's x_{2^r+s} corresponds to Walsh's $\phi_r^{(s)}$ (up to an immaterial shift of index).

While Walsh, as already mentioned, obtains his convergence results from the analogous ones for the Haar functions, Rademacher considers directly the kernels $K_n(x,y) = \sum_{\nu=1}^{n} x_\nu(x)x_\nu(y)$. He uses only the most elementary, geometric reasonings in the (x,y)-plane (in fact, on the square $0 \leq x \leq 1$, $0 \leq y \leq 1$). Both authors observe that the convergence of the formal series for F(x) at a point x, depends only on the behaviour of F(x) in a neighborhood of that point. Rademacher states his results for (locally) monotonic functions; Walsh, equivalently, formulates the corresponding statements for functions of (locally) bounded variation. Both authors mention the special importance of the dyadic rationals, but Walsh stresses it more. Both authors realize the particular interest of the (C,1)-summability (by first Cesaro means), for which one has the theorem that if F(x) is (at least piecewise) continuous on (0,1), then its series in Rademacher-Walsh functions is uniformly (C,1)-summable.

While the condition of monotonicity (equivalently, of bounded variation) used in the proof of convergence in Rademacher's work implies that continuity alone is insufficient to guarantee said convergence, Walsh actually states this as a theorem and gives a construction for a continuous function, whose formal series diverges at a dyadic irrational (however, the partial sums s_{2^n} of order 2^n do converge to the function!).

Rademacher-Walsh Functions

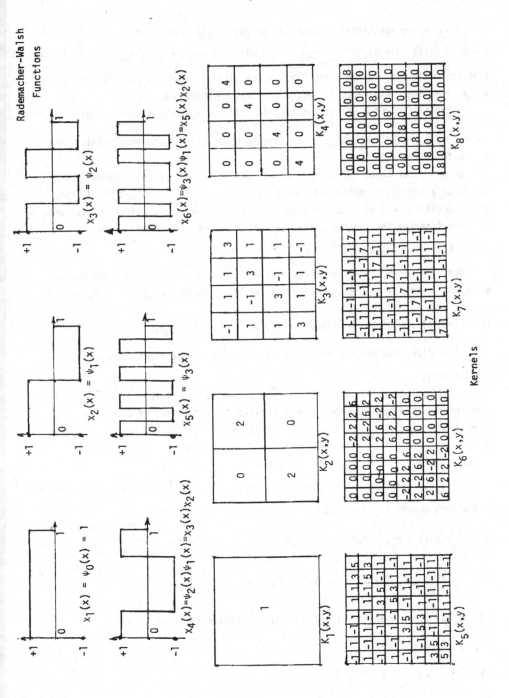

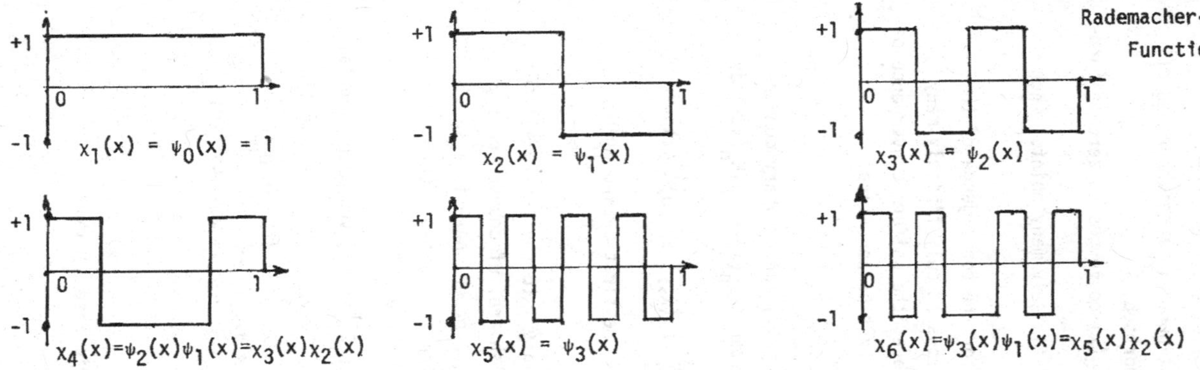

1

$K_1(x,y)$

0	2
2	0

$K_2(x,y)$

-1	1	1	3
1	-1	3	1
1	3	-1	1
3	1	1	-1

$K_3(x,y)$

0	0	0	4
0	0	4	0
0	4	0	0
4	0	0	0

$K_4(x,y)$

-1	1	-1	1	1	1	3	5
1	-1	1	-1	1	-1	5	3
-1	1	-1	1	3	5	-1	1
1	-1	1	-1	5	3	1	-1
-1	1	3	5	-1	1	-1	1
1	-1	5	3	1	-1	1	-1
3	5	-1	1	-1	1	-1	1
5	3	1	-1	1	-1	1	-1

$K_5(x,y)$

0	0	0	0	-2	2	2	6
0	0	0	0	2	-2	6	2
0	0	0	0	2	6	-2	2
0	0	0	0	6	2	2	-2
-2	2	2	6	0	0	0	0
2	-2	6	2	0	0	0	0
2	6	-2	2	0	0	0	0
6	2	2	-2	0	0	0	0

$K_6(x,y)$

1	-1	-1	1	-1	1	1	7
-1	1	1	-1	1	-1	7	1
-1	1	1	-1	1	7	-1	1
1	-1	-1	1	7	1	1	-1
-1	1	1	7	1	-1	-1	1
1	-1	7	1	-1	1	1	-1
1	7	-1	1	-1	1	1	-1
7	1	1	-1	1	-1	-1	1

$K_7(x,y)$

0	0	0	0	0	0	0	8
0	0	0	0	0	0	8	0
0	0	0	0	0	8	0	0
0	0	0	0	8	0	0	0
0	0	0	8	0	0	0	0
0	0	8	0	0	0	0	0
0	8	0	0	0	0	0	0
8	0	0	0	0	0	0	0

$K_8(x,y)$

Kernels

Walsh gives additional theorems about the convergence of the formal series at $x = a$ to $\frac{1}{2}\{F(a+0)+F(a-0)\}$, if $F(x)$ is (locally) of bounded variation and a is a dyadic rational. If, in addition, $F(x)$ is actually continuous in a neighborhood of $x = a$, then the convergence is uniform in that neighborhood.

Walsh also gives a uniqueness theorem: If $\sum_{\nu} a_\nu \phi_\nu(x)$ converges to zero uniformly on (0,1), except, perhaps, in the neighborhoods of finitely many points, then $a_\nu = 0$ for all ν.

On the other hand, Rademacher states conditions for the convergence of the series of a piecewise continuous function,left continuous at only finitely many discontinuities, all at dyadic rationals. He also proves the uniform convergence of the formal series for functions that satisfy a Lipschitz type condition: $|F(x_0)-F(x)| < A|\log(x_0-x)|^{-\alpha}$, $\alpha > 1$.

Next, he makes a detailed study of the kernels $K_n(x,y)$ for ordinary convergence and investigates the corresponding Lebesgue constants $\rho_n(x) = \int_0^1 |K_n(x,y)|dy$.

At the end of the paper, he discusses $K_n^{(1)}(x,y) = \frac{1}{n}\sum_{\nu=1}^{n} K_\nu(x,y)$ and $\rho_n^{(1)}(x) = \int_0^1 |K_n^{(1)}(x,y)|dy$, the kernels and Lebesgue constants of the first Cesaro means, respectively.

It is not possible - and may not even be appropriate - to quote here all the theorems obtained by Rademacher (those obtained by Walsh are, of course, easily accessible in [9]). But just to give the flavour of the paper, here is a sample of the results. In all cases it should be kept in mind that $n = 2^r+s$, with $0 < s \leq 2^r$ (observe the equal sign in the last, rather than the first equality).

Let D_r be the union of the set of squares of sides 2^{-r} along the main diagonal of the unit square $0 \leq x \leq 1$, $0 \leq y \leq 1$; then

(1) $K_{2^r}(x,y) = K_{2^{r-1}}(x,y)(1+\psi_r(x)\psi_r(y)) = \prod_{j=1}^{r} (1+\psi_j(x)\psi_j(y)) = \begin{cases} 2^r \text{ in } D_r, \\ 0 \text{ elsewhere;} \end{cases}$

(2) $K_n(x,y) = K_{2^r}(x,y)+\psi_{r+1}(x)\psi_{r+1}(y)K_s(x,y)$;

(3) $\rho_n(x)$ is independent of x (as in the trigonometric case !) and

(4) $\rho_n = 1+\rho_s - s/2^r$;

(5) $\rho_{2^r+s} = \rho_{2^r-s} + s/2^r$;

(6) $\rho_{2n} = \rho_n$ (so that each value of a Lebesgue constant occurs infinitely often);

(7) $\rho_{2^{r+1}+s} = \rho_{2^r+s} + s/2^{r+1}$.

Next, set $s_r = 2^{r-1}-2^{r-2}+2^{r-3}- \dots + (-1)^{r-1} = \frac{1}{3}(2^r-(-1)^r)$;

then

(8) $\rho_{2^r+s_r}$ and $\rho_{2^r+s_{r+1}}$ $(= \rho_{2^{r+1}-s_r})$ are the largest Lebesgue constants for a given r; also, if $0 \leq \tau < s_r$, then

(9) $\rho_{2^r+s_r} - \rho_{2^r+\tau} > \dfrac{s_r-\tau}{2^{r+1}}$.

Furthermore,

(10) $\rho_{2^r+s_r} = \rho_{2^r+s_{r+1}} = \frac{10}{9} + \frac{r}{3} - \frac{1}{9}(-\frac{1}{2})^r$,

so that

(11) $\lim\limits_{r \to \infty} (\rho_{2^r+s_r} - (\frac{10}{9} + \frac{r}{3})) = 0$.

From this follows rather easily

(12) $\overline{\lim\limits_{n \to \infty}} (\rho_n - (\frac{4}{9} + \frac{\log 3n}{\log 8})) = 0$,

and

(13) $\underline{\lim\limits_{n \to \infty}} \rho_n = 1$.

For the first Cesaro mean, the Lebesgue constants $\rho_n^{(1)}(x)$ are bounded. In fact

(14) $\rho_{2^r}^{(1)}(x) = 1$ identically in x, and, more generally

(15) $\rho_n^{(1)}(x) < 2$.

In order to put this work into perspective, it may be appropriate to make a brief mention of some further developments of this topic.

Kaczmarz [4] and Kaczmarz and Steinhaus [5] investigated the system of Rademacher-Walsh functions. (They call them Rademacher, or Rademacher-Kaczmarz functions, although they know, and in fact, quote Walsh). They study in detail properties and conditions on sets of constants $\{c_k\}$ that serve as coefficients for series $\sum c_k \chi_k(x)$, by taking as their model similar theorems on the coefficients of (trigonometric) Fourier series.

Paley [6] wrote a long paper in two parts on these functions, which he calls Walsh-Kaczmarz functions. He changes slightly the indexing of these functions. Like Walsh, he takes as his starting point the Haar functions. Among his numerous results, perhaps the most interesting ones deal with partial sums of order 2^n of the formal series of a function f(x) and with certain norm-type inequalities. The following are rather typical: For each $k > 1$, there exists a constant B_k, that depends only on k, such that, if $n = n(t)$ is an arbitrary, integral valued function of t, $\int_0^1 |s_{2^{n(t)}}(t)|^k dt \leq B_k \int_0^1 |f(t)|^k\, dt$; while for $k = 1$

$\int_0^1 |s_{2^{n(t)}}(t)|dt \leq B \int_0^1 |f(t)|\log^+|f(t)|dt+B$; for fixed n, one has

$\int_0^1 |s_{2^n}(t)|dt \leq \int_0^1 |f(t)|dt$.

Perhaps the most important further developments are due to N. Fine [1]. He introduced into the study of these functions the dyadic group G and interprets the Rademacher-Walsh functions as characters on G. He computes the expansions of the fractional part x-[x] of a real x and of the "smoothed out" function $J_k(x) = \int_0^x x_k(u)du$ and uses these most effectively. He obtains bounds for the "Fourier"-coefficients of expansions for different categories of functions, finds again many of Rademacher's results on Lebesgue constants (in fact, he states that some were communicated to him by Rademacher), often with different, very elegant proofs, etc.

He also computes the average value of these constants, $\frac{1}{x}\sum_{\nu=1}^{[x]} \rho_\nu = \frac{\log x}{\log 16} + O(1)$.

He proves analogues of the theorems of Dini and of Lipschitz for the present system, considers also the (C,1)-summability and gives some results on the Abel kernel $\sum_{j=o}^{n-1} x_j(u)r^j$, apparently not discussed anywhere else. He defines the functions

$\mu_n(t) = 2^n t-[2^n t] - \frac{1}{2}$ and $\lambda_n = 2^n t - [2^n t + \frac{1}{2}]$ and, in connection with a theorem

of Kac [3], contrasts the behaviour of $\sum_{n=0}^{N} \mu_n(t)$ with that of $\sum_{n=0}^{N} \lambda_n(t)$. Finally, Fine proves a number of uniqueness and of localization theorems.

Although the elegance of Rademacher's manuscript and a feel for historic justice militate in favor of a publication in full of the manuscript, other considerations militate against it. Among these is the unfinished state of the manuscript(s) and the fact that, during his lifetime, Rademacher himself did not want to have it published. Instead, still following the suggestions of Professor König and Lamprecht, xeroxed copies of the original manuscript will be deposited in an easily accessible library, perhaps together with the previously mentioned edited and completed translation into English. Also, if a need should be perceived, an information, somewhat similar to the present note, could be made available in German.

BIBLIOGRAPHY

1. N.J. Fine - On the Walsh Functions - Transactions of the Amer. Math. Soc., vol. 65 (1949), pp. 372-414.

2. A. Haar - Zur Theorie der orthogonalen Funktionensysteme - Diss. Göttingen (1909) and Mathem. Annalen.,vol. 69 (1910), pp. 331-371.

3. M. Kac - Sur les fonctions $2^n t-[2^n t]-\frac{1}{2}$ - Journal London Math. Soc., vol. 13 (1938), 131-134.

4. S. Kaczmarz- Über ein Orthogonalsystem - C.R. du premier Congrès des Mathematiciens des pays Slaves, Warsaw, 1929, pp. 189-192.

5. S. Kaczmarz and H. Steinhaus - Le systeme orthogonal de M. Rademacher - Studia Mathematica, vol. 2 (1930), pp. 231-247.

6. R.E.A.C. Paley - A remarkable series of orthogonal functions - Proc. London Math. Soc., vol. 34 (1932), pp. 241-279.

7. H. Rademacher - Einige Sätze über Reihen von allgemeinen Orthogonal - funktionen-Mathem. Annalen, vol. 87 (1922), pp. 112-138.

8. H. Rademacher - Collected Papers, 2 volumes (edited by E. Grosswald), The MIT Press (G.-C. Rota series editor), Cambridge, Mass., 1974.

9. J.L. Walsh - A closed set of normal, orthogonal functions - Amer. J. Math., vol. 55 (1923), pp. 5-24.

MORDELL INTEGRALS AND RAMANUJAN'S "LOST" NOTEBOOK

George E. Andrews[1]

Dedicated to an outstanding mathematician and human being, Emil Grosswald

1. Introduction. In previous papers [4], [5], [6], [7], [8], [9], I have presented treatments of several collections of related results from Ramanujan's "Lost" Notebook. The purpose of this paper is two-fold: First we shall be considering another closely related set of results. Second and most important we shall see the way of these results fit into a general study of Mordell integrals.

Let $Re(a) < 0$, the Mordell integral $M(a,b,c,d)$ is defined by

$$M(a,b,c,d) = \int_{-\infty}^{\infty} \frac{e^{ax^2+bx}}{e^{cx}+d}\, dx,$$

where the line of integration is deformed to avoid singularities of the integrand. The Mordell integrals have played an important role in various problems. Mordell describes a number of these in [16] and [17]. While special cases of such integrals actually arose in the work of Kronecker and Riemann (c.f. Siegel [20]), Mordell is the first one to analyze their behaviour relative to modular transformations (see Sections 3 and 4 of [17]). Consequently following R. Bellman's lead [11], we shall refer to these integrals as Mordell integrals.

Perhaps their greatest importance to date lies in the use Riemann (as described by Siegel [20]) made of them in proving the approximate functional equation for the Riemann zeta function. This work was utilized by Levinson [15] to prove that at least 1/3 of the complex zeros of the Riemann zeta function have real part equal to 1/2.

In Section 2 we shall consider three functions:

(1) Partially supported by National Science Foundation Grant MCS-7901754.

(1.1) $$M_1(q) = \sum_{n=-\infty}^{\infty} \frac{q^{2n^2+n}}{1+q^{2n}}, \quad |q| < 1;$$

(1.2) $$M_2(q) = \sum_{n=-\infty}^{\infty} \frac{q^{2n^2-n}}{1+q^{2n-1}}, \quad |q| < 1;$$

(1.3) $$M_3(q) = \sum_{n=-\infty}^{\infty} \frac{(-1)^n q^{2n^2+2n}}{1-q^{2n+1}}, \quad |q| < 1.$$

The first two of these functions appear explicitly in the "Lost" Notebook, and the third arises naturally in our analysis. The functions are closely related to squares of classical theta functions since

(1.4) $$M_1(q) - M_2(q) = \frac{1}{2} (\theta_4(0,q))^2 = \frac{1}{2} \Big(\sum_{n=-\infty}^{\infty} (-1)^n q^{n^2} \Big)^2$$ [14; p. 238, eq. (14)],

and

(1.5) $$M_3(q) + M_3(-q) = \frac{1}{4} q^{-1/2} (\theta_2(0,q))^2,$$ [14; p. 238, eq. (13)].

We shall see that these functions are related to Mordell integrals through identities like (Theorem 1 below)

(1.6) $$M_1(q) = \frac{2\pi}{\alpha} q_1^{1/2} M_3(q_1) + \Big(\sum_{k=-\infty}^{\infty} (-1)^k q_1^{2k^2} \Big) \int_{-\infty}^{\infty} \frac{e^{-2\alpha x^2+\alpha x}}{e^{2\alpha x}+1} dx,$$

where $q = e^{-\alpha}$, $q_1 = e^{-\pi^2/\alpha}$, $\operatorname{Re} \alpha > 0$.

In Ramanujan's "Lost" Notebook, a number of identities (see eq's. (3.1), (3.4), (3.8), (3.9), (3.11) and (3.13) in Section 3 with a = 1) involve

(1.7) $$\mu(q) = \sum_{n=0}^{\infty} \frac{(-1)^n q^{n^2} (1-q)(1-q^3)\dots(1-q^{2n-1})}{(1+q^2)^2(1+q^4)^2\dots(1+q^{2n})^2}.$$

Let us define also

(1.8) $$\beta(q) = \sum_{n=0}^{\infty} \frac{q^{n^2+n}(1+q^2)(1+q^4)\dots(1+q^{2n})}{(1-q)^2(1-q^2)(1-q^3)^2(1-q^4)\dots(1-q^{2n})(1-q^{2n+1})^2}.$$

This function does not appear in the "Lost" Notebook; however $\beta(q)$ is related to $\mu(q)$ and Mordell integrals through identities such as

$$(1.9)\qquad q^{-1/8}\sqrt{\frac{\pi}{2\alpha}}\ \mu(q) = \frac{2\pi}{\alpha}\ q_1^{1/2}\ \beta(q_1) + \int_{-\infty}^{\infty} \frac{e^{-2\alpha x^2+\alpha x}}{1+e^{2\alpha x}}\ dx,$$

where $q = e^{-\alpha}$, $q_1 = e^{-\beta}$, $\alpha\beta = \pi^2$.

The main body of related results that Ramanujan presents in the "Lost" Notebook entails Eulerian series (or q-series) identities for these functions and will be treated in Section 3. The developments in Section 3 will center on a single analytic result (Lemma 2) that can be utilized to prove a very large number of Ramanujan's results. The transformations of Section 2 together with the resulting discovery of $M_3(q)$ allows us to find a number of results that apparently eluded Ramanujan. Section 4 will provide this extension of some of Ramanujan's results. As in [7], [8], and [9] we shall use the subscript "R" on each equation that appears in Ramanujan's "Lost" Notebook.

2. The Mordell integrals. Here we derive the basic modular type transformations for $M_1(q)$, $M_2(q)$ and $M_3(q)$.

THEOREM 1. For $q = e^{-\alpha}$, $q_1 = e^{-\pi^2/\alpha}$, $\mathrm{Re}\ \alpha > 0$,

$$(2.1)\qquad M_1(q) = \frac{2\pi}{\alpha}\, q_1^{1/2}\ M_3(q_1) + \Big(\sum_{k=-\infty}^{\infty} (-1)^k q_1^{2k^2}\Big)\int_{-\infty}^{\infty} \frac{e^{-2\alpha x^2+\alpha x}}{e^{2\alpha x}+1}\ dx.$$

REMARK. To simplify our proof we shall assume that α is real with $0 < \alpha < 1$. The full theorem then follows by analytic continuation.

PROOF. We apply the Poisson summation formula [11; pp. 7-9]:

$$(2.2)\qquad M_1(e^{-\alpha}) = \sum_{n=-\infty}^{\infty} \frac{e^{-2\alpha n^2}}{e^{-\alpha n}+e^{\alpha n}}$$

$$= \frac{1}{2}\sum_{n=-\infty}^{\infty} \frac{e^{-2\alpha n^2}}{\cosh \alpha n}$$

$$= \frac{1}{2} \sum_{k=-\infty}^{\infty} \int_{-\infty}^{\infty} \frac{e^{-2\alpha x^2+2\pi ikx}}{\cosh \alpha x} dx.$$

$$= \sum_{k=0}^{\infty}{}^{*} \int_{-\infty}^{\infty} \frac{e^{-2\alpha x^2+2\pi ikx}}{\cosh \alpha x} dx,$$

where $\sum^* a_k = \frac{1}{2} a_0 + a_1 + a_2 + a_3 + \ldots$.

We now shift the line of integration from the real x-axis to the line passing through the stationary point of the function

$$e^{-2\alpha x^2+2\pi ikx}$$

following the method of steepest descent.

Hence with $x_k = \frac{i\pi k}{2\alpha}$,

$$(2.3) \qquad M_1(e^{-\alpha}) = \Big(\sum_{k=0}^{\infty}{}^{*} P \int_{-\infty+x_k}^{\infty+x_k} \frac{e^{-2\alpha x^2+2\pi ikx}}{\cosh \alpha x} dx\Big)$$

$$+ \Big(\sum_{k=0}^{\infty}{}^{*} 2\pi i \sum_{0 \leq j \leq (k-1)/2}^{**} \lambda_{k,j}\Big)$$

$$= I + R,$$

where P denotes the "principal value" of the integral, $\lambda_{k,j}$ is the residue of the integrand at $x = \frac{\pi i}{\alpha}(j+1/2)$, and $\sum^{**}$ means that the factor $\frac{1}{2}$ is introduced for the term $j = \frac{k-1}{2}$ (which occurs only when k is odd).

Now

$$(2.4) \qquad I = \sum_{k=0}^{\infty}{}^{*} P \int_{-\infty}^{\infty} \frac{e^{-2\alpha(x+x_k)^2+2\pi ik(x+x_k)}}{\cosh \alpha(x+x_k)} dx$$

$$= \sum_{k=0}^{\infty}{}^{*} e^{\frac{-\pi^2 k^2}{2\alpha}} P \int_{-\infty}^{\infty} \frac{e^{-2\alpha x^2}}{\cosh \alpha(x+x_k)} dx$$

$$= \sum_{k=0}^{\infty}{}^{*} e^{\frac{-2\pi^2k^2}{\alpha}} (-1)^k \int_{-\infty}^{\infty} \frac{e^{-2\alpha x^2}}{\cosh \alpha x} dx$$

$$-i \sum_{k=0}^{\infty} e^{\frac{-\pi^2(2k+1)^2}{2\alpha}} (-1)^k \, P \int_{-\infty}^{\infty} \frac{e^{-2\alpha x^2}}{\sinh \alpha x} dx$$

$$= \frac{1}{2} \left(\sum_{k=-\infty}^{\infty} (-1)^k e^{-\frac{2\pi^2 k^2}{\alpha}} \right) \int_{-\infty}^{\infty} \frac{e^{-2\alpha x^2}}{\cosh \alpha x} dx.$$

On the other hand,

$$R = 2\pi i \sum_{k=0}^{\infty}{}^{*} \sum_{0 \leq j \leq (k-1)/2}^{**} \lambda_{k,j} \tag{2.5}$$

$$= 2\pi i \sum_{j=0}^{\infty} \left(\frac{1}{2} \lambda_{2j+1,j} + \lambda_{2j+2,j} + \lambda_{2j+3,j} + \ldots\right).$$

Since $\lambda_{k,j}$ is the residue of the integrand at $\rho_j = \pi i(j+1/2)/\alpha$, we see that

$$\lambda_{k,j} = \lim_{x \to \rho_j} (x-\rho_j) \frac{e^{-2\alpha x^2+2\pi ikx}}{\cosh \alpha x} \tag{2.6}$$

$$= \frac{e^{-2\alpha\rho_j^2+2\pi ik\rho_j}}{i\alpha(-1)^j} ,$$

and so

$$\lambda_{2j+1+\ell,j} = \lambda_{2j+1,j}\, e^{2\pi i\ell\rho_j} . \tag{2.7}$$

Therefore

$$R = 2\pi i \sum_{j=0}^{\infty} \lambda_{2j+1,j} \frac{1+e^{2\pi i\rho_j}}{1-e^{2\pi i\rho_j}} . \tag{2.8}$$

$$= \frac{2\pi}{\alpha} \sum_{j=0}^{\infty} (-1)^j e^{-2(\frac{\pi^2}{\alpha})(j+1/2)^2} \frac{1+e^{-(\frac{\pi^2}{\alpha})(2j+1)}}{1-e^{-(\frac{\pi^2}{\alpha})(2j+1)}} .$$

Substituting (2.4) and (2.8) into (2.3), we obtain (2.1). Thus Theorem 1 is proved.

The three remaining theorems of this section are proved similarly so we omit many of the details.

THEOREM 2. For $q = e^{-\alpha}$, $q_1 = e^{-\pi^2/\alpha}$, Re $\alpha > 0$,

$$\text{(2.9)}\quad M_2(q) = -\frac{2\pi}{\alpha} q_1^{1/2} M_3(-q_1) + \Big(\sum_{k=-\infty}^{\infty} (-1)^k q_1^{2k^2}\Big) \int_{\infty}^{\infty} \frac{e^{-2\alpha x^2+\alpha x}}{e^{2\alpha x}+1}\, dx.$$

PROOF. As in Theorem 1,

$$M_2(q) = \frac{1}{2} \sum_{k=-\infty}^{\infty} (-1)^k \int_{-\infty}^{\infty} \frac{e^{-2\alpha x^2+2\pi ikx}}{\cosh x\alpha}\, dx.$$

Thus the only change from Theorem 1 is the introduction of $(-1)^k$, and if we trace its effect throughout the argument we wind up with (2.9).

It is now a simple matter to obtain the corresponding results for $M_3(q)$ and $M_3(-q)$. The main tools required are the transformation formulas for the theta functions $\theta_2(0,q)$ and $\theta_4(0,q)$ as well as the following simple result:

LEMMA 1. Let Re $\alpha > 0$, $\beta = \pi^2/\alpha$. Then

$$\int_0^{\infty} \frac{e^{-2\alpha x^2}}{\cosh \alpha x}\, dx = \frac{1}{2} \left(\frac{2\beta}{\pi}\right)^{3/2} \int_0^{\infty} \frac{e^{-2\beta x^2}}{\cosh 2\beta x}\, dx.$$

PROOF. An equivalent form of this result was actually given by Ramanujan in an issue of the Journal of the Indian Mathematical Society [18; Question 295, pp. 324-325].

THEOREM 3. For $q = e^{-\alpha}$, $q_1 = e^{-\pi^2/\alpha} = e^{-\beta}$, Re $\alpha > 0$,

(2.10) $$q^{1/2}M_3(q) = \frac{\pi}{2\alpha} M_1(q_1) - q_1^{1/8}\left(\sum_{k=0}^{\infty} q_1^{k(k+1)/2}\right) \int_0^{\infty} \frac{e^{-2\alpha x^2}}{\cosh 2\alpha x}\,dx.$$

PROOF. In equation (2.1) of Theorem 1, replace α by π^2/α (i.e. interchange α and $\beta = \pi^2/\alpha$). Hence

$$M_1(q_1) = \frac{2\alpha}{\pi} q^{1/2}M_3(q) + \left(\sum_{k=-\infty}^{\infty} (-1)^k q^{2k^2}\right) \int_0^{\infty} \frac{e^{-2\beta x^2}}{\cosh \beta x}\,dx$$

$$= \frac{2\alpha}{\pi} q^{1/2}M_3(q) + \left(\sum_{k=-\infty}^{\infty} (-1)^k q^{2k^2}\right) \frac{1}{2}\left(\frac{2\alpha}{\pi}\right)^{3/2} \int_0^{\infty} \frac{e^{-2\alpha x^2}}{\cosh 2\alpha x}\,dx \text{ (by Lemma 1)}$$

$$= \frac{2\alpha}{\pi} q^{1/2}M_3(q) + \left(\frac{\pi}{2\alpha}\right)^{1/2} \sum_{k=0}^{\infty} q_1^{\frac{1}{2}(k+1/2)^2} \left(\frac{2\alpha}{\pi}\right)^{3/2} \int_0^{\infty} \frac{e^{-2\alpha x^2}}{\cosh 2\alpha x}\,dx$$

(by eq. (6), Table XLIII of [22; p. 263]).

(2.11) $$= \frac{2\alpha}{\pi} q^{1/2}M_3(q) + \frac{2\alpha}{\pi} q_1^{\frac{1}{8}} \sum_{k=0}^{\infty} q_1^{k(k+1)/2} \int_0^{\infty} \frac{e^{-2\alpha x^2}}{\cosh 2\alpha x}\,dx.$$

Now multiplying (2.11) by $\pi/2\alpha$ we obtain (2.10).

THEOREM 4. For $q = e^{-\alpha}$, $q_1 = e^{-\pi^2/\alpha} = e^{-\beta}$, $\mathrm{Re}\ \alpha > 0$,

(2.12) $$q^{1/2}M_3(-q) = -\frac{\pi}{2\alpha} M_2(q_1) + q_1^{1/8}\left(\sum_{k=0}^{\infty} q_1^{k(k+1)/2}\right) \int_0^{\infty} \frac{e^{-2\alpha x^2}}{\cosh 2\alpha x}\,dx.$$

PROOF. In equation (2.9) of Theorem 2, replace α by π^2/α (i.e. interchange α and $\beta = \pi^2/\alpha$). Theorem 4 follows in precisely the way that Theorem 3 did.

We observe that the fundamental modular transformation for the squares of $\theta_2(0,q)$ and $\theta_4(0,q)$ easily follows from Theorems 1 and 2 (or alternatively Theorems 3 and 4).

COROLLARY 1. For $q = e^{-\alpha}$, $q_1 = e^{-\pi^2/\alpha} = e^{-\beta}$, $\mathrm{Re}\ \alpha > 0$,

(2.13) $$(\theta_4(0,q))^2 = \frac{\pi}{\alpha}(\theta_2(0,q_1))^2.$$

PROOF. By (1.4)

$$(\theta_4(0,q))^2 = 2(M_1(q)-M_2(q))$$

$$= 2\cdot\frac{2\pi}{\alpha}\, q_1^{1/2}\,(M_3(q_1)+M_3(-q_1)) \quad \text{(by (2.1) and (2.9))}$$

$$= \frac{4\pi}{\alpha}\, q_1^{1/2}\,\frac{1}{4}\, q_1^{-1/2}\,(\theta_2(0,q_1))^2 \quad \text{(by (1.5))}$$

$$= \frac{\pi}{\alpha}(\theta_2(0,q_1))^2.$$

3. Ramanujan's identities. Section 2 has shown us the behavior of $M_1(q)$, $M_2(q)$ and $M_3(q)$ under the fundamental modular substitution. We shall now see how these functions fit in with the following set of results from Ramanujan's "Lost" Notebook.

The identities in question are:

$(3.1)_R$ $$\left(\prod_{n=1}^{\infty}(1+aq^n)(1+a^{-1}q^n)(1-q^n)\right)\sum_{n=0}^{\infty}\frac{(-1)^n q^{n^2}(1-q)(1-q^3)\dots(1-q^{2n-1})}{\prod_{j=1}^{n}(1+aq^{2j})(1+a^{-1}q^{2j})}$$

$$= 1+\sum_{n=1}^{\infty}(a^n+a^{-n}+2(-1)^n)\,\frac{q^{n(n+1)/2}}{1+q^n};$$

$(3.2)_R$ $$(1+a^{-1})\sum_{n=0}^{\infty}\frac{q^{n+1}(1+q)(1+q^2)\dots(1+q^{2n})}{\prod_{j=0}^{n}(1-aq^{2j+1})(1-a^{-1}q^{2j+1})}$$

$$= \frac{\sum_{n=1}^{\infty}(-1)^{n-1}\left(\frac{q^{n^2}}{1-aq^{2n-1}}+\frac{q^{n^2}}{a-q^{2n-1}}\right)}{\sum_{n=-\infty}^{\infty}(-1)^n q^{n^2}};$$

$$(3.3)_R \quad = \frac{\sum_{n=1}^{\infty} (-1)^{n-1}(a^{n-1}+a^{-n}) \dfrac{q^{n^2}}{1-q^{2n-1}}}{\sum_{n=-\infty}^{\infty} (-1)^n a^n q^{n^2}} ;$$

$$(3.4)_R \quad 1 + (1+a) \sum_{n=1}^{\infty} \left(\frac{q^{n(2n+1)}}{1+aq^{2n}} + \frac{q^{n(2n+1)}}{a+q^{2n}}\right)$$

$$= \left(\sum_{n=0}^{\infty} q^{n(n+1)/2}\right) \sum_{n=0}^{\infty} \frac{(-1)^n q^{n^2}(1-q)(1-q^3)\dots(1-q^{2n-1})}{\prod_{j=1}^{n} (1+aq^{2j})(1+a^{-1}q^{2j})} ;$$

$$(3.5)_R \quad \frac{1}{1+a} + \sum_{n=1}^{\infty} \left(\frac{q^{n(n+1)/2}}{1+aq^n} + \frac{q^{n(n+1)/2}}{a+q^n}\right)$$

$$= \frac{\prod_{n=1}^{\infty} (1-q^n)^3}{\sum_{n=-\infty}^{\infty} a^n q^{n(n-1)/2}} ;$$

$$(3.6)_R \quad \sum_{n=0}^{\infty} \frac{q^{n+1}(1+q^2)(1+q^4)\dots(1+q^{2n})}{(1-q)(1-q^3)\dots(1-q^{2n+1})} = \sum_{n=0}^{\infty} \frac{q^{(n+1)^2}(1+q)(1+q^3)\dots(1+q^{2n-1})}{(1-q)^2(1-q^3)^2\dots(1-q^{2n+1})^2} ;$$

$$(3.7)_R \quad = \frac{-\frac{1}{2} M_2(-q)}{\sum_{n=0}^{\infty} (-q)^{n(n+1)/2}} \qquad \text{(see (1.2) for } M_2(q)\text{)};$$

$$(3.8)_R \quad \sum_{n=0}^{\infty} \frac{(-1)^n q^{n^2}(1-q)(1-q^3)\dots(1-q^{2n-1})}{\prod_{j=1}^{n} (1+aq^{2j})(1+a^{-1}q^{2j})}$$

$$- (1+a^{-1})(1+a) \sum_{n=0}^{\infty} \frac{(-1)^n q^{(n+1)^2}(1-q)(1-q^3)\ldots(1-q^{2n-1})}{\prod_{j=0}^{n} (1+aq^{2j+1})(1+a^{-1}q^{2j+1})}$$

$$= \frac{\left(\sum_{n=-\infty}^{\infty} (-1)^n q^{n^2}\right) \prod_{n=1}^{\infty} (1-q^{2n-1})}{\prod_{j=1}^{\infty} (1+aq^j)(1+a^{-1}q^j)} ;$$

$(3.9)_R$
$$\sum_{n=0}^{\infty} \frac{(-1)^n q^{2n^2}(1-q^2)(1-q^6)\ldots(1-q^{4n-2})}{\prod_{j=1}^{n} (1+aq^{4j}(1+a^{-1}q^{4j})}$$

$$+ (1+a)(1+a^{-1}) \sum_{n=0}^{\infty} \frac{q^{n+1}(1+q)(1+q^2)\ldots(1+q^{2n})}{\prod_{j=0}^{n} (1-aq^{2j+1})(1-a^{-1}q^{2j+1})}$$

$$= \frac{\left(\sum_{n=0}^{\infty} q^{n(n+1)/2}\right) \prod_{j=0}^{\infty} (1+aq^{4j+2})(1+a^{-1}q^{4j+2})}{\prod_{j=1}^{\infty} (1-q^{2j-1})(1+aq^{4j})(1+a^{-1}q^{4j})(1-aq^{2j-1})(1-aq^{2j-1})}$$

$(3.10)_R$
$$\sum_{n=0}^{\infty} \frac{q^{2n^2}(1+q^2)(1+q^6)\ldots(1+q^{4n-2})}{\prod_{j=1}^{n} (1+q^{4j})^2}$$

$$+ 4 \sum_{n=0}^{\infty} \frac{q^{n+1}(1+q)(1+q^2)\ldots(1+q^{2n})}{(1+q^2)(1+q^6)\ldots(1+q^{4n+2})}$$

$$= \frac{\left(\sum_{n=-\infty}^{\infty} q^{n^2}\right)^2}{\sum_{n=0}^{\infty} (-1)^{n(n+1)/2} q^{n^2+n}} ;$$

$$(3.11)_R \qquad \sum_{n=0}^{\infty} \frac{(-1)q^{n^2}(1-q)(1-q^3)\dots(1-q^{2n-1})}{\prod_{j=1}^{n}(1+aq^{2j})(1+a^{-1}q^{2j})}$$

$$=^{*} \sum_{n=0}^{\infty} \frac{(-1)^n(1-q)(1-q^2)\dots(1-q^{2n-1})}{\prod_{j=1}^{n}(1+aq^{j})(1+a^{-1}q^{j})}$$

$$+ \frac{1}{2} \frac{\left(\sum_{n=-\infty}^{\infty} (-1)^n q^{n^2}\right)(1-q)(1-q^3)(1-q^5)\dots}{\prod_{j=1}^{\infty}(1+aq^{j})(1+a^{-1}q^{j})} ;$$

($=^{*}$ indicates that some "summability" convention is required)

$$(3.12)_R \qquad \sum_{n=0}^{\infty} \frac{(-1)^n q^{n^2}(1-q)(1-q^3)\dots(1-q^{2n-1})}{(1-q^2)^2(1-q^4)^2\dots(1-q^{2n})^2}$$

$$= \frac{1}{1+q+q^3+q^6+q^{10}+\dots} ;$$

$$(3.13)_R \qquad \sum_{n=0}^{\infty} \frac{(-1)^n q^{n^2}(1-q)(1-q^3)\dots(1-q^{2n-1})}{\prod_{n=1}^{n}(1+aq^{2j})(1+a^{-1}q^{2j})}$$

$$= (1+a) \sum_{n=0}^{\infty} \frac{q^{2n+1}(1-q)(1-q^3)\dots(1-q^{2n-1})}{(1+aq)(1+aq^2)\dots(1+aq^{2n+1})}$$

$$+ \left(\prod_{n=1}^{\infty} \frac{(1-q^{2n-1})}{(1+aq^n)}\right) \sum_{n=0}^{\infty} \frac{(-1)^n q^{n^2}}{(1+a^{-1}q^2)(1+a^{-1}q^4)\dots(1+q^{-1}q^{2n})} .$$

The multiplicity and similarity of these results makes it difficult to see what is really going on here. Indeed apart from (3.7) there seems to be no obvious relationship of these results to the work in Section 2. For the remainder of this section we shall prove the above identities. The next section will consider their implications.

Our approach to proving these results will be through the classical theory of basic hypergeometric series. Indeed it may well be that all of the results given above may be established via this method; however we shall also introduce a method which we term "q-partial fractions". A prototype of this technique (which has its theoretical foundation in Lemma 2 below) is our proof of identity (3.1).

<u>LEMMA 2</u>. Let $f(a)$ be analytic in a except possibly at $a = 0, \infty$. Let $f(a)$ satisfy:

(i) $$f(a) = f(a^{-1}), \quad a \neq 0, \infty;$$

(ii) There exists a sequence of positive real numbers $1 = \rho_1, \rho_2, \rho_3, \ldots \to +\infty$ and a sequence of nonnegative real numbers M_n, such that $\lim_{n \to \infty} M_n = 0$, and

$$M_n \geqq \max_{|a| = \rho_n} |f(a)| .$$

Then $f(a)$ is identically zero.

<u>PROOF</u>. Since $\lim_{n \to \infty} M_n = 0$, there exists a nonnegative real number μ defined by

$$\mu = \max_{n \geqq 1} M_n .$$

Then by the maximum modulus principle for closed annuli [12; p. 163], we see that for $|a| \geq 1$, $|f(a)| \leqq \mu$. Furthermore by (i) we see that for all nonzero but finite values of a, we have $|f(a)| \leqq \mu$. Next by (ii) and (i) we see that for any $\varepsilon > 0$, we may choose n_0 sufficiently large that $0 \leqq M_n < \varepsilon$ for $n \geqq n_0$, and therefore again by the maximum modulus principle $|f(a)| = |f(a^{-1})| < \varepsilon$ for $0 < |a| < |\rho_{n_0}|^{-1}$. Consequently $f(a)$ is in fact analytic at $a = 0$. This is because 0 is at worst some sort of isolated singularity of $f(a)$; however, boundedness in a neighborhood of 0 implies that this singularity is neither essential nor a pole and so must be removable [12; pp. 78-81]. Indeed $f(0) = 0 \leq \mu$. Therefore we may invoke Liouville's theorem and conclude that $f(a)$ is a constant. Since $f(0) = 0$ we see that $f(a) = 0$ for all a.

To facilitate our treatment we introduce the following notation from basic hypergeometric series [21; pp. 88-89]:

$$(3.14)\qquad (A;q)_\infty = (A)_\infty = \prod_{n=0}^{\infty} (1-Aq^n);$$

$$(3.15)\qquad (A;q)_n = (A)_n = \prod_{m=0}^{\infty} \frac{(1-Aq^m)}{(1-Aq^{m+n})}.$$

Note that when n is a nonnegative integer

$$(3.16)\qquad (A;q)_n = (A)_n = (1-A)(1-Aq)\dots(1-Aq^{n-1}).$$

PROOF OF (3.1). The right hand side of (3.1) may clearly be written as

$$\sum_{n=-\infty}^{\infty} \frac{a^n q^{n(n+1)/2}}{1+q^n 1+q^n} + \sum_{n=-\infty}^{\infty} \frac{(-1)^n q^{n(n+1)/2}}{1+q^n}.$$

$$= \sum_{n=-\infty}^{\infty} \frac{a^n q^{n(n+1)/2}}{1+q^n} + \frac{1}{2}\,\frac{(q)_\infty^2}{(-q)_\infty^2} \quad \text{(by [14; p. 238, eq. (14)]).}$$

$$= \frac{1}{2}\,\frac{(-a)_\infty(-q/a)_\infty}{(-q)_\infty^2} \sum_{n=-\infty}^{\infty} \frac{q^{n(n+1)/2}}{1+aq^n} + \frac{1}{2}\,\frac{(q)_\infty^2}{(-q)_\infty^2}$$

(by [10; p. 196, eq. (2.4)] with $\alpha = -q/\tau$, $\beta = -1$, $\gamma = \tau$, $\delta = -q$, $\tau \to 0$).

Hence after replacing a by a^2 and multiplying both sides by an appropriate factor, we may rewrite (3.1) as

$$(3.17)\qquad a \sum_{n=0}^{\infty} \frac{(-1)^n q^{n^2} (q;q^2)_n}{(-a^2;q^2)_{n+1}(-\frac{q^2}{a^2};q^2)_n}$$

$$-\frac{1}{2(-q)_\infty(q^2;q^2)_\infty}\sum_{n=-\infty}^{\infty}\frac{q^{n^2/2}}{a^{-1}q^{n/2}+aq^{n/2}}$$

$$-\frac{a(q)_\infty}{2(-q)_\infty^2(-a^2)_\infty(-q/a^2)_\infty}=0.$$

Let us denote the left side of (3.17) by $f(a)$; our object now is to fulfill the conditions of Lemma 2. Inspection shows that $f(a) = f(a^{-1})$. However it appears that $f(a)$ has simple poles at $a^2 = -q^N$ for all integers N. These are only apparent singularities as we now establish by showing that the residue at each such singularity is 0.

CASE 1. $a^2 = -q^{-2N+1}$ (where without loss of generality we may assume $N > 0$). In the following the choice of sign in $\mp$ is fixed throughout with $\mp = -(\pm)$.

$$\lim_{a\to\mp iq^{-N+1/2}}(1+a^2q^{2N-1})f(a)$$

$$=\frac{q^{N(2N-1)}(\pm iq^{-N+1/2})}{2(-q)_\infty(q^2;q^2)_\infty}$$

$$+\frac{(\pm iq^{-N+1/2})(q)_\infty}{2(-q)_\infty^2(q^{-2N+1})_{2N-1}(q)_\infty(q^{2N})_\infty}$$

$$=\frac{q^{N(2N-1)}(\pm iq^{-N+1/2})}{2(-q)_\infty(q^2;q^2)_\infty}-\frac{q^{N(2N-1)}(\pm iq^{-N+1/2})}{2(-q)_\infty(q^2;q^2)_\infty}=0.$$

CASE 2. $a^2 = -q^{-2N}$ (where without loss of generality we may assume $N \geq 0$).

$$\lim_{a\to\mp iq^{-N}}(1+a^2q^{2N})f(a)$$

$$=(\mp iq^{-N})\sum_{n=N}^{\infty}\frac{(-1)^n q^{n^2}(q;q^2)_n}{(q^{-2N};q^2)_N(q^2;q^2)_{n-N}(q^{2N+2};q^2)_n}$$

$$- \frac{(\mp iq^{-N})q^{N(2N+1)}}{2(-q)_\infty(q^2;q^2)_\infty} - \frac{(\mp iq^{-N})(q)_\infty}{(-q)_\infty^2(q^{-2N})_{2N}(q)_\infty(q^{2N+1})_\infty}$$

$$= (\mp iq^{-N}) \sum_{n=0}^{\infty} \frac{(-1)^n q^{n^2+2nN+2N^2+N}(q;q^2)_{n+N}}{(q^2;q^2)_n(q^2;q^2)_{n+2N}} - \frac{(\mp iq^{-N})q^{N(2N+1)}}{(-q)_\infty(q^2;q^2)_\infty}$$

$$= \frac{q^{2N^2+N}(q;q^2)_N(\mp iq^{-N})}{(q^2;q^2)_{2N}} \lim_{\tau \to 0} \sum_{j=0}^{\infty} \frac{(q/\tau;q^2)_j(q^{2N+1};q^2)_j\tau^j q^{2Nj}}{(q^2;q^2)_j(q^{4N+2};q^2)_j}$$

$$- \frac{(\mp iq^{-N})q^{N(2N+1)}}{(-q)_\infty(q^2;q^2)_\infty}$$

$$= \frac{q^{2N^2+N}(q;q^2)_N(\mp iq^{-N})(q^{2N+1};q^2)_\infty}{(q^2;q^2)_{2N}(q^{4N+2};q^2)_\infty} - \frac{(\mp iq^{-N})q^{N(2N+1)}}{(-q)_\infty(q^2;q^2)_\infty} \quad \text{(by [3; p.20, Cor. 2.4])}$$

$$= \frac{q^{2N^2+N}(\mp iq^{-N})(q;q^2)_\infty}{(q^2;q^2)_\infty} - \frac{(\mp iq^{-N})q^{N(2N+1)}}{(-q)_\infty(q^2;q^2)_\infty} = 0 \quad \text{(by [3; p.5, eq. (1.2.5)]).}$$

Thus we have established that $f(a)$ is analytic for every a except possibly $a = 0,\infty$. All we need to do now is show that condition (ii) of Lemma 2 is fulfilled. To do this we choose for the ρ_n numbers that are in effect far away from the apparent singularities that required the above special treatment. Namely take $\rho_N = |q|^{-N-\frac{1}{4}}$, $n \geqq 2$; so that for $|a| = \rho_N$ we have $|a^2| = |q|^{-2N-\frac{1}{2}}$. Hence on the circle $|a| = \rho_N$, $N \geq 2$:

$$|f(a)| \leqq |q|^{-N-\frac{1}{4}} \sum_{n \geqq 0} \frac{|q|^{n^2}(1+|q|)(1+|q|^3)\dots(1+|q|^{2n-1})}{(|q|^{-2N-\frac{1}{2}}-1)\dots(|q|^{-\frac{1}{2}}-1)\prod_{m=0}^{\infty}(1-|q|^{\frac{1}{2}+2m})^2}$$

$$+ \frac{|q|^{-N-1/4}}{2\prod_{n=1}^{\infty}(1-|q|^n)(1-|q|^{2n})} \frac{\sum_{n=-N/2}^{N/2} |q|^{(n^2+n)/2}}{|q|^{-2N-1/2+N/4}-1} + \frac{\sum_{|n|>\frac{N}{2}} |q|^{\frac{n^2}{2}+\frac{n}{2}}}{|q|^{-1/2}-1}$$

$$+ \frac{|q|^{-N-1/4}\prod_{n=1}^{\infty}(1+|q|^n)}{2\prod_{n=1}^{\infty}(1-|q|^n)^4(|q|^{-2N-1/2}-1)\dots(|q|^{-1/2}-1)}$$

$$= |q|^{2N^2-1/4+N} \sum_{n=0}^{\infty} \frac{|q|^{n^2}(1+|q|)(1+|q|^3)\dots(1+|q|^{2n-1})}{(1-|q|^{1/2})(1-|q|^{3/2})\dots(1-|q|^{2N+1/2})\prod_{m=0}^{\infty}(1-|q|^{1/2+2m})^2}$$

$$+ \frac{1}{2\prod_{n=1}^{\infty}(1-|q|^n)(1-|q|^{2n})} \left\{ \frac{|q|^{(3N+1)/4}}{1-|q|^{3N/4+1/2}}(N+1) + \frac{2|q|^{N^2/8-N-1/4}|q|^{N/4}}{(|q|^{-1/2}-1)(1-|q|^{1/2})} \right\}$$

$$+ \frac{|q|^{2N^2-\frac{1}{4}+N}}{2\prod_{n=1}^{\infty}(1-|q|^n)^4(1-|q|^{1/2})(1-|q|^{3/2})\dots(1-q^{2N+1/2})} = M_N \to 0 \text{ as } N \to \infty$$

under the assumption used throughout that $|q| < 1$.

Thus we have fulfilled all the conditions of Lemma 2 for $f(a)$. Hence $f(a) = 0$ thereby establishing (3.17) and with it (3.1).

PROOF OF (3.2). Again we invoke Lemma 2; however since all the steps are easier than with the proof of (3.1), we merely sketch the treatment. First if we replace a by a^2 and then multiply both sides by a, we see that (3.2) is equivalent to

$$(3.18) \qquad (a+a^{-1}) \sum_{n=0}^{\infty} \frac{q^{n+1}(-q)_{2n}}{(a^2q;q^2)_{n+1}(a^{-2}q;q^2)_{n+1}} - \frac{\sum_{n=1}^{\infty}(-1)^{n-1}\left(\frac{q^{n^2}}{a^{-1}-aq^{2n-1}} + \frac{q^{n^2}}{a-a^{-1}q^{2n-1}}\right)}{\sum_{n=-\infty}^{\infty}(-1)^n q^{n^2}}$$

$$= 0.$$

As before denote the left side of (3.18) by $f(a)$. Again clearly $f(a) = f(a^{-1})$. Furthermore condition (ii) of Lemma 2 is fulfilled if (following the lead of our proof of (3.1)) we take $\rho_N = |q|^{-N+1/4}$. The only remaining obstacle to the utilization of Lemma 2 is the possibility of simple poles at $a^2 = q^{-2N+1}$. Without loss of generality (since $f(a) = f(a^{-1})$) we may consider for $N \gtrless 0$:

$$\lim_{a \to \pm q^{-N+1/2}} (1-a^2q^{2N-1})f(a)$$

$$= (\pm q^{-N+1/2})(1+q^{2N-1}) \sum_{n=N-1}^{\infty} \frac{q^{n+1}(-q)_{2n}}{(q^{-2N+2};q^2)_{N-1}(q^2;q^2)_{n-N+1}(q^{2N};q^2)_{n+1}}$$

$$- \frac{(\pm q^{-N+1/2})q^{N^2}(-1)^{N-1}}{\sum_{n=-\infty}^{\infty}(-1)^n q^{n^2}}$$

$$= \frac{(\pm q^{-N+1/2})q^{N^2-N}(-1)^{N-1}(1+q^{2N-1})}{(q^2;q^2)_{N-1}} \sum_{n=0}^{\infty} \frac{q^{n+N}(-q)_{2n+2N-2}}{(q^2;q^2)_n(q^{2N};q^2)_{n+N}}$$

$$- \frac{(\pm q^{-N+1/2})q^{N^2}(-1)^{N-1}(-q)_\infty}{(q)_\infty}$$

(by [3; p.23, eq. (2.2.12)])

$$= \frac{(\pm q^{-N+1/2})q^{N^2}(-1)^{N-1}(-q)_{2N-1}}{(q^2;q^2)_{2N-1}} \sum_{n=0}^{\infty} \frac{(-q^{2N-1};q^2)_n(-q^{2N};q^2)_n q^n}{(q^2;q^2)_n(q^{4N};q^2)_n}$$

$$-\frac{(\pm q^{-N+1/2})q^{N^2}(-1)^{N-1}(-q)_\infty}{(q)_\infty}$$

$$=\frac{(\pm q^{-N+1/2})q^{N^2}(-1)^{N-1}(-q)_{2N-1}}{(q^2;q^2)_{2N-1}}\quad\frac{(-q^{2N+1};q^2)_\infty(-q^{2N};q^2)_\infty}{(q^{4N};q^2)_\infty(q;q^2)_\infty}$$

$$-\frac{(\pm q^{-N+1/2})q^{N^2}(-1)^{N-1}(-q)_\infty}{(q)_\infty}$$

(by [3; p. 20, Cor. 2.4])

$$=\frac{(\pm q^{-N+1/2})q^{N^2}(-1)^{N-1}(-q)_\infty}{(q)_\infty}-\frac{(\pm q^{-N+1/2})q^{N^2}(-1)^{N-1}(-q)_\infty}{(q)_\infty}=0.$$

Thus (3.18) (and consequently (3.2)) is established.

PROOF OF (3.3). The expression here is immediately identified with the right hand side of (3.2) under W.N. Bailey's transformation [10; p. 196, eq. (2.4)] wherein q is replaced by q^2 then $\alpha = q/\tau$, $\beta = aq^{-1}$, $\gamma = \tau q^{-1}$, $\delta = aq$, $z = \tau$ and $\tau \to 0$.

PROOF OF (3.4). If we subtract the right side of (3.4) from the left side and call the result $f(a)$, then we are again faced with the task of invoking Lemma 2. The proof that $f(a) = f(a^{-1})$ and the establishment of condition (ii) with $\rho_n = |q|^{-2n+1}$ are routine. The only obstacle again is the verification that the possible simple poles are spurious. This follows from the fact that for $N \geq 0$

$$\lim_{a\to -q^{-2N}} (1+aq^{2N})f(a)$$

$$= q^{2N^2+N}-\left(\sum_{n=0}^{\infty} q^{n(n+1)/2}\right)\sum_{n\geq N}\frac{(-1)^n q^{n^2}(q;q^2)_n}{(q^{-2N};q^2)_N(q^2;q^2)_{n-N}(q^{2N+2};q^2)_n}$$

$$= q^{2N^2+N} - \left(\sum_{n=0}^{\infty} q^{n(n+1)/2}\right)(-1)^N q^{N^2+N} \sum_{n=0}^{\infty} \frac{(-1)^{n+N} q^{n^2+2nN+N^2}(q;q^2)_{n+N}}{(q^2;q^2)_n (q^2;q^2)_{n+2N}}$$

$$= q^{2N^2+N} - \left(\sum_{n=0}^{\infty} q^{n(n+1)/2}\right) \frac{q^{2N^2+N}(q;q^2)_N}{(q^2;q^2)_{2N}} \lim_{\tau \to 0} \sum_{n=0}^{\infty} \frac{(q/\tau;q^2)_n (q^{2N+1};a^2)_n \tau^n q^{2Nn}}{(q^2;q^2)_n (q^{4N+2};q^2)_n}$$

$$= q^{2N^2+N} - \frac{\left(\sum_{n=0}^{\infty} q^{n(n+1)/2}\right) q^{2N^2+N}(q;q^2)_\infty}{(q^2;q^2)_\infty}$$

(by [3; p. 20, Cor. 2.4])

$$= q^{2N^2+N} - q^{2N^2+N}$$

(by [3; p. 23, eq. (2.2.13)])

$$= 0.$$

Hence (3.4) is proved.

<u>PROOF OF (3.5)</u>. This is a classical formula for the expansion of the reciprocal of a theta function in partial fractions. It is completely equivalent to the expansion for $\frac{1}{2\pi} \frac{\theta_1(0)}{\theta_4(\nu)}$ given at the end of Section 486 in [22; p. 136] and is originally due to Jacobi.

<u>PROOF OF (3.6)</u>.

$$\sum_{n=0}^{\infty} \frac{q^{(n+1)^2}(1+q)(1+q^3)\cdots(1+q^{2n-1})}{(1-q)^2(1-q^3)^2\cdots(1-q^{2n+1})^2}$$

$$= \lim_{\tau \to 0} \frac{q}{(1-q)^2} \sum_{n \geq 0} \frac{(-q^3/\tau;q^2)_n (-q;q^2)_n (q^2;q^2)_n \tau^n}{(q^2;q^2)_n (q^3;q^2)_n (q^3;q^2)_n}$$

$$= \frac{q}{(1-q)^2}(1-q) \sum_{n=0}^{\infty} \frac{(-q^2;q^2)_n q^n}{(q^3;q^2)_n}$$

(by [18; p. 174, eq. (10.1)] with $q \to q^2$, $a = -q^3/\tau$, $b = -q$, $c = q^2$, $e = f = q^3$)

$$= \sum_{n=0}^{\infty} \frac{(-q^2;q^2)_n q^{n+1}}{(q;q^2)_{n+1}} .$$

Thus (3.6) is proved.

PROOF OF (3.7). Here we require Watson's q-analog of Whipple's theorem [21; p.100, eq. (3.4.1.5)] wherein we replace q by q^2 and then set $e = -q^3/\tau$, $f = -q$, $a = q^2$, $c = d = q$, $\tau \to 0$. As a result we obtain

$$\sum_{n=0}^{\infty} \frac{q^{(n+1)^2}(1+q)(1+q^3)\dots(1+q^{2n-1})}{(1-q)^2(1-q^3)^2\dots(1-q^{2n+1})^2}$$

$$= \frac{q}{(1-q)^2} \frac{(-q^3;q^2)_\infty}{(q^4;q^2)_\infty} \sum_{n=0}^{\infty} \frac{(1-q^{4n+2})(q;q^2)_n(-q;q^2)_n(-1)^n q^{2n^2+3n}}{(1-q^2)(q^3;q^2)_n(-q^3;q^2)_n}$$

$$= \frac{(-q;q^2)_\infty}{(q^2;q^2)_\infty} \sum_{n=0}^{\infty} \frac{q^{(n+1)(2n+1)}(-1)^n}{1-q^{2n+1}}$$

$$= \frac{1}{\sum_{n=0}^{\infty} (-q)^{n(n+1)/2}} \sum_{n=1}^{\infty} \frac{(-1)^{n-1} q^{n(2n-1)}}{1-q^{2n-1}} \qquad \text{(by [3; p. 23, eq. (2.2.13)])}$$

$$= \frac{-\frac{1}{2} M_2(-q)}{\sum_{n=0}^{\infty} (-q)^{n(n+1)/2}} \qquad \text{(by (1.2))},$$

and so we have (3.7).

PROOF OF (3.8). Here we return to an application of Lemma 2. If on both sides of (3.8), we replace a by a^2 and then multiply throughout by $(a+a^{-1})^{-1}$, we find that we are asked to prove

$$f(a) = 0$$

where

$$(3.19) \qquad f(a) = a \sum_{n=0}^{\infty} \frac{(-1)^n q^{n^2} (q;q^2)_n}{(-a^{-2}q^2;q^2)_n(-a^2;q^2)_{n+1}}$$

$$- (a+a^{-1}) \sum_{n=0}^{\infty} \frac{(-1)^n q^{(n+1)^2} (1-q)(1-q^3)\dots(1-q^{2n-1})}{(-a^2q;q^2)_{n+1}(-a^{-2}q;q^2)_{n+1}}$$

$$- \frac{a(q)_\infty}{(-q)_\infty^2(-a^2)_\infty(-a^{-2}q)_\infty} .$$

The proof now completely parallels the proof of (3.1) via Lemma 2, and so it is omitted.

PROOF OF (3.9). This is another intricate utilization of Lemma 2 in that two cases are required to deal with the various possible simple poles. Also the application of condition (ii) in Lemma requires some care. After replacing a by a^2 throughout and multiplying by $(a+a^{-1})^{-1}$, we find that (3.9) is equivalent to the assertion that

$$f(a) = 0$$

where

$$(3.20) \qquad f(a) = a \sum_{n=0}^{\infty} \frac{(-1)^n q^{2n^2} (q^2;q^4)_n}{(-a^2;q^4)_{n+1}(-a^{-2}q^4;q^4)_n}$$

$$+ (a+a^{-1}) \sum_{n=0}^{\infty} \frac{q^{n+1}(-q)_{2n}}{(a^2q;q^2)_{n+1}(a^{-2}q;q^2)_{n+1}}$$

$$- \frac{a(q^2;q^2)_\infty(-a^2q^2;q^4)_\infty(-a^{-2}q^2;q^4)_\infty}{(q;q^2)_\infty^2(-a^2;q^4)_\infty(-a^{-2}q^4;q^4)_\infty(a^2q;q^2)_\infty(a^{-2}q;q^2)_\infty}$$

(we have utilized [3; p. 23, eq. (2.2.13)] to rewrite the third term in f(a)). As always the fact that $f(a) = f(a^{-1})$ is clear.

For condition (ii) of Lemma 2 we take $\rho_N = |q|^{-2N-1}$, $N \geqq 2$. With this choice of ρ_N we find that the first two terms in f(a) with $|a| = \rho_N$ tend to zero as $N \to \infty$ just as before; the third term is also appropriately behaved since with $|a| = \rho_N$

$$\left|\frac{a(q^2;q^2)_\infty(-a^2q^2;q^4)_\infty(-a^{-2}q^2;q^4)_\infty}{(q;q^2)^2_\infty(-a^2;q^4)_\infty(-a^{-2}q^4;q^4)_\infty(a^2q;q^2)_\infty(a^{-2}q;q^2)_\infty}\right|$$

$$\leqq \frac{|q|^{-2N-1}(-|q|^2;|q|^2)_\infty(-|q|^{-4N};q^4)_\infty(-|q|^{4N+4};|q|^4)_\infty}{(|q|;|q|^2)^2(|q|^{-4N-2}-1)\dots(|q|^{-2}-1)(|q|^2;|q|^4)_\infty(|q|^{4N+6};|q|^4)_\infty} \times$$

$$\frac{1}{(|q|^{-4N-1}-1)(|q|^{-4N+1}-1)\dots(|q|^{-1}-1)(|q|;|q|^2)_\infty(|q|^{4N+3};|q|^2)_\infty}$$

$$= \frac{|q|^{(2N+1)^2+1}(-|q|^2;|q|^2)_\infty 2(-|q|^4;|q|^4)^2_\infty}{(|q|;|q|^2)^2_\infty(|q|^2;|q|^4)^2_\infty(|q|;|q|^2)^2_\infty} \to 0 \text{ as } N \to +\infty.$$

The proof that the f(a) of (3.20) is identically zero will now follow directly from Lemma 2 once we show that the various apparent simple poles are in fact removable singularities.

<u>CASE 1</u>. $a \to \pm iq^{-N-1/2}$, i.e. $a^2 \to q^{-2N-1}$ (where without loss of generality we may assume $N \geqq 0$). Thus (assuming as always a fixed choice of sign in $\pm$) we see that

$$\lim_{a \to \pm iq^{-N-1/2}} (1-a^2q^{2N+1})f(a)$$

$$= (\pm iq^{-N-1/2})(1+q^{2N+1}) \sum_{n=N}^{\infty} \frac{q^{n+1}(-q)_{2n}}{(q^{-2N};q^2)_N(q^2;q^2)_{n-N}(q^{2N+2};q^2)_{n+1}}$$

$$- \frac{(\pm iq^{-N-1/2})(q^2;q^2)_\infty(-q^{-2N+1};q^4)_\infty(-q^{2N+3};q^4)_\infty}{(q;q^2)^2_\infty(-q^{-2N-1};q^4)_\infty(-q^{2N+5};q^4)_\infty(q^{-2N};q^2)_N(q^2;q^2)_\infty(q^{2N+2};q^2)_\infty} .$$

$$= \frac{(\pm iq^{-N-1/2})(1+q^{2N+1})(-1)^N q^{(N+1)^2}(-q)_{2N}}{(q^2;q^2)_{2N+1}} \sum_{n=0}^{\infty} \frac{q^n(-q^{2N+1};q^2)_n(-q^{2N+2};q^2)_n}{(q^2;q^2)_n(q^{4N+4};q^2)_n}$$

$$- \frac{(\pm iq^{-N-1/2})(q^2;q^2)_\infty(-q^{-2N+1};q^4)_{N+1}(-1)^N q^{N^2+N}}{(q;q^2)^2_\infty\ (-q^{-2N-1};q^4)_{N+1}(q^2;q^2)^2_\infty}$$

$$= \frac{(\pm iq^{-N-1/2})(-1)^N q^{(N+1)^2}(-q)_{2N+1}}{(q^2;q^2)_{2N+1}} \cdot \frac{(-q^{2N+3};q^2)_\infty(-q^{2N+2};q^2)_\infty}{(q^{4N+4};q^2)_\infty(q;q^2)_\infty}$$

$$- \frac{(\pm iq^{-N-1/2})(q^2;q^2)_\infty(-1)^N q^{(N+1)^2}}{(q;q^2)^2_\infty(q^2;q^2)^2_\infty}$$

$$= \frac{(\pm iq^{-N-1/2})(-1)^N q^{(N+1)^2}}{(q)_\infty(q;q^2)_\infty} - \frac{(\pm iq^{-N-1/2})(-1)^N q^{(N+1)^2}}{(q)_\infty(q;q^2)_\infty} = 0.$$

CASE 2. $a \to \pm iq^{-2N}$, i.e. $a^2 \to -q^{-4N}$

(where without loss of generality we may assume $N \geqq 0$). Hence

$$\lim_{a \to \pm iq^{-2N}} (1+a^2q^{4N})f(a)$$

$$= (\pm iq^{-2N}) \sum_{n=N}^{\infty} \frac{(-1)^n q^{2n^2}(q^2;q^4)_n}{(q^{-4N};q^4)_N(q^4;q^4)_{n-N}(q^{4N+4};q^4)_n}$$

$$- \frac{(\pm iq^{-2N})(q^2;q^2)_\infty(q^{-4N+2};q^4)_\infty(q^{4N+2};q^4)_\infty}{(q;q^2)^2_\infty(q^{-4N};q^4)_N(q^4;q^4)_\infty(q^{4N+4};q^4)_\infty(-q^{-4N+1};q^2)_\infty(-q^{4N+1};q^2)_\infty}$$

$$= \frac{(\pm iq^{-2N})q^{4N^2+2N}(q^2;q^4)_N}{(q^4;q^4)_{2N}} \sum_{n=0}^{\infty} \frac{(-1)^n q^{2n^2+4nN}(q^{4N+2};q^4)_n}{(q^4;q^4)_n(q^{8N+4};q^4)_n}$$

$$- \frac{(\pm iq^{-2N})q^{4N^2+2N}(q^2;q^4)_\infty^2(q^2;q^2)_\infty}{(q;q^2)_\infty^2(q^4;q^4)_\infty^2(-q;q^2)_\infty^2}$$

$$= \frac{(\pm iq^{-2N}q^{4N^2+2N}(q^2;q^4)_N}{(q^4;q^4)_{2N}} \cdot \frac{(q^{4N+2};q^4)_\infty}{(q^{8N+4};q^4)_\infty}$$

$$- \frac{(\pm iq^{-2N})q^{4N^2+2N}(q^2;q^2)_\infty}{(q^4;q^4)_\infty^2}$$

$$= \frac{(\pm iq^{-2N})q^{4N^2+2N}(q^2;q^4)_\infty}{(q^4;q^4)_\infty} - \frac{(\pm iq^{-2N})q^{4N^2+2N}(q^2;q^4)_\infty}{(q^4;q^4)_\infty} = 0.$$

Hence (3.9) is established.

PROOF OF (3.10). This identity requires some of the previous work plus a rather extended analysis of the series involved. We define

(3.21)
$$\lambda(q) = \sum_{n=0}^{\infty} \frac{q^{n+1}(-q)_{2n}}{(-q^2;q^4)_{n+1}},$$

and

(3.22)
$$\mu(q) = \sum_{n=0}^{\infty} \frac{(-1)^n q^{n^2}(q;q^2)_n}{(-q^2;q^2)_n^2}.$$

Then (3.10) is the assertion that

(3.23)
$$\mu(-q^2) + 4\lambda(q) = \Big(\sum_{n=-\infty}^{\infty} q^{n^2}\Big)^2 \sum_{n=0}^{\infty} (-q^2)^{n(n+1)/2}.$$

Now

$$\lambda(q) = \sum_{n=0}^{\infty} \frac{q^{n+1}(-q;q^2)_n(-q^2;q^2)_n(q^2;q^2)_n}{(q^2;q^2)_n(-q^2;q^4)_{n+1}}$$

$$= \sum_{n=0}^{\infty} \frac{q^{n+1}(-q;q^2)_n(q^4;q^4)_n}{(q^2;q^2)_n(-q^2;q^4)_{n+1}}$$

$$= \frac{(q^4;q^4)_\infty}{(q^2;q^4)_\infty} \sum_{n=0}^{\infty} \frac{q^{n+1}(-q;q^2)_n(-q^{4n+6};q^4)_\infty}{(q^2;q^2)_n(q^{4n+4};q^4)_\infty}$$

$$= \frac{(q^4;q^4)}{(-q^2;q^4)} \sum_{n=0}^{\infty} \frac{q^{n+1}(-q;q^2)_n}{(q^2;q^2)_n} \sum_{m=0}^{\infty} \frac{(-q^2;q^4)_m q^{4nm+4m}}{(q^4;q^4)_m}$$

(by [3; p. 17, eq. (2.2.1)])

$$= \frac{(q^4;q^4)_\infty}{(-q^2;q^4)_\infty} \sum_{m=0}^{\infty} \frac{q^{4m+1}(-q^2;q^4)_m}{(q^4;q^4)_m} \frac{(-q^{4m+2};q^2)_\infty}{(q^{4m+1};q^2)_\infty}$$

(by [3; p. 17, eq. (2.2.1)])

$$= \frac{q(q^4;q^4)_\infty(-q^2;q^2)_\infty}{(-q^2;q^4)\ (q;q^2)_\infty} \sum_{m=0}^{\infty} \frac{(q;q^2)_{2m}(-q^2;q^4)_m q^{4m}}{(-q^2;q^2)_{2m}(q^4;q^4)_m}$$

$$= \frac{q(q^8;q^8)}{(q;q^2)} \sum_{m=0}^{\infty} \frac{(q;q^2)_{2m} q^{4m}}{(q^8;q^8)_m}$$

$$= q(q^8,q^8) \sum_{m=0}^{\infty} \frac{q^{4m}}{(q^8;q^8)_m} \sum_{n=0}^{\infty} \frac{q^{n(4m+1)}}{(q^2;q^2)_n}$$

(by [3; p. 19, eq. (2.2.5)])

$$= q(q^8;q^8)_\infty \sum_{n=0}^{\infty} \frac{q^{n}}{(q^2;q^2)_n(q^{4n+4};q^8)_\infty}$$

$$= q \sum_{n=0}^{\infty} \frac{q^{2n+1}(q^8;q^8)_n}{(q^2;q^2)_{2n+1}}$$

$$+ q \frac{(q^8;q^8)_\infty}{(q^4;q^8)_\infty} \sum_{n=0}^{\infty} \frac{q^{2n}(q^4;q^8)_{2n}}{(q^2;q^2)_{2n}}$$

(3.24) $$= \sum_{n=0}^{\infty} \frac{q^{2(n+1)}(-q^4;q^4)_n}{(q^2;q^4)_{n+1}} + \frac{q(q^8;q^8)_\infty(-q^4;q^4)_\infty}{(q^4;q^8)_\infty(q^2;q^4)_\infty}$$

(by [3; p. 17, eq. (2.2.1)]).

If we now define

(3.25) $$\alpha(q) = \sum_{n=0}^{\infty} \frac{q^{n+1}(-q^2;q^2)_n}{(q;q^2)_{n+1}} ,$$

then we may rewrite (3.24) as

(3.26) $$\lambda(q) = \alpha(q^2) + \frac{q(q^8;q^8)_\infty(-q^4;q^4)_\infty}{(q^4;q^8)_\infty(q^2;q^4)_\infty} .$$

Now by (3.6)

(3.27) $$\alpha(q) = \sum_{n=0}^{\infty} \frac{q^{(n+1)^2}(-q;q^2)_n}{(q;q^2)_{n+1}^2} ,$$

and so by (3.8) with a = 1 and q replaced by -q, we find

$$(3.28)\qquad \mu(-q) + 4\alpha(q) = \frac{\left(\sum_{n=-\infty}^{\infty} q^{n^2}\right)(-q;q^2)_\infty}{(q;q^2)_\infty^2(-q^2;q^2)_\infty^2}.$$

Hence if we replace q by q^2 in (3.28) and then use the resulting equation to eliminate $\alpha(q^2)$ from (3.26), we see that

$$(3.29)\qquad \mu(-q^2) + 4\lambda(q) = \frac{(q^4;q^4)_\infty(-q^2;q^4)_\infty^3}{(q^2;q^4)_\infty^2(-q^4;q^4)_\infty^2} + \frac{2q(q^8;q^8)_\infty(-1;q^4)_\infty}{(q^4;q^8)_\infty(q^2;q^4)_\infty}$$

$$= \frac{(-q^2;q^4)_\infty}{(q^4;q^4)_\infty}\left((q^4;q^4)_\infty^2(-q^2;q^4)_\infty^4 + 4q(q^4;q^4)_\infty^2(-q^4;q^4)_\infty^4\right)$$

$$= \frac{(-q^2;q^4)_\infty}{(q^4;q^4)_\infty}\left(\left(\sum_{n=-\infty}^{\infty} q^{2n^2}\right)^2 + q\left(\sum_{n=-\infty}^{\infty} q^{2n^2+2n}\right)^2\right)$$

$$= \frac{(-q^2;q^4)_\infty}{(q^4;q^4)_\infty}\left(\sum_{n=-\infty}^{\infty} q^{n^2}\right)^2$$

(by [22; p. 269, eq. (4), 5th line])

$$= \left(\sum_{n=-\infty}^{\infty} q^{n^2}\right)^2 \Big/ \left(\sum_{n=0}^{\infty} (-q^2)^{n(n+1)/2}\right)$$

(by [3; p. 23, eq. (2.2.13)]).

Thus (3.23) (and with it (3.10)) is proved.

PROOF OF (3.11). In light of the divergence of the series in the first term on the right hand side of (3.11), we must find a suitable representation of this function. To effect this we consider the q-analog of Whipple's theorem [21; p. 100, eq. (3.4.1.5)] wherein we let $N \to \infty$ (i.e. $g \to \infty$), $e = -f = q^{1/2}$, $d = c^{-1}$ and $a \to 1$. This yields

$$\lim_{a\to 1} \sum_{n=0}^{\infty} \frac{(aq)_n(q;q^2)_n(-a)^n}{(q)_n(-cq)_n(-c^{-1}q)_n}$$

$$= \frac{(q;q^2)_\infty}{(q)_\infty(-1)_\infty} \left(1+ \sum_{n=1}^{\infty} \frac{(1+q^n)(1+c^{-1})(1+c)q^{n(n+1)/2}}{(1+cq^n)(1+c^{-1}q^n)}\right)$$

$$= \frac{1}{2} \frac{(q;q^2)_\infty}{(q^2;q^2)_\infty} \sum_{n=-\infty}^{\infty} \frac{(1+c^{-1})(1+c)q^{n(n+1)/2}}{(1+cq^n)(1+c^{-1}q^n)} .$$

Now the limiting expression on the left above formally converges to Ramanujan's divergent series in (3.11) (once a is replaced by c in (3.11)). Hence we shall replace Ramanujan's divergent series by the righthand side of the above expression. Once this is done, we replace a by a^2, then multiply throughout by $(a+a^{-1})^{-1}$ and as a result we find that (3.11) is equivalent to the assertion that

$$f(a) = 0$$

where

(3.30) $$f(a) = a \sum_{n=0}^{\infty} \frac{(-1)^n q^{n^2} (q;q^2)_n}{(-a^2;q^2)_{n+1}(-a^{-2}q^2;q^2)_n}$$

$$- \frac{1}{2} \frac{(q;q^2)}{(q^2;q^2)} \sum_{n=-\infty}^{\infty} \frac{(a+a^{-1})q^{n(n+1)/2}}{(1+a^2q^n)(1+a^{-2}q^n)}$$

$$- \frac{1}{2} \frac{(q)_\infty(q;q^2)_\infty a}{(-q)_\infty(-a^2)_\infty(-a^{-2}q)_\infty} .$$

The application of Lemma 2 to (3.30) is now routine. One choose $\rho_N = |q|^{-N/2-1/4}$, $N \geq 2$, for condition (ii). The identity $f(a) = f(a^{-1})$ is immediate. As usual the only potential problem comes from the possibility of simple poles.

CASE 1. $a \to \pm iq^{-N-1/2}$, (i.e. $a^2 \to -q^{-2N-1}$), and without loss of generality we assume $N \geqq 0$.

$$\lim_{a \to \pm iq^{-N-1/2}} (1+a^2q^{2N+1})f(a) =$$

$$= -\frac{1}{2}\ \frac{(q;q^2)_\infty}{(q^2;q^2)_\infty}\ \frac{(\pm iq^{-N-1/2})(1-q^{2N+1})q^{(N+1)(2N+1)}(1+q^{2N+1})}{(1-q^{4N+2})}$$

$$-\frac{1}{2}\ \frac{(q)_\infty(q;q^2)_\infty(\pm iq^{-N-1/2})}{(-q)_\infty(q^{-2N-1})_{2N+1}(q)_\infty(q^{2N+2})_\infty}$$

$$= -\frac{1}{2}\ \frac{(q;q^2)_\infty(\pm iq^{-N-1/2})q^{(N+1)(2N+1)}}{(-q)_\infty(q)_\infty}$$

$$+\frac{1}{2}\ \frac{(q)_\infty(q;q^2)_\infty(\pm iq^{-N-1/2})q^{(N+1)(2N+1)}}{(q)_\infty^2(-q)_\infty} = 0.$$

CASE 2. $a \to \pm iq^{-N}$, (i.e. $a^2 \to -q^{-2N}$), and without loss of generality we assume $N > 0$.

$$\lim_{a \to \pm iq^{-N}} (1+a^2q^{2N})f(a) =$$

$$= (\pm iq^{-N}) \sum_{n=N}^{\infty} \frac{(-1)^n q^{n^2}(q;q^2)_n}{(q^{-2N};q^2)_N(q^2;q^2)_{n-N}(q^{2N+2};q^2)_n}$$

$$-\frac{1}{2}\ \frac{(\pm iq^{-N})(q;q^2)_\infty(1-q^{2N})q^{N(2N+1)}(1+q^{2N})}{(q^2;q^2)_\infty(1-q^{4N})}$$

$$-\frac{1}{2}\ \frac{(q)_\infty(q;q^2)_\infty(\pm iq^{-N})}{(-q)_\infty(q^{-2N})_{2N}(q)_\infty(q^{2N+1})_\infty}$$

$$= (\pm iq^{-N})(-1)^N q^{N^2+N} \sum_{n=0}^{\infty} \frac{(-1)^{n+N} q^{n^2+2nN+N^2}(q;q^2)_{n+N}}{(q^2;q^2)_n(q^2;q^2)_{n+2N}}$$

$$-\frac{(\pm iq^{-N})(q;q^2)_\infty q^{N(2N+1)}}{(q^2;q^2)_\infty}$$

$$= \frac{(\pm iq^{-N})q^{2N^2+N}(q;q^2)_N}{(q^2;q^2)_{2N}} \sum_{n=0}^{\infty} \frac{(-1)^n q^{n^2+2nN}(q^{2N+1};q^2)_n}{(q^2;q^2)_n(q^{4N+2};q^2)_n}$$

$$-\frac{(\pm iq^{-N})q^{2N^2+N}(q;q^2)_\infty}{(q^2;q^2)_\infty}$$

$$= \frac{(\pm iq^{-N})q^{2N^2+N}(q;q^2)_\infty}{(q^2;q^2)_\infty} - \frac{(\pm iq^{-N})q^{2N^2+N}(q;q^2)_\infty}{(q^2;q^2)_\infty} = 0.$$

CASE 3. $a \to \pm i$, (i.e. $a^2 \to -1$).

$$\lim_{a \to \pm i} (1+a^2)f(a)$$

$$= (\pm i) \sum_{n=0}^{\infty} \frac{(-1)^n q^{n^2}(q;q^2)_n}{(q^2;q^2)_n^2}$$

$$-\frac{1}{2}\ \frac{(\pm i)(q;q^2)_\infty}{(q^2;q^2)_\infty} - \frac{1}{2}\ \frac{(q)_\infty(q;q^2)_\infty(\pm i)}{(-q)\ (q)_\infty^2}$$

$$= \frac{(\pm i)(q;q^2)_\infty}{(q^2;q^2)_\infty} - \frac{1}{2}\frac{(\pm i)(q;q^2)_\infty}{(q^2;q^2)_\infty} - \frac{1}{2}\frac{(\pm i)(q;q^2)_\infty}{(q^2;q^2)_\infty}$$

(by [3; p. 20, Col. 2.4])

$$= 0.$$

Hence we see by Lemma 2 that the $f(a)$ in (3.30) is identically zero, and so, therefore, is (3.11) proved valid.

<u>PROOF OF (3.12)</u>. This identity (which incidentally was used in Case 3 of (3.11) above) is an immediate corollary of the q-analog of Gauss's theorem [3; p. 20, Cor. 2.4]. It is also the case $a = -1$ of (3.8) once one observes that

$$\frac{\left(\sum_{n=-\infty}^{\infty} (-1)^n q^{n^2}\right)(q;q^2)_\infty}{(q)_\infty^2} = \frac{(q)_\infty (q;q^2)_\infty}{(-q)(q)_\infty^2}$$

$$= \frac{(q;q^2)_\infty}{(q^2;q^2)_\infty}$$

$$= \frac{1}{\sum_{n=0}^{\infty} q^{n(n+1)/2}},$$

by [3; p. 23, eq. (2.2.13)].

<u>PROOF OF (3.13)</u>. While this result is included here because of the relationship of the series therein to the other identities in this paper, it is a direct corollary of the main theorem of [7] (see [1] for an elegant alternative proof). Indeed if we replace q by q^2 in Theorem 1 of [7] and then set $B = q$, $a = bq$ and let $A \to 0$, we find after multiplication by $q(1+b)(1+bq)^{-1}$ that

$$(1+b)\sum_{n=0}^{\infty} \frac{(q;q^2)_n q^{2n+1}}{(-bq)_{2n+1}}$$

$$= -\frac{(q;q^2)_\infty}{(-bq)_\infty} \sum_{m=0}^{\infty} \frac{(-1)^m q^{m^2}}{(-q^2b^{-1};q^2)_m}$$

$$+ (1+b) \sum_{m=0}^{\infty} \frac{(-q/b;q^2)_m(-b)^m}{(-b^{-1}q^2;q^2)_m}$$

$$(3.31) \qquad = -\frac{(q;q^2)_\infty}{(-bq)_\infty} \sum_{m=0}^{\infty} \frac{(-1)^m q^{m^2}}{(-q^2b^{-1};q^2)_m}$$

$$+ \sum_{m=0}^{\infty} \frac{(-1)^m q^{m^2}(q;q^2)_m}{(-bq^2;q^2)_m(-b^{-1}q^2;q^2)_m}$$

(by [18; p. 174, eq. (10.1)] with q replaced by q^2, then $a = q$, $b = q/\tau$, $c = q^2$, $e = -q^2/b$, $f = -bq^2$, $\tau \to 0$),

and this is just (3.13) with a replaced by b.

Now that we have seen the intricacy of these results of Ramanujan we proceed to view them in terms of their interrelationship with the transformations presented in Section 2.

4. Extensions and modular transformations of Ramanujan's functions.

While the approach of the last section provides uniformity, it fails to place these results in the basic hypergeometric hierarchy. In fact many of the results of Section 3 may be viewed as specializations of Watson's q-analog of Whipple's theorem [21; p. 100, eq. (3.4.1.5)]: (N is a nonnegative integer)

$$(4.1) \qquad {}_8\phi_7 \left[\begin{matrix} \alpha, & q\sqrt{\alpha}, & -q\sqrt{\alpha}, & b, & c, & d, & e, & q^{-N} & ; q, \frac{\alpha^2 q^{N+2}}{bcde} \\ \sqrt{\alpha}, & -\sqrt{\alpha}, & \frac{\alpha q}{b}, & \frac{\alpha q}{c}, & \frac{\alpha q}{d}, & \frac{\alpha q}{e}, & \alpha q^{N+1} & \end{matrix} \right]$$

$$= \frac{(\alpha q)_N (\frac{\alpha q}{de})_N}{(\frac{\alpha q}{e})_N (\frac{\alpha q}{d})_N} \; {}_4\phi_3 \left[\begin{matrix} \frac{\alpha q}{bc}, & d, & e, & q^{-N} & ; q,q \\ \frac{\alpha q}{b}, & \frac{\alpha q}{c}, & \frac{deq^{-N}}{\alpha} & & \end{matrix} \right]$$

where

(4.2) $${}_r\phi_s \left[\begin{matrix} a_1, a_2, \ldots, a_r; \; q, t \\ b_1, b_2, \ldots, b_s \end{matrix} \right]$$

$$= \sum_{n=0}^{\infty} \frac{(a_1)_n (a_2)_n \ldots (a_r)_n \, t^n}{(q)_n (b_1)_n \ldots (b_s)_n} .$$

If in (4.1) we replace q by q^2 then set $\alpha = 1$, $b = a$, $c = a^{-1}$, $d = q$, $e \to \infty$, $N \to \infty$, we obtain a result easily seen to be equivalent to (3.4).

Next in (4.1), we may replace q by q^2 then set $\alpha = q^2$, $b = qa$, $c = qa^{-1}$, $d = -q$, $e \to \infty$, $N \to \infty$ we obtain an extension of (3.2) which, in fact, explicitly specializes to (3.7) when $a = 1$.

Most noteworthy in all these observations however is the fact that neither $M_3(q)$ nor a generalization appears in any of Ramanujan's work. This is of course easily overcome once the role of (4.1) in this work has been established. Actually we find a generalization of $M_3(q)$ if in (4.1) we replace q by q^2 and then set $\alpha = q^2$, $b = aq$, $c = a^{-1}q$, $d = -q^2$, $e \to \infty$, $N \to \infty$. The case $a = 1$ is thus

(4.3) $$M_3(q) = \frac{(q^2;q^2)_\infty}{(-q^2;q^2)_\infty} \sum_{n=0}^{\infty} \frac{(-q^2;q^2)_n q^{n^2+n}}{(q)_{2n+1}(q;q^2)_{n+1}}$$

$$\equiv \frac{(q^2;q^2)_\infty}{(-q^2;q^2)_\infty} \beta(q),$$

where $\beta(q)$ was originally given by (1.8).

Once we have found $\beta(q)$ from (4.3), we see immediately that the four modular transformations (2.1), (2.9), (2.10) and (2.12) are now easily translated into transformations connecting $\mu(q)$, $\beta(q)$ and $\alpha(q)$. Indeed the transformations from Section 2 yield the following four results immediately once we recall that for

(4.4) $$\theta_4(0,q) = \sum_{n=-\infty}^{\infty} (-1)^n q^{n^2} = \frac{(q)_\infty}{(-q)_\infty},$$

and

(4.5) $$\frac{1}{2} q^{-1/8}\theta_2(0,\sqrt{q}) = \sum_{n=0}^{\infty} q^{n(n+1)/2} = \frac{(q^2;q^2)_\infty}{(q;q^2)_\infty},$$

we have [22; p. 263, eq. (8)]

(4.6) $$\theta_4(0,q_1^2) = \sqrt{\frac{\alpha}{2\pi}}\,\theta_2(0,\sqrt{q}) = q^{1/8}\sqrt{\frac{2\alpha}{\pi}} \sum_{n=0}^{\infty} q^{n(n+1)/2},$$

where $q = e^{-\alpha}$, $q_1 = e^{-\beta}$, $\alpha\beta = \pi^2$.

Hence from (2.1) we see that

(4.7) $$q^{-1/8}\mu(q) = 2\sqrt{\frac{2\pi}{\alpha}}\, q^{1/2}\, \beta(q_1) + \sqrt{\frac{2\alpha}{\pi}} \int_{-\infty}^{\infty} \frac{e^{-2\alpha x^2+\alpha x}}{e^{2\alpha x}+1}\, dx,$$

essentially a restatement of (1.9).

From (2.9), we obtain that

(4.8) $$q^{-1/8}\alpha(q) = \sqrt{\frac{2\pi}{\alpha}}\, q_1^{1/2}\, \beta(-q_1) - \sqrt{\frac{\alpha}{2\pi}} \int_{-\infty}^{\infty} \frac{e^{-2\alpha x^2+\alpha x}}{e^{2\alpha x}+1}\, dx.$$

From (2.10), it follows that

(4.9) $$q^{1/2}\beta(q) = \frac{1}{2}\sqrt{\frac{\pi}{2\alpha}}\, q_1^{-1/8}\, \mu(q_1) - 2\sqrt{\frac{\alpha}{2\pi}} \int_{0}^{\infty} \frac{e^{-2\alpha x^2+2\alpha x}}{e^{4\alpha x}+1}\, dx.$$

Finally, from (2.12), we see that

$$(4.10)\qquad q^{1/2}\beta(-q) = \sqrt{\frac{\pi}{2\alpha}}\, q_1^{-1/8}\,\alpha(q_1) + 2\sqrt{\frac{\alpha}{2\pi}} \int_0^\infty \frac{e^{-2\alpha x^2+2\alpha x}}{e^{4\alpha x}+1}\,dx$$

5. Numerical aspects. One of the most important ways of spotting where a particular function arises in an arithmetic or combinatorial setting is to examine the first few coefficients in its power series expansion. These types of functions have been found to have combinatorial applications. For example, the third order mock theta function

$$f(q) = \sum_{n=0}^{\infty} \frac{q^{n^2}}{(-q)_n^2}$$

is the generating function for the excess of the number of partitions of n with even rank (rank = largest part minus number of parts) over those with odd rank. This may be easily verified by following the Durfee square analysis of partitions [3; pp. 27-28] and keeping track of the rank as well. In order to facilitate possible subsequent applications of Ramanujan's functions we list the first 36 coefficients in the MacLauren series expansions of $\alpha(q)$ defined in equation (3.25), $\beta(q)$ defined in equation (1.8) and $\mu(q)$ defined in equation (1.7):

n	n^{th} coefficient in $\alpha(q)$	n^{th} coefficient in $\beta(q)$	n^{th} coefficient in $\mu(q)$
0	0	1	1
1	1	2	-1
2	2	4	1
3	3	6	2
4	5	10	-1
5	8	16	-4
6	11	25	1
7	16	38	6
8	23	57	-3
9	31	80	-7
10	43	113	6
11	58	156	8
12	74	210	-5
13	95	278	-11

n	n^{th} coefficient in $\alpha(q)$	n^{th} coefficient in $\beta(q)$	n^{th} coefficient in $\mu(q)$
14	122	362	3
15	151	462	15
16	186	586	-8
17	229	732	-18
18	274	904	13
19	329	1106	20
20	394	1344	-9
21	460	1616	-24
22	537	1931	8
23	626	2288	32
24	722	2690	-17
25	832	3150	-38
26	953	3671	23
27	1080	4248	41
28	1223	4896	-21
29	1383	5612	-50
30	1552	6407	20
31	1737	7290	62
32	1940	8267	-29
33	2153	9332	-71
34	2389	10500	41
35	2648	11776	81

6. Conclusion. Our object in this paper has been to show the relationship of one particular set of identities from Ramanujan's "Lost" Notebook to the Mordell integrals. Obviously the study begun in Section 4 can easily be extended to include a number of other functions just as G.N. Watson [24] did for the third order mock theta functions. Indeed we have primarily restricted our attention to the case a = -1 when specializing the identities of Section 3. However it is clear that interesting real results arise whenever a is an algebraic unit and at most a quadratic irrationality. Thus other interesting choices for a are i, $e^{2\pi i/3}$, $-e^{2\pi i/3}$. Certainly Ramanujan must have recognized this in some sense, since he often specialized identities of this nature with these particular values (see [7]).

The work in Section 4 clearly indicates that one could treat these functions and their identities via the modular function techniques utilized by Mordell in his fundamental papers on Mordell integrals [16], [17]. We have not chosen this route since it is unlikely to have been the one that Ramanujan followed, and the technique followed in Section 3 wherein Lemma 2 played a central role allows the establishment of a diverse collection of results prior to a subsequent careful classification via related Mordell integrals.

In closing, we must emphasize the debt Sections 2 and 4 of this paper owe to G.N. Watson's paper on the third order mock theta functions [24]. Indeed the functions introduced here could be subjected to examination under the full modular group in the way that Watson's work was extended for the third order mock theta functions [2], [13]. Subsequent papers in the series [7], [8], [9] will make even clearer the role that Mordell integrals play in a complete treatment of Ramanujan's "Lost" Notebook.

REFERENCES

1. R.P. Agarwal, On the paper "A 'lost' notebook of Ramanujan I" of G.E. Andrews, Advances in Math., (to appear).
2. G.E. Andrews, On the theorems of Watson and Dragonette for Ramanujan's mock theta functions, Amer. J. Math., 88 (1966), 454-490.
3. G.E. Andrews, The Theory of Partitions, Encyclopedia of Mathematics and Its Applications, Vol. 2, G.-C. Rota ed., Addison-Wesley, Reading, 1976.
4. G.E. Andrews, An introduction to Ramanujan's "lost" notebook, Amer. Math. Monthly, 86 (1979), 89-108.
5. G.E. Andrews, Ramanujan and his "lost" notebook, Vinculum, 16 (1979), 91-94.
6. G.E. Andrews, Partitions: Yesterday and Today, New Zealand Math. Soc., Wellington, 1979.
7. G.E. Andrews, Ramanujan's "lost" notebook I: partial theta functions, Advances in Math., (to appear).
8. G.E. Andrews, Ramanujan's "lost" notebook II: θ-function expansions, Advances in Math., (to appear).

9. G.E. Andrews, Ramanujan's "lost" notebook III: the Rogers-Ramanujan continued fraction, Advances in Math., (to appear).

10. W.N. Bailey, On the basic bilateral hypergeometric series ${}_2\psi_2$, Quart. J. Math., Oxford Ser., 1 (1950), 194-198.

11. R. Bellman, A Brief Introduction to Theta Functions, Holt, Rinehart and Winston, New York, 1961.

12. E.T. Copson, An Introduction to the Theory of Functions of a Complex Variable, Oxford University Press, London, 1935.

13. L.A. Dragonette, Some asymptotic formulae for the mock theta series of Ramanujan, Amer. J. Math., 72 (1952), 474-500.

14. C.G.J. Jacobi, Fundamenta Nova Theoriae Functionum Ellipticarum, Regiomonti, fratrum Bornträger, 1829 (Reprinted: in vol. 1, 49-239 of Jacobi's, Gesammelte Werke, Reimer, Berlin 1881-1891, now by Chelsea, New York, 1969).

15. N. Levinson, More than one third of zeros of Riemann's zeta-function are on $\sigma = 1/2$, Advances in Math., 13 (1974), 383-436.

16. L.J. Mordell, The value of the definite integral $\int_{-\infty}^{\infty} e^{at^2+bt}dt/(e^{ct}+d)$, Quarterly Journal of Math., 68 (1920), 329-342.

17. L.J. Mordell, The definite integral $\int_{-\infty}^{\infty} e^{ax^2+bx}dx/(e^{cx}+d)$ and the analytic theory of numbers, Acta Math. Stockholm, 61 (1933), 323-360.

18. S. Ramanujan, Collected Papers, Cambridge University Press, London and New York, 1927; reprinted Chelsea, New York.

19. D.B. Sears, On the transformation theory of basic hypergeometric functions, Proc. London Math. Soc. (2), 53 (1951), 158-180.

20. C.L. Siegel, Über Riemanns Nachlass zur analytischen Zahlentheorie, Quellen und Studien zur Geschichte der Mathematik, Astronomie und Physik, 2 (1933), 45-80.

21. L.J. Slater, Generalized Hypergeometric Functions, Cambridge University Press, London, 1966.

22. J. Tannery and J. Molk, Éléments de la Théorie des Fonctions Elliptiques, Vol. II., Gauthier-Villars, Paris, 1896 (Reprinted: Chelsea, New York, 1972).

23. J. Tannery and J. Molk, Éléments de la Théorie des Fonctions Elliptiques, Vol. III, Gauthier-Villars, Paris 1898 (Reprinted: Chelsea, New York, 1972).

24. G.N. Watson, The final problem: an account of the mock theta functions, J. London Math. Soc., 11 (1936), 55-80.

The Pennsylvania State University
University Park, Pennsylvania 16802

CHAPTER 5 OF RAMANUJAN'S SECOND NOTEBOOK

Bruce C. Berndt* and B.M. Wilson+

Dedicated to Emil Grosswald, with respect and admiration.

Chapter 5 of Ramanujan's second notebook contains more number theory than any of the remaining 20 chapters. Of the 94 formulas or statements of theorems in Chapter 5, the great majority pertain to Bernoulli numbers, Euler numbers, Eulerian polynomials and numbers, and the Riemann zeta-function. As is to be expected, most of these results are not new. The geneses of Ramanujan's first published paper [28] (on Bernoulli numbers) and fourth published paper [29] (on sums connected with the Riemann zeta-function) are found in Chapter 5.

We shall not follow the convention used by Ramanujan for the Bernoulli numbers B_n, $0 \le n < \infty$, but instead we employ the contemporary definition

$$(0.1) \qquad \frac{x}{e^x-1} = \sum_{n=0}^{\infty} \frac{B_n}{n!} x^n, \qquad |x| < 2\pi.$$

Since most of Ramanujan's results involving Bernoulli numbers are well known, we shall generally refrain from supplying proofs. The books of Bromwich [11], Nielsen [24], Nörlund [26], and Uspensky and Heaslet [35], for example, are standard references which contain most of Ramanujan's discoveries about Bernoulli numbers.

We also employ the current convention for the Euler numbers E_n, $0 \le n < \infty$; thus, $E_{2n+1} = 0$, $n \ge 0$, while E_{2n}, $n \ge 0$, is defined by

$$(0.2) \qquad \sec x = \sum_{n=0}^{\infty} \frac{(-1)^n E_{2n}}{(2n)!} x^{2n}, \qquad |x| < \pi/2,$$

which again differs from that of Ramanujan. We shall omit proofs of those formulas involving Euler numbers that are familiar.

The notations for the Eulerian polynomials and numbers are not particularly standard, and so we shall employ Ramanujan's notations. Define the Eulerian polynomials $\psi_n(p)$, $0 \le n < \infty$, $p \ne -1$, by

---

* Research partially supported by National Science Foundation grant no. MCS-7903359.

+ Deceased, March 18, 1935.

(0.3) $$\frac{1}{e^x+p} = \sum_{n=0}^{\infty} \frac{(-1)^n \psi_n(p) x^n}{n!(p+1)^{n+1}}, \quad |x| < |\mathrm{Log}(-p)|.$$

It will be shown in the sequel that, indeed, $\psi_n(p)$ is a polynomial in p of degree n-1. In the notation of Carlitz's paper [12], which is perhaps the most extensive source of information about Eulerian polynomials and numbers, $R_n[-p] = \psi_n(p)$. The Eulerian numbers A_{nk}, $1 \le k \le n$, are generally defined by [12]

$$\psi_n(p) = \sum_{k=1}^{n} A_{nk}(-p)^{k-1}.$$

In Ramanujan's notation $A_{nk} = F_k(n)$; see (6.1) below. The Eulerian polynomials and numbers were first introduced by Euler [19] in 1755. Carlitz and his colleagues [12-18] have written extensively about Eulerian polynomials and numbers and certain generalizations thereof. See also a paper of Frobenius [20], [21, pp. 809-847], for much historical information, and Riordan's book [32], which contains combinatorial applications. In particular, A_{nk} is equal to the number of permutations of $\{1,2,\ldots,n\}$ with exactly k rises including a conventional one at the left. Some of Ramanujan's theorems on Eulerian polynomials and numbers appear to be new. Also, since most results in this area are not well known and the proofs are very short, we shall give most proofs.

The content of this paper is based partially upon notes left by the second author upon his death in 1935.

In Entries 1(i), 1(ii), and 3 it is assumed that, for $|h|$ sufficiently small, f can be expanded in the form

(1.1) $$f(x) = \sum_{n=0}^{\infty} a_n h^n \phi^{(n)}(x),$$

where a_n, $0 \le n < \infty$, is independent of ϕ.

ENTRY 1(i). Let $f(x+h)-f(x) = h\phi'(x)$. Then, in the notation of (1.1), $a_n = B_n/n!$, $0 \le n < \infty$, where B_n denotes the nth Bernoulli number.

PROOF. Since a_n, $0 \le n < \infty$, is independent of ϕ, let $\phi(x) = e^x$. Then, by (1.1),

$$he^x = f(x+h)-f(x) = e^x(e^h-1) \sum_{n=0}^{\infty} a_n h^n,$$

i.e.,

$$\sum_{n=0}^{\infty} a_n h^n = \frac{h}{e^h-1}. \tag{1.2}$$

Comparing (1.2) with (0.1), we complete the proof.

ENTRY 1(ii). Let $f(x+h)+f(x) = h\phi'(x)$. Then $a_n = (1-2^n)B_n/n!$, $0 \le n < \infty$.

PROOF. Again, without loss of generality, we take $\phi(x) = e^x$. Then, from (1.1),

$$he^x = f(x+h)+f(x) = e^x(e^h+1) \sum_{n=0}^{\infty} a_n h^n.$$

Thus, from (0.1),

$$\sum_{n=0}^{\infty} a_n h^n = \frac{h}{e^h+1} = \frac{h}{e^h-1} - \frac{2h}{e^{2h}-1} = \sum_{n=0}^{\infty} \frac{(1-2^n)B_n h^n}{n!}, \tag{1.3}$$

and the desired result follows.

Suppose that f is a solution of either the difference equation of Entry 1(i) or of Entry 1(ii). Then, in general, the series on the right side of (1.1) diverges. However, under suitable conditions [26, pp. 58-60], the series on the right side of (1.1) represents the function asymptotically as h tends to 0.

Let ϕ be any function defined on $(-\infty,\infty)$. In anticipation of Entries 2(i) and 2(ii), define

$$F_{2n+1}(x) = \phi(x) + \sum_{k=1}^{n} (-1)^k \frac{(n!)^2}{(n+k)!(n-k)!} \{\phi(x+k)+\phi(x-k)\}, \tag{2.1}$$

where n is a nonnegative integer.

ENTRY 2(i). Let $f(x+1)-f(x-1) = 2\phi'(x)$, where ϕ is a polynomial. Then there exists a polynomial solution f of the form

$$f(x) = \sum_{n=0}^{\infty} \frac{F_{2n+1}(x)}{2n+1}, \tag{2.2}$$

where F_{2n+1} is defined by (2.1).

ENTRY 2(ii). Let $f(x+1)+f(x-1) = 2\phi(x)$, where ϕ is a polynomial. Then there exists a polynomial solution f of the form

(2.3) $$f(x) = \sum_{n=0}^{\infty} 2^{-n}\binom{2n}{n} F_{2n+1}(x),$$

where F_{2n+1} is defined by (2.1).

Before embarking upon the proofs of these theorems we shall make several comments.

First, the series on the right sides of (2.2) and (2.3) are, in fact, finite series. This will be made evident in the proofs of (2.2) and (2.3). In some cases, the restrictions that f and ϕ be polynomials may be lifted. However, in general, if f is a solution of the difference equation in Entry 2(i) or Entry 2(ii), the series on the right side of (2.2) or (2.3), respectively, does not converge.

Ramanujan actually considers the seemingly more general difference equations $f(x+h)-f(x-h) = 2h\phi'(x)$ and $f(x+h)+f(x-h) = 2\phi(x)$ in Entries 2(i) and 2(ii), respectively. However, it is no loss of generality to assume that $h = 1$. For suppose, for example, that Entry 2(i) has been established. Put $y = xh$, $f(x) = g(y)$, and $\phi(x) = \psi(y)$. Then the difference equation of Entry 2(i) becomes $g(y+h)-g(y-h) = 2h\psi'(y)$.

We are very grateful to Doron Zeilberger for suggesting the following method of proof for Entries 2(i) and 2(ii). Since the proofs use operator calculus, we need to define a couple of operators. As customary, let D denote the differential operator and define E by $Ef(x) = f(x+1)$. Note that $E = e^D$.

<u>PROOF OF ENTRY 2(i)</u>. We shall first derive another formulation of $F_{2n+1}(x)$. From (2.1),

(2.4) $$F_{2n+1}(x) = \sum_{k=-n}^{n} \frac{(-1)^k(n!)^2E^k\phi}{(n+k)!(n-k)!}$$

$$= \frac{(n!)^2(-E)^n}{(2n)!} \sum_{j=0}^{2n} \binom{2n}{j} (-E^{-1})^{2n-j}\phi$$

$$= \frac{(n!)^2(-1)^n}{(2n)!} e^{nD}(1-e^{-D})^{2n}\phi$$

$$= \frac{(-1)^n (n!)^2}{(2n)!} (2 \sinh(D/2))^{2n} \phi \quad .$$

In operator notation, the given difference equation is $(E-E^{-1})f = 2D\phi$, or $f = 2D\phi/(e^D - e^{-D})$. Hence, in operator notation, the proposed identity may be written as

$$(2.5) \qquad \sum_{n=0}^{\infty} \frac{(-1)^n (n!)^2}{(2n+1)!} (2 \sinh(D/2))^{2n+1} \phi = \frac{2D\phi}{e^{D/2}+e^{-D/2}} .$$

Set $y = \sinh(D/2) = (\sqrt{E} - 1/\sqrt{E})/2$. (Note that $(\sqrt{E} - 1/\sqrt{E})^n \phi = 0$ if n exceeds the degree of ϕ , and so the series in (2.5), indeed, does terminate.) A short calculation shows that $\sqrt{y^2+1} = (e^{D/2}+e^{-D/2})/2$. Hence, it suffices to prove that

$$(2.6) \qquad \sqrt{y^2+1} \sum_{n=0}^{\infty} \frac{(-1)^n (n!)^2 (2y)^{2n+1} \phi}{(2n+1)!} = 2(\sinh^{-1} y)\phi \quad .$$

Obviously, (2.6) is valid for $y = 0$. Thus, it suffices to show that the derivatives of both sides of (2.6) are equal. After taking derivatives in (2.6) and multiplying both sides by $\sqrt{y^2+1}/2$, we find that it is sufficient to prove that

$$(2.7) \qquad \sum_{n=0}^{\infty} \frac{(-1)^n (n!)^2 2^{2n} y^{2n+2} \phi}{(2n+1)!} + (y^2+1) \sum_{n=0}^{\infty} \frac{(-1)^n (n!)^2 (2y)^{2n} \phi}{(2n)!} = \phi \; .$$

Combining the series together on the left side of (2.7), we easily obtain ϕ after a short calculation. This completes the proof.

PROOF OF ENTRY 2(ii). Using (2.4), we find that the left side of (2.3) may be written as

$$\sum_{n=0}^{\infty} (-2 \sinh^2(D/2))^n \phi = \frac{1}{1+2 \sinh^2(D/2)} \phi$$

$$= \frac{1}{\cosh D} \phi \; .$$

Since the given difference equation in operator notation is $(E+E^{-1})f = 2\phi$ or $f = \phi/\cosh D$, the desired result follows.

PROOF 3. Let $f(x+h)+pf(x) = \phi(x)$, where $p \neq -1$. Then, in the notation of (1.1),

$$a_n = \frac{(-1)^n \psi_n(p)}{n!(p+1)^{n+1}}, \quad 0 \le n < \infty,$$

where $\psi_n(p)$ is defined by (0.3).

PROOF. As before, since a_n, $0 \le n < \infty$, is independent of ϕ, we shall let $\phi(x) = e^x$. Using (1.1) and proceeding in the same fashion as in the proof of Entry 1(i), we find that

$$\sum_{n=0}^{\infty} a_n h^n = \frac{1}{e^h+p}. \tag{3.1}$$

Comparing (3.1) with (0.3), we deduce the desired result.

Appell [3], Brodén [9], [10], and Picard [27] have studied periodic solutions of $f(x+h)+pf(x) = \phi(x)$. For a discussion of solutions to many types of difference equations and for numerous references, see Nörlund's article in the Encyklopädie [25].

ENTRY 4. For $|p| < 1$ and $n \ge 0$,

$$(p+1)^{-n-1}\psi_n(p) = \sum_{k=0}^{\infty} (k+1)^n(-p)^k. \tag{4.1}$$

PROOF. for $|e^{-x}p| < 1$,

$$\frac{1}{e^x+p} = \sum_{k=0}^{\infty} e^{-(k+1)x}(-p)^k = \sum_{k=0}^{\infty} (-p)^k \sum_{n=0}^{\infty} \frac{(k+1)^n(-x)^n}{n!} \tag{4.2}$$

$$= \sum_{n=0}^{\infty} \frac{(-x)^n}{n!} \sum_{k=0}^{\infty} (k+1)^n(-p)^k.$$

If we now equate coefficients of x^n, $n \ge 0$, in (0.3) and (4.2), we deduce (4.1). However, this procedure is valid only when the series in (0.3) and (4.2) have a common domain of convergence in the complex x-plane. Put $p = re^{i\alpha}$, $0 < r < 1$, $-\pi < \alpha \le \pi$. Then the series in (0.3) converges if and only if $|x|^2 < (\text{Log } r)^2+\alpha^2$. The double series in (4.2) are absolutely convergent when $\text{Re } x > \text{Log } r$. Thus, there is a common domain of convergence, and the proof is complete.

ENTRY 5. We have $\psi_0(p) = 1$, while for $n \ge 1$,

(5.1) $$(-1)^{n+1}p(p+1)^{-n}\psi_n(p) = \sum_{k=0}^{n} (-1)^k\binom{n}{k}(p+1)^{-k}\psi_k(p).$$

PROOF. From (0.3),

$$1 = \left\{p + \sum_{n=0}^{\infty} \frac{x^n}{n!}\right\} \sum_{n=0}^{\infty} \frac{(-1)^n\psi_n(p)x^n}{n!(p+1)^{n+1}} .$$

Equating coefficients on both sides, we find that $\psi_0(p) = 1$ and that (5.1) holds for $n \geq 1$.

For $n \geq 1$, the recursion formula (5.1) may be written in the form

(5.2) $$\psi_n(p) = \sum_{j=0}^{n-1} (-1)^j\binom{n}{j+1} (p+1)^j\psi_{n-j-1}(p),$$

where we have set $j = n-k-1$ in (5.1). Note that $\psi_1(p) = 1$. By inducting on n, we see that $\psi_n(p)$ is a polynomial in p of degree n-1. Thus, after Ramanujan, we write

(6.1) $$\psi_n(p) = \sum_{k=0}^{n-1} F_{k+1}(n)(-p)^k.$$

ENTRY 6. Let $1 \leq r \leq n$. Then

(i) $F_r(n) = F_{n-r+1}(n)$,

(ii) $\sum_{k=0}^{r-1} \binom{n+k}{k}F_{r-k}(n) = r^n$,

and

(iii) $F_r(n) = \sum_{k=0}^{r-1} (-1)^k\binom{n+1}{k}(r-k)^n$.

PROOF OF (i). Since $\psi_0(p) = 1$, (0.3) yields

$$\frac{e^x-1}{(p+1)(e^x+p)} = \frac{1}{p+1} - \frac{1}{e^x+p} = \sum_{n=1}^{\infty} \frac{(-1)^{n-1}\psi_n(p)x^n}{n!(p+1)^{n+1}} .$$

Replacing x by -x and p by 1/p, we obtain

$$\sum_{n=1}^{\infty} \frac{(-1)^{n-1}\psi_n(p)x^n}{n!(p+1)^n} = \frac{e^x-1}{e^x+p} = \sum_{n=1}^{\infty} \frac{p^{n-1}\psi_n(1/p)x^n}{n!(p+1)^n} .$$

Equating coefficients of x^n, we find that

(6.2) $$(-1)^{n-1}\psi_n(p) = p^{n-1}\psi_n(1/p), \quad n \geq 1.$$

Using (6.1) in (6.2) and equating coefficients of p^{r-1} on both sides, we complete the proof of (i).

PROOF OF (ii). By (4.1) and (6.1), if $|p| < 1$,

$$\sum_{k=1}^{\infty} k^n(-p)^{k-1} = \sum_{j=1}^{n} F_j(n)(-p)^{j-1} \sum_{k=0}^{\infty} \binom{n+k}{k}(-p)^k.$$

Equating the coefficients of p^{r-1} on both sides, we deduce (ii).

PROOF OF (iii). Again, by (4.1) and (6.1), for $|p| < 1$,

$$\sum_{k=0}^{n-1} F_{k+1}(n)(-p)^k = \sum_{k=0}^{\infty} (k+1)^n(-p)^k \sum_{j=0}^{n+1} \binom{n+1}{j} p^j.$$

Equate coefficients of p^{r-1} on both sides to deduce (iii).

The statement of (ii) in the notebooks [31, vol. II, p. 48] is incorrect; replace n by n+1 on the left side of (ii) in [31]. Entry 6(ii) is due to Worpitzky [38] while (iii) is due to Euler [19].

ENTRY 7. If $n \geq 0$ and $0 < x < 1-e^{-2\pi}$, then

(7.1) $$\psi_n(x-1) = \frac{x^{n+1}}{1-x}\left\{\frac{n!}{(\text{Log}\,\frac{1}{1-x})^{n+1}} + (-1)^n \sum_{k=n+1}^{\infty} \frac{B_k}{k(k-n-1)!}\left(\text{Log}\,\frac{1}{1-x}\right)^{k-n-1}\right\},$$

where B_k denotes the kth Bernoulli number.

This formulation of Entry 7 is not the same as Ramanujan's version. Ramanujan claims that $\psi_n(x-1)$ is the "integral part" of

(7.2) $$\frac{x^{n+1}}{1-x}\left\{\frac{n!}{(\text{Log}\,\frac{1}{1-x})^{n+1}} + (-1)^n \frac{B_{n+1}}{n+1}\right\}.$$

Since $\psi_n(x-1)$ is generally not an integer, we are not sure what Ramanujan intends. Perhaps Ramanujan is indicating that the primary contribution to $\psi_n(x-1)$ in (7.1) is (7.2), especially if x is small.

PROOF. By (0.1), if $0 < x < 2\pi$,

$$\sum_{k=1}^{\infty} e^{-kx} = \frac{e^{-x}}{1-e^{-x}} = \frac{1}{e^{x}-1} = \sum_{k=0}^{\infty} \frac{B_k}{k!} x^{k-1}. \tag{7.3}$$

Differentiate both sides of (7.3) n times with respect to x and multiply both sides by $(-1)^n$ to get

$$\sum_{k=1}^{\infty} k^n e^{-kx} = \frac{n!}{x^{n+1}} + (-1)^n \sum_{k=n+1}^{\infty} \frac{B_k}{k(k-n-1)!} x^{k-n-1}. \tag{7.4}$$

Now replace x by $-\mathrm{Log}(1-x)$ in (7.4). We then observe that the left side of (7.4) is $(1-x)x^{-n-1}\psi_n(x-1)$ by Entry 4. This completes the proof of (7.1).

ENTRY 8. For $n \geq 1$, $\psi_n(-1) = n!$ and $\psi_n(1) = 2^{n+1}(2^{n+1}-1)B_{n+1}/(n+1)$. Furthermore,

$$\psi_0(p) = \psi_1(p) = 1,$$
$$\psi_2(p) = 1-p,$$
$$\psi_3(p) = 1-4p+p^2,$$
$$\psi_4(p) = 1-11p+11p^2-p^3,$$
$$\psi_5(p) = 1-26p+66p^2-26p^3+p^4,$$
$$\psi_6(p) = 1-57p+302p^2-302p^3+57p^4-p^5,$$

and

$$\psi_7(p) = 1-120p+1191p^2-2416p^3+1191p^4-120p^5+p^6.$$

PROOF. Since $\psi_0(-1) = 1$, the values for $\psi_n(-1)$, $n \geq 1$, follow immediately from (5.2) and induction on n.

Next, by (0.1), for $|x| < \pi$,

$$\sum_{k=0}^{\infty} \frac{(1-2^k)B_k x^{k-1}}{k!} = \frac{1}{e^x-1} - \frac{2}{e^{2x}-1}$$

$$= \frac{1}{e^x+1} = \sum_{n=0}^{\infty} \frac{(-1)^n \psi_n(1) x^n}{2^{n+1} n!}.$$

The formula for $\psi_n(1)$ now readily follows from comparing coefficients of x^n above.

By Entry 5 and (5.2), we previously had shown that $\psi_0(p) = \psi_1(p) = 1$. To calculate the remaining polynomials, we employ Entry 6(i) and the recursion formula

$$F_k(n) = kF_k(n-1)+(n-k+1)F_{k-1}(n-1), \tag{8.1}$$

where $2 \leq k \leq n$. Ramanujan does not state (8.1), but he indicates that he was in possession of such a formula.

To prove (8.1), we employ Entry 6(iii) to get

$$\begin{aligned} &kF_k(n-1)+(n-k+1)F_{k-1}(n-1) \\ &= k\sum_{j=0}^{k-1} (-1)^j \binom{n}{j} (k-j)^{n-1}+(n-k+1)\sum_{j=0}^{k-2} (-1)^j \binom{n}{j} (k-j-1)^{n-1} \\ &= k^n+\sum_{j=1}^{k-1} (-1)^j(k-j)^{n-1}\left\{k\binom{n}{j} -(n-k+1)\binom{n}{j-1}\right\} \\ &= \sum_{j=0}^{k-1} (-1)^j \binom{n+1}{j} (k-j)^n = F_k(n). \end{aligned}$$

COROLLARY 1. Let $f(x)$ denote the solution found in Entry 3. Then $f(x)$ is the term independent of n in

$$\frac{\sum_{k=0}^{\infty} \phi^{(k)}(x)/n^k}{e^{nh}+p}, \tag{8.2}$$

where it is understood that the series in the numerator above does, indeed, converge.

PROOF. Expand $1/(e^{nh}+p)$ by (0.3). Upon the multiplication of the two series in (8.2), the proposed result readily follows.

Corollary 2 is a complete triviality and not worth recording here.

COROLLARY 3. If n is even and positive, then $\psi_n(p)$ is divisible by $1-p$.

PROOF. The result follows readily from (6.1) and Entry 6(i).

COROLLARY 4. For $|x| < |\mathrm{Log}(-p)|$,

$$\frac{\cos x + p}{p^2+2p\cos x + 1} = \sum_{n=0}^{\infty} \frac{(-1)^n \psi_{2n}(p) x^{2n}}{(2n)!(p+1)^{2n+1}}.$$

COROLLARY 5. For $|x| < |\mathrm{Log}(-p)|$,

$$\frac{\sin x}{p^2+2p\cos x + 1} = \sum_{n=0}^{\infty} \frac{(-1)^n \psi_{2n+1}(p) x^{2n+1}}{(2n+1)!(p+1)^{2n+2}}.$$

4 AND 5. Replacing x by ix in (0.3), we have

$$\frac{1}{e^{ix}+p} = \frac{e^{-ix}+p}{p^2+2p\cos x + 1} = \sum_{n=0}^{\infty} \frac{\psi_n(p)(-ix)^n}{n!(p+1)^{n+1}}.$$

Equating real and imaginary parts on both sides above, we deduce Corollaries 4 and 5, respectively.

In the hypothesis of Corollary 6, which consists of four parts, Ramanujan attempts to define a sequence of numbers $\{A_n\}$. However, these numbers are not uniquely determined. It is preferable to define A_n by Corollary 6(iv) and then deduce the equality of the hypothesis. Hence, we have taken the liberty of inverting (iv) and the hypothesis below. Thus, put

$$\psi_n(p-1) = \sum_{k=0}^{n-1} A_{n-k}(-p)^k. \tag{8.3}$$

Ramanujan's notation is unfortunate because A_k depends upon n.

For Re $s > 1$, the Riemann zeta-function $\zeta(s)$ is defined by $\zeta(s) = \sum_{n=1}^{\infty} n^{-s}$; in Ramanujan's notation, $\zeta(k) = S_k$.

COROLLARY 6. Let $1 \le r \le n$. Then

(i) $\sum_{k=1}^{r} \binom{r}{k} A_k = r^n$,

(ii) $A_r = \sum_{k=0}^{r-1} (-1)^k \binom{r}{k} (r-k)^n$,

(iii) $A_r/n!$ is the coefficient of x^n in $(e^x-1)^r$,

and

(iv) $$\sum_{k=1}^{\infty} (-1)^{k+1} k^n (\zeta(k+1)-1)$$

$$= (-1)^n + (-1)^n 2^{-n-1} \psi_n(1) + \sum_{k=1}^{n} (-1)^{k+1} A_k \zeta(k+1),$$

where $\psi_n(1)$ is determined in Entry 8.

PROOF OF (i). By (6.1),

(8.4) $$\psi_n(p-1) = \sum_{k=0}^{n-1} F_{k+1}(n)(1-p)^k.$$

Equating coefficients of p^{n-j} in (8.3) and (8.4), we find that, for $1 \leq j \leq n$,

$$A_j = \sum_{k=n-j}^{n-1} \binom{k}{n-j} F_{k+1}(n).$$

Thus, using Vandermonde's theorem below, we find that

$$\sum_{j=1}^{r} \binom{r}{j} A_j = \sum_{j=1}^{r} \binom{r}{j} \sum_{k=n-j}^{n-1} \binom{k}{n-j} F_{k+1}(n)$$

$$= \sum_{k=n-r}^{n-1} F_{k+1}(n) \sum_{j=n-k}^{r} \binom{r}{j}\binom{k}{n-j}$$

$$= \sum_{k=n-r}^{n-1} \binom{r+k}{n} F_{k+1}(n)$$

$$= \sum_{j=0}^{r-1} \binom{n+j}{n} F_{n+j-r+1}(n)$$

$$= \sum_{j=0}^{r-1} \binom{n+j}{j} F_{r-j}(n) = r^n,$$

where we have employed Entries 6(i) and 6(ii).

PROOF OF (ii). The proposed formula follows from the inversion of (i) [33, pp. 43-44].

From (ii) it is seen that $A_r = A_r(n) = r!S(n,r)$, where the numbers $S(n,r)$ are Stirling numbers of the second kind [1, p. 824]. Because (iii) is such a familiar property of the Stirling numbers of the second kind, we shall omit the proof.

PROOF OF (iv). Using successively Entry 4, (6.2), and (8.3) below, we find that

$$\sum_{k=1}^{\infty} (-1)^{k+1}k^n(\zeta(k+1)-1) = \sum_{j=2}^{\infty} \sum_{k=1}^{\infty} \frac{(-1)^{k+1}k^n}{j^{k+1}}$$

$$= \sum_{j=2}^{\infty} \frac{\psi_n(1/j)}{j^2(1+1/j)^{n+1}}$$

$$= (-1)^{n-1} \sum_{j=2}^{\infty} \frac{\psi_n(j)}{(j+1)^{n+1}}$$

$$= (-1)^{n-1} \sum_{j=2}^{\infty} \frac{1}{(j+1)^{n+1}} \sum_{k=0}^{n-1} (-1)^k A_{n-k}(j+1)^k$$

$$= (-1)^{n-1} \sum_{k=0}^{n-1} (-1)^k A_{n-k}\{\zeta(n-k+1)-1-2^{k-n-1}\},$$

from which the desired formula follows with the use of (8.3).

EXAMPLES 1 AND 2. We have

$$\sum_{k=1}^{\infty} k^5/2^k = 1082, \qquad \text{and} \qquad \sum_{k=1}^{\infty} k^5/3^k = 273/4.$$

PROOF. By Entry 4 and (6.2),

$$\sum_{k=1}^{\infty} k^n/p^k = \frac{\psi_n(-1/p)}{p(1-1/p)^{n+1}} = \frac{p\psi_n(-p)}{(p-1)^{n+1}} . \tag{8.5}$$

From Entry 8, $\psi_5(-2) = 541$ and $\psi_5(-3) = 1456$. Putting these values in (8.5), we achieve the two given evaluations.

Entry 9 is simply the definition (0.1) of the Bernoulli numbers, and Entry 10 is the expansion given on the right side of (1.3).

ENTRY 11. For $|x| < 2\pi$,

$$\operatorname{Log}\left(\frac{x}{e^x-1}\right) = \sum_{n=1}^{\infty} \frac{(-1)^{n-1}B_n x^n}{n\, n!} .$$

PROOF. Using (0.1), we have, for $|x| < 2\pi$,

$$\text{Log}\left(\frac{x}{e^x-1}\right) = \int_0^x \left(\frac{1}{t} - \frac{e^t}{e^t-1}\right)dt = \int_0^x \frac{1}{t}\left(1 - \frac{-t}{e^{-t}-1}\right)dt$$

$$= \int_0^x \sum_{n=1}^{\infty} \frac{B_n}{n!}(-t)^{n-1}dt = \sum_{n=1}^{\infty} \frac{(-1)^{n-1}B_n x^n}{n\, n!}.$$

ENTRY 12. For $|x| < \pi$,

$$\text{Log}\left(\frac{2}{e^x+1}\right) = \sum_{n=1}^{\infty} \frac{(-1)^{n-1}(2^n-1)B_n x^n}{n\, n!}.$$

PROOF. Observe that

$$\text{Log}\left(\frac{2}{e^x+1}\right) = \text{Log}\left(\frac{2x}{e^{2x}-1}\right) - \text{Log}\left(\frac{x}{e^x-1}\right)$$

and then use Entry 11.

The conclusions of Examples 1-3 below are written as equalities in the notebooks, but Ramanujan clearly realizes that his results are approximations. Put $e^P = 1+p$, $e^Q = 1+q$, $e^R = 1+r$, $e^S = 1+s$, and $e^T = 1+t$, where p,q,r,s, and t are to be regarded as small. In the first example below, Ramanujan has incorrectly written -1/2 instead of 1/2 on the right side.

EXAMPLE 1. If $e^P+e^Q+e^R = 2+e^{P+Q+R}$, then

$$\frac{1}{P} + \frac{1}{Q} + \frac{1}{R} + \frac{1}{12}(P+Q+R) \approx \frac{1}{2}.$$

PROOF. In terms of p,q, and r, we are given that $3+p+q+r = 2+(1+p)(1+q)(1+r)$, which may be written as

$$-1 = \frac{1}{p} + \frac{1}{q} + \frac{1}{r} = \frac{1}{e^P-1} + \frac{1}{e^Q-1} + \frac{1}{e^R-1}$$

$$\approx \frac{1}{P} - \frac{1}{2} + \frac{P}{12} + \frac{1}{Q} - \frac{1}{2} + \frac{Q}{12} + \frac{1}{R} - \frac{1}{2} + \frac{R}{12},$$

where we have used (0.1) and ignored all terms with powers of P,Q, and R greater than the first. The desired approximation now follows.

EXAMPLE 2. If

$$e^{P+Q+R+S} = \frac{e^P+e^Q+e^R+e^S-2}{e^{-P}+e^{-Q}+e^{-R}+e^{-S}-2},$$

then

$$\frac{1}{P}+\frac{1}{Q}+\frac{1}{R}+\frac{1}{S}+\frac{1}{12}(P+Q+R+S) \approx 0.$$

PROOF. The given equality is equivalent to

$$(1+p)(1+q)(1+r)(1+s) = \frac{2+p+q+r+s}{2-p-q-r-s}.$$

Now cross multiply and ignore all products involving p^2, q^2, r^2, and s^2. After a tedious calculation and much simplification, we find that

$$-2 \approx \frac{1}{p}+\frac{1}{q}+\frac{1}{r}+\frac{1}{s}.$$

Proceeding as in Example 1, we achieve the sought approximation.

EXAMPLE 3. If

$$2e^{P+Q+R+S+T} = \frac{(e^P+e^Q+e^R+e^S+e^T-2)^2-(e^{2P}+e^{2Q}+e^{2R}+e^{2S}+e^{2T}-2)}{e^{-P}+e^{-Q}+e^{-R}+e^{-S}+e^{-T}-2},$$

then

$$\frac{1}{P}+\frac{1}{Q}+\frac{1}{R}+\frac{1}{S}+\frac{1}{T}+\frac{1}{12}(P+Q+R+S+T) \approx \frac{1}{2}.$$

PROOF. The proof is straightforward, very laborious, and along the lines of the proofs of Examples 1 and 2. Rewrite the given equality in terms of p,q,r,s, and t. Cross multiply and ignore all terms involving p^2,q^2,r^2,s^2, and t^2. After a lengthy calculation and considerable cancellation, we arrive at

$$\frac{1}{p}+\frac{1}{q}+\frac{1}{r}+\frac{1}{s}+\frac{1}{t} \approx -2.$$

Now proceed as in Example 1.

ENTRY 13. For $|x| < \pi$,

$$x \cot x = \sum_{n=0}^{\infty} \frac{(-1)^n B_{2n}(2x)^{2n}}{(2n)!}.$$

ENTRY 14. For $|x| < \pi$,

$$x \csc x = \sum_{n=0}^{\infty} \frac{(-1)^n(2-2^{2n})B_{2n}x^{2n}}{(2n)!}.$$

ENTRY 15. For $|x| < \pi/2$,

$$x \tan x = \sum_{n=1}^{\infty} \frac{(-1)^n 2^{2n}(1-2^{2n})B_{2n}x^{2n}}{(2n)!}.$$

Entries 13-15 are very familiar, and so there is no point in supplying proofs here.

ENTRY 16. For $|x| < \pi$,

$$\operatorname{Log}\left(\frac{x}{\sin x}\right) = \sum_{n=1}^{\infty} \frac{(-1)^{n+1}B_{2n}(2x)^{2n}}{(2n)(2n)!}.$$

PROOF. For $|x| < \pi$,

$$\operatorname{Log}\left(\frac{x}{\sin x}\right) = \int_0^x \frac{1}{t}(1-t \cot t)dt.$$

Now employ Entry 13.

ENTRY 17. For $|x| < \pi/2$,

$$\operatorname{Log}(\sec x) = \sum_{n=1}^{\infty} \frac{(-1)^n 2^{2n}(1-2^{2n})B_{2n}x^{2n}}{(2n)(2n)!}.$$

PROOF. For $|x| < \pi/2$,

$$\operatorname{Log}(\sec x) = \int_0^x \tan t \, dt.$$

Now employ Entry 15.

Ramanujan now makes three remarks, the first of which is trivial and the second of which is a special case of the third,

$$\frac{B_{2n}}{B_{2n-2h}} \sim \frac{(-1)^h(2n)!}{(2\pi)^{2h}(2n-2h)!}$$

as n tends to ∞, where $0 \le h \le n-1$. This asymptotic formula is a simple consequence of Euler's formula for $\zeta(2n)$; see Entry 25(i).

The following recurrence relation is due to Euler [19].

ENTRY 18. Let n be an integer exceeding 1. Then

$$-(2n+1)B_{2n} = \sum_{k=1}^{n-1} \binom{2n}{2k} B_{2k}B_{2n-2k}.$$

Ramanujan's first published paper [28], [30, pp. 1-14] is on Bernoulli numbers, and Section 2 of that paper contains a proof of Entry 18. After Entry 18, Ramanujan lists all of the Bernoulli numbers with index ≤ 38. All of the values are correct and agree with the table found in [1, p. 810]. The next result is contained in (16) of [28].

ENTRY 19(i). For each positive integer n, $2(2^{2n}-1)B_{2n}$ is an integer.

PROOF. We give a somewhat easier proof than that in [28]. By the von Staudt-Clausen theorem, the denominator of B_{2n} is the product of all those primes p such that $(p-1)|2n$. Let p be such an odd prime. Then by Fermat's theorem, $p|(2^{p-1}-1)$. But since $(p-1)|2n$, we have $p|(2^{2n}-1)$, which completes the proof.

ENTRY 19(ii). The numerator of B_{2n} is divisible by the largest factor of 2n which is relatively prime to the denominator of B_{2n}.

Entry 19(ii) is contained in (18) of [28] and is originally due to J.C. Adams [35, p. 261]. In fact, in both Entry 19(ii) and (18) of [28], Ramanujan claims a stronger result, viz., the implied quotient is a prime number. However, this is false; for example, the numerator of B_{22} is $854513 = 11\cdot131\cdot593$ [36, p. 589].

Entry 19(iii) is a statement of the von Staudt-Clausen theorem, which we mentioned above and which is (19) of [28].

ENTRY 20. For each nonnegative integer n, $(-1)^{n-1}B_{2n}+(-1)^n(1-F_{2n})$ is an integer I_{2n}, where F_{2n} is the sum of the reciprocals of those primes p such that $(p-1)|2n$.

Of course, Entry 20 is another version of the von Staudt-Clausen theorem. Ramanujan next lists the following values for I_{2n}: $I_{2n} = 0$, $0 \leq n \leq 6$, $I_{14} = 1$, $I_{16} = 7$, $I_{18} = 55$, $I_{20} = 529$, $I_{22} = 6192$, $I_{24} = 86580$, and $I_{26} = 1425517$. All of these values are correct.

In Example 1, Ramanujan calculates B_{22}, partly with the aid of the von Staudt-Clausen theorem. However, his reasoning is fallacious because Ramanujan thought that the numerator of B_{22} divided by 11 is prime. We pointed out above that this is false.

EXAMPLE 2. The fractional part of B_{200} is 216641/1366530.

PROOF. The given result is a direct consequence of the von Staudt-Clausen theorem.

Several of the results in Ramanujan's first paper [28] are not completely proved or are false. Wagstaff [37] has made a thorough examination of Ramanujan's paper and has given complete proofs of all the correct results in [28].

Entry 21 consists of two tables. The first is a table of primes up to 211. In constructing the table, Ramanujan makes use of the fact that any prime other than 2,3, and 5 is of the form $p+30n$, where $n \geq 0$ and p is either 7,11,13,17,19, 23,29, or 31. The second table lists all primes up to 4969. The numbers at the extreme left of the table give the number of hundreds in the primes immediately following; and the inked vertical strokes divide the hundreds. The table was presumably constructed as follows. First insert the primes (constructed in the first table) up to 211. Then use the fact that any prime other than 2,3,5, and 7 is of the form $q+210n$, where $n \geq 0$ and q is one of the numbers 11,13,17,...,199, 211. Thus, increase each of these primes already in the table by multiples of 210 until we reach the prime $2 \cdot 3 \cdot 5 \cdot 7 \cdot 11+1 = 2311$. Then proceed in a similar fashion with the arithmetic progressions $r+2310n$, where $n \geq 0$ and r is any of the numbers 13,17,19,...,2309,2311.

ENTRY 22. If n is a natural number, then

$$2^{2n}(2^{2n}-1)B_{2n} = 2n \sum_{k=0}^{n-1} \binom{2n-2}{2k} E_{2k}E_{2n-2k-2},$$

where E_j denotes the jth Euler number.

PROOF. By (0.2) and Entry 15,

$$\left\{\sum_{k=0}^{\infty} \frac{(-1)^k E_{2k}x^{2k}}{(2k)!}\right\}^2 = \sec^2 x = \frac{d}{dx} \tan x$$

$$= \sum_{n=1}^{\infty} \frac{(-1)^n 2^{2n}(1-2^{2n})B_{2n}x^{2n-2}}{(2n)(2n-2)!}.$$

Now equate coefficients of x^{2n-2} on both sides to achieve the proposed formula.

Next, Ramanujan records the Euler numbers E_{2n}, $0 \le n \le 7$. All values are correct.

Entries 23(i)-(iv) give the well-known partial fraction decompositions of $\cot(\pi x)$, $\tan(\pi x/2)$, $\csc(\pi x)$, and $\sec(\pi x/2)$. Ramanujan has also derived these expansions in Chapter 2, Entry 10, Corollaries 1-3 [7]. Entries 24(i)-(iv) offer the familiar partial fraction expansions of $\coth(\pi x)$, $\tanh(\pi x/2)$, $\operatorname{csch}(\pi x)$, and $\operatorname{sech}(\pi x/2)$.

After Entry 24, Ramanujan makes three remarks. In the first he claims that the last digit of E_{4n} is 5 and that $E_{4n-2}+1$ is divisible by 4, $n \ge 1$. Ramanujan probably discovered these results empirically. The latter is due to Sylvester. Moreover, they are, respectively, special cases of the following congruences

$$E_{4n} \equiv 5 (\text{mod } 60) \qquad \text{and} \qquad E_{4n-2} \equiv -1 (\text{mod } 60),$$

normally attributed to Stern [24, p. 261].

The second remark is a special case of the third, namely,

$$\frac{E_{2n+2h}}{E_{2n}} \sim \frac{(-1)^h 2^{2h}(2n+2h)!}{\pi^{2h}(2n)!}$$

as n tends to ∞, where h is a nonnegative integer. This asymptotic formula is an easy consequence of Entry 25(iv) below.

For Re s > 1, let $\lambda(s) = \sum_{k=0}^{\infty} (2k+1)^{-s}$, and for Re s > 0, define

$\eta(s) = \sum_{k=1}^{\infty} (-1)^{k+1} k^{-s}$ and $L(s) = \sum_{k=0}^{\infty} (-1)^k (2k+1)^{-s}$.

<u>   25</u>. If n is a positive integer, then

(i) $$\zeta(2n) = \frac{(-1)^{n-1}(2\pi)^{2n}}{2(2n)!} B_{2n},$$

(ii) $$\lambda(2n) = \frac{(-1)^{n-1}(1-2^{-2n})(2\pi)^{2n}}{2(2n)!} B_{2n},$$

(iii) $$\eta(2n) = \frac{(-1)^{n-1}(1-2^{1-2n})(2\pi)^{2n}}{2(2n)!} B_{2n},$$

and

(iv) $$L(2n-1) = \frac{(-1)^{n-1}(\pi/2)^{2n-1}}{2(2n-2)!} E_{2n-2}.$$

The first equality is Euler's very famous formula for $\zeta(2n)$. For an interesting account of Euler's discovery of this formula, see [5]. A proof of (i) that uses only elementary calculus is found in [6], which also contains references to several other proofs. Observe that $\lambda(s) = (1-2^{-s})\zeta(s)$ and that $\eta(s) = (1-2^{1-s})\zeta(s)$, and so (ii) and (iii) both follow from (i). Formula (iv) is also well known and is an illustration of the following fact. If χ is an odd character, then the associated Dirichlet L-function can be explicitly evaluated at odd, positive integral values; if χ is even, then the associated L-function can be determined at even, positive integral values. See [8] for a verification of this remark.

Ramanujan next uses (i) and (iii) to define Bernoulli numbers for any index. Thus, for any real number s, he defines

$$B_s = \frac{2\Gamma(s+1)}{(2\pi)^s} \zeta(s). \tag{25.1}$$

(We now employ Ramanujan's convention for the ordinary Bernoulli numbers.) Note that $B_{2n+1} \neq 0$, $n \geq 1$, in Ramanujan's definition, which conflicts with (0.1). Similarly, (iv) can be used to extend the definition of E_s to all real s. The next two corollaries give numerical examples.

COROLLARY 1. $B_1 = \infty$, $B_{3/2} = \dfrac{3\zeta(3/2)}{4\pi\sqrt{2}}$, and $B_3 = \dfrac{3\zeta(3)}{2\pi^3}$.

COROLLARY 2. $B_0 = -1$, $B_{1/2} = -(1+1/\sqrt{2})\eta(1/2)$, $E_{-1} = \infty$, $E_{-1/2} = 2\sqrt{2}\,L(1/2)$, and $E_1 = 8L(2)/\pi^2$.

PROOF. The value for B_0 arises from the extension (25.1) and the fact that $\zeta(0) = -1/2$. The value for E_{-1} arises from the fact that $\Gamma(s)$ has a simple pole at $s = 0$. All other tabulated values are easily verified.

Let

$$f(a) = \sum_{k=1}^{\infty} (a+bk)^{-n},$$

where $a,b > 0$ and $n > 1$. Note that $f(a) = b^{-n}\zeta(n,a/b)-a^{-n}$, where $\zeta(s,\alpha)$ denotes Hurwitz's zeta-function.

ENTRY 26. As b/a tends to 0,

$$a^n f(a) \sim \frac{a}{b(n-1)} - \frac{1}{2} + \sum_{k=1}^{\infty} \frac{B_{2k}}{n-1}\binom{n+2k-2}{2k}\left(\frac{b}{a}\right)^{2k-1}. \tag{26.1}$$

PROOF. A simple calculation shows that $f(a-b)-f(a) = a^{-n}$. In the notation of Entry 1(i), $h = -b$ and $\phi(a) = 1/\{a^{n-1}b(n-1)\}$. If Entry 1(i) were applicable, we could readily deduce (26.1) with the asymptotic sign $\sim$ replaced by an equality sign. But, the series on the right side of (26.1) does not converge. However, by appealing to the theorem in Nörlund's text [26, pp. 58-60] that we mentioned after the proof of Entry 1(ii), we can conclude that the right side of (26.1) represents the function $a^n f(a)$ asymptotically as b/a tends to 0.

In the notebooks [31, vol. II, p. 58] the asymptotic sign in (26.1) is replaced by an equality sign. It appears that Ramanujan thought that the series in (26.1) converges.

EXAMPLE: For $n > 1$, as the positive integer r tends to ∞,

(26.2)
$$\zeta(n) \sim \sum_{k=1}^{r-1} k^{-n} + \frac{1}{(n-1)r^{n-1}} + \frac{1}{2r^n} + \sum_{k=1}^{\infty} \frac{B_{2k}}{n-1} \binom{n+2k-2}{2k} r^{-n-2k+1}.$$

PROOF. Apply Entry 26 with $a = r$ and $b = 1$. Thus, as r tends to ∞,

$$\sum_{k=1}^{\infty} (r+k)^{-n} \sim \frac{1}{(n-1)r^{n-1}} - \frac{1}{2r^n} + \sum_{k=1}^{\infty} \frac{B_{2k}}{n-1} \binom{n+2k-2}{2k} r^{-n-2k+1}.$$

Adding $\sum_{k=1}^{r} k^{-n}$ to both sides above, we obtain (26.2).

Asymptotic series like those in Entry 26 and the Example above were initially found by Euler. See Bromwich's book [11, pp. 324-329] for a very complete discussion of such asymptotic series and their applications to numerical calculation. Indeed, like Euler, Ramanujan employs (26.2) to calculate $\zeta(n)$, where n is a positive integer with $2 \leq n \leq 10$, to the tenth decimal place and provides the following table.

n	$\zeta(n)$	$1/B_n$
2	1.6449340668	6
3	1.2020569031	17.19624
4	1.0823232337	30
5	1.0369277551	39.34953
6	1.0173430620	42
7	1.0083492774	38.03538
8	1.0040773562	30
9	1.0020083928	20.98719
10	1.0009945781	13.2

For $\zeta(3)$, the tenth decimal place should be 2; for $\zeta(10)$, the ninth decimal place should be 5. Euler used this same method to calculate $\zeta(n)$, $2 \leq n \leq 16$, to 18 decimal places [11, p. 326]. A different method was used by Legendre [23, p. 432] to calculate $\zeta(n)$, $2 \leq n \leq 35$, to 16 decimal places. Stieltjes

[34], using Legendre's method, calculated $\zeta(n)$, $2 \leq n \leq 70$, to 32 decimal places. For even n, the values of $1/B_n$ above are determined from Ramanujan's table of Bernoulli numbers. For odd n, Ramanujan employs (25.1) and his previously determined values of $\zeta(n)$. In the far right column above, the last recorded digit for $1/B_3$ should be 3; the last digit for $1/B_7$ should be 6; and the last two digits for $1/B_9$ should be 20.

COROLLARY 1. The Riemann zeta-function has a simple pole at $s = 1$ with residue 1, and the constant term in the Laurent expansion of $\zeta(s)$ about $s = 1$ is Euler's constant γ.

Of course, Corollary 1 is very well known [22, p. 164]. Ramanujan's wording for Corollary 1 is characteristically distinct: "$n\ S_{n+1} = 1$ if $n = 0$ and $S_{n+1} - 1/n = .577$ nearly." In the sketch following Corollary 1, Ramanujan gives the first three terms of the asymptotic series

$$\gamma - \frac{1}{2} + \sum_{n=1}^{\infty} \frac{B_{2n}}{2n}. \tag{26.3}$$

In fact, this asymptotic series for γ was first discovered by Euler and used by him to calculate γ [11, pp. 324-325]. It is curious that Ramanujan's approximation .577 to γ is better than any that can be gotten from Euler's series. Bromwich [11, p. 325] points out that the best approximation .5790 is achieved by taking four terms on the right side of (26.3). If we take just three terms and average the two approximations, we get the mean approximation .5770. Perhaps this is how Ramanujan reasoned, or possibly he calculated γ by using the Euler-Maclaurin summation formula to approximate a partial sum of the harmonic series.

Corollary 2 is merely a reformulation of the latter part of Corollary 1 in terms of Ramanujan's generalization (25.1) of Bernoulli numbers.

ENTRY 27. Suppose that $|a_p| \leq p^{-c}$ for each prime p and for some constant $c > 1$. Then

$$\prod_p (1-a_p)^{-1} = 1 + \sum_{\substack{n=2 \\ n=p_1\cdots p_k}}^{\infty} a_{p_1}\cdots a_{p_k},$$

where the product on the left is over all primes p, and where the suffixes on the right side are the (not necessarily distinct) primes in the canonical factorization of n.

PROOF. Expand the product on the left side and use the unique factorization theorem.

ENTRY 28. For Re s > 1,

$$\zeta(s) = \prod_p (1-p^{-s})^{-1}.$$

Entry 28 is the familiar Euler product for $\zeta(s)$ and, in fact, is a special instance of Entry 27. The next two corollaries are simple consequences of Entry 28.

COROLLARY 1. For Re s > 1,

$$\prod_p (1+p^{-s}) = \zeta(s)/\zeta(2s).$$

COROLLARY 2. For Re s > 1,

$$\prod_p \left(\frac{1+p^{-s}}{1-p^{-s}}\right) = \zeta^2(s)/\zeta(2s).$$

COROLLARY 3. For Re s > 1,

$$T(s) \equiv \sum_n n^{-s} = \frac{\zeta^2(s)-\zeta(2s)}{2\zeta(s)},$$

where the sum on the left is over all positive integers which have an odd number of prime factors in their canonical factorizations.

PROOF. By Entry 28 and Corollary 1,

$$\zeta(s) - \zeta(2s)/\zeta(s) = \prod_p (1-p^{-s})^{-1} - \prod_p (1+p^{-s})^{-1},$$

from which the desired result follows.

Examples 1(i), (ii), and (iii) record the familiar results $\zeta(2) = \pi^2/6$, $L(3) = \pi^3/32$, and $\lambda(4) = \pi^4/96$, respectively, deducible from Entry 25.

EXAMPLE 2. We have

$$\text{(i)} \quad \prod_p \left(\frac{1+p^{-2}}{1-p^{-2}}\right) = 5/2 \qquad \text{and} \qquad \text{(ii)} \quad \prod_p (1+p^{-4}) = 105/\pi^4.$$

PROOF. Equality (i) follows from Corollary 2, and (ii) follows from Corollary 1.

EXAMPLE 3. We have $T(2) = \pi^2/20$ and $T(4) = \pi^4/1260$, where $T(s)$ is given in Corollary 3.

PROOF. The proposed values follow from Corollary 3 and Entry 25(i).

COROLLARY 4. For Re $s > 1$,

$$L(s) = \prod_p (1-\sin(\pi p/2)p^{-s})^{-1}.$$

Corollary 4 is simply the Euler product for $L(s)$, and is an instance of Entry 27. Corollary 5 below is a well-known result arising from the logarithmic derivative of the Euler product in Entry 28.

COROLLARY 5. For Re $s > 1$,

$$\sum_{k=1}^{\infty} k^{-s} \operatorname{Log} k = \zeta(s) \sum_p \frac{\operatorname{Log} p}{p^s-1}.$$

EXAMPLE. The series $\sum_p \frac{\sin(\pi p/2)}{p}$ converges.

Although the result above is a special case of a well-known theorem in the theory of L-functions [22, pp. 446-449], its proof is considerably deeper than the other results in Section 28. Ramanujan supplies no hint of how he deduced this result.

ENTRY 29. For $|x| < 1$,

$$\prod_{k=1}^{\infty} (1-x^{p_k})^{-1} = 1 + \sum_{k=1}^{\infty} \frac{x^{p_1+p_2+\ldots+p_k}}{(1-x)(1-x^2)\ldots(1-x^k)}, \tag{29.1}$$

where $p_1, p_2, \ldots$ denote the primes in ascending order.

Entry 29, in fact, is cancelled by Ramanujan. Let a_n and b_n, $2 \le n < \infty$, denote the coefficients of x^n on the left and right sides, respectively, of (29.1). Then, quite amazingly, $a_n = b_n$ for $2 \le n \le 20$. But $a_{21} = 30$ and $b_{21} = 31$. Thus, as indicated by Ramanujan, (29.1) is false. Andrews [2] has thoroughly discussed (29.1) and has also formulated some related conjectures.

ENTRY 30. Suppose that $|a_p| \le p^{-c}$ for each prime p and for some constant $c > 1$. Then

$$\prod_p (1+a_p) = 1 + \sum_{\substack{n \\ n=p_1\cdots p_k}} a_{p_1}\cdots a_{p_k}.$$

The sum on the right side is over all squarefree integers $n = p_1 \cdots p_k$, where $p_1, \ldots, p_k$ are distinct primes.

PROOF. Expand the product on the left side above.

COROLLARY 1. For Re $s > 1$,

$$\sum_{\substack{n=1 \\ n \text{ squarefree}}}^{\infty} n^{-s} = \zeta(s)/\zeta(2s).$$

PROOF. Let $a_p = p^{-s}$ in Entry 30 and apply Corollary 1 of Entry 28.

COROLLARY 2. For Re $s > 1$,

$$\sum_n n^{-s} = \frac{\zeta^2(s)-\zeta(2s)}{2\zeta(s)\zeta(2s)},$$

where the sum on the left is over all squarefree integers n which contain an odd number of prime factors.

PROOF. By Entry 28 and Corollary 1 of Entry 28,

$$\zeta(s)/\zeta(2s) - 1/\zeta(s) = \prod_p (1+p^{-s}) - \prod_p (1-p^{-s}),$$

from which the desired equality follows.

COROLLARY 3. For Re s > 1,

$$\sum_{\substack{n=1 \\ n \text{ not squarefree}}}^{\infty} n^{-s} = \frac{\zeta(s)(\zeta(2s)-1)}{\zeta(2s)} .$$

PROOF. By Corollary 1, the series on the left side is $\zeta(s)-\zeta(s)/\zeta(2s)$.

Entry 27, Entry 28, Corollaries 1 and 3 and Examples 2(ii) and 3 in Section 28, and Entry 30 and its first three corollaries are found in Ramanujan's fourth published paper [29], [30, pp. 20-21].

COROLLARY 1. The sum of the reciprocals of the primes diverges.

As is well known, Corollary 1 is due to Euler, and Ramanujan's proof is similar to that of Euler [4, p. 5].

COROLLARY 2. $\lim_{s \to 1+} \{\text{Log}(s-1) + \sum_p p^{-s}\}$ exists.

PROOF. By Entry 28, for s > 1,

$$\text{Log } \zeta(s) = \sum_p \frac{1}{p^s} + \sum_p \sum_{k=2}^{\infty} \frac{1}{kp^{ks}} .$$

But by Corollary 1 of Entry 26, Log $\zeta(s) \sim -\text{Log}(s-1)$ as s tends to 1 from the right. The sought result now follows.

COROLLARY 3. If p_n denotes the nth prime, then $p_n/n - \text{Log } n$ tends to a limit as n tends to ∞.

Ramanujan has rightly struck Corollary 3 out, for [22, p. 215]

$$p_n/n = \text{Log } n + \text{Log Log } n + O(1)$$

as n tends to ∞.

REFERENCES

1. M. Abramowitz and I.A. Stegun, editors, Handbook of mathematical functions, Dover, New York, 1965.
2. G.E. Andrews, An incredible formula of Ramanujan, Australian Math. Soc. Gazette 6 (1979), 80-89.
3. P. Appell, Sur une classe de fonctions analogues aux fonctions Eulériennes, Math. Ann. 19 (1882), 84-102.
4. R. Ayoub, An introduction to the analytic theory of numbers, American Mathematical Society, Providence, 1963.
5. R. Ayoub, Euler and the zeta function, Amer. Math. Monthly 81 (1974), 1067-1086.
6. B.C. Berndt, Elementary evaluation of $\zeta(2n)$, Math. Mag. 48 (1975), 148-154.
7. B.C. Berndt, P.T. Joshi, and B.M. Wilson, Chapter 2 of Ramanujan's second notebook, Glasgow Math. J. (to appear).
8. B.C. Berndt and L. Schoenfeld, Periodic analogues of the Euler-Maclaurin and Poisson summation formulas with applications to number theory, Acta Arith. 28 (1975), 23-68.
9. T.B.N. Brodén, Bemerkungen über sogenannte finite Integration, Arkiv för Mat. Astr. och Fys. (Stockholm) 7 (1911), 34pp.
10. T.B.N. Brodén, Einige Anwendungen diskontinuierlicher Integrale auf Fragen der Differenzenrechnung, Acta Univ. Lund. (2) 8 (1912), 17pp.
11. T.J. I'A. Bromwich, An introduction to the theory of infinite series, second ed., Macmillan, London, 1926.
12. L. Carlitz, Eulerian numbers and polynomials, Math. Mag. 33 (1959), 247-260.
13. L. Carlitz, Eulerian numbers and operators, Coll. Math. 24 (1973), 175-200.
14. L. Carlitz, Some remarks on the Eulerian function, Univ. Beograd. Publ. Elektrotehn. Fak. Ser. Mat. Fiz., No. 611, 1978, 79-91.
15. L. Carlitz, D.C. Kurtz, R. Scoville, and O.P. Stackelberg, Asymptotic properties of Eulerian numbers, Z. Wahrscheinlichkeitstheorie 23 (1972), 47-54.

16. L. Carlitz and J. Riordan, Congruences for Eulerian numbers, Duke Math. J. 20 (1953), 339-343.

17. L. Carlitz and R. Scoville, Generalized Eulerian numbers: combinatorial applications, J. Reine Angew. Math. 265 (1974), 110-137. Corrigendum 288 (1976), 218-219.

18. L. Carlitz and R. Scoville, Eulerian numbers and operators, Fibonacci Quart. 13 (1975), 71-83.

19. L. Euler, Institutiones calculi differentialis, Acad. Imperialis Sci., Petrograd, 1755.

20. F.G. Frobenius, Über die Bernoullischen Zahlen und die Eulerschen Polynome, Sitz. d. K. Preuss. Akad. Wiss. Berlin 1910, 809-847.

21. F.G. Frobenius, Gesammelte Abhandlungen, Band III, Springer-Verlag, Berlin, 1968.

22. E. Landau, Primzahlen, Chelsea, New York, 1953.

23. A.M. Legendre, Traité des fonctions elliptiques et des intégrales Eulériennes, tome II, Huzard-Courcier, Paris, 1826.

24. N. Nielsen, Traité élémentaire des nombres de Bernoulli, Gauthier-Villars, Paris, 1923.

25. N.E. Nörlund, IIC7. Neure Untersuchungen über Differenzengleichungen, Encyklopädie der mathematischen Wissenschaften, Band 2, Teil 3, B.G. Teubner, Leipzig, 1923, pp. 675-721.

26. N.E. Nörlund, Vorlesungen über Differenzenrechnung, Chelsea, New York, 1954.

27. E. Picard, Sur une classe de transcendantes nouvelles, Acta Math. 18 (1894), 133-154.

28. S. Ramanujan, Some properties of Bernoulli's numbers, J. Indian Math. Soc. 3 (1911), 219-234.

29. S. Ramanujan, Irregular numbers, J. Indian Math. Soc. 5 (1913), 105-106.

30. S. Ramanujan, Collected papers, Chelsea, New York, 1962.

31. S. Ramanujan, Notebooks (2 volumes), Tata Institute of Fundamental Research, Bombay, 1957.

32. J. Riordan, An introduction to combinatorial analysis, John Wiley, New York, 1958.

33. J. Riordan, Combinatorial identities, John Wiley, New York, 1968.

34. T.J. Stieltjes, Table des valeurs des sommes $S_k = \sum_1^\infty n^{-k}$, Acta Math. 10 (1887), 299-302.

35. J.V. Uspensky and M.A. Heaslet, Elementary number theory, McGraw-Hill, New York, 1939.

36. S.S. Wagstaff, Jr., The irregular primes to 125000, Math. Comp. 32 (1978), 583-591.

37. S.S. Wagstaff, Jr., Ramanujan's paper on Bernoulli numbers, submitted for publication.

38. J. Worpitzky, Studien über die Bernoullischen und Eulerschen Zahlen, J. Reine Angew. Math. 94 (1883), 203-232.

OSCILLATION THEOREMS

Robert J. Anderson and H.M. Stark

To Emil Grosswald, colleague and friend

1. Introduction. Given a sequence of numbers $a_1, a_2, \ldots$, number theorists will naturally investigate the sum,

$$A(x) = \sum_{n \leq x} a_n. \tag{1}$$

Examples of such functions are

$$M(x) = \sum_{n \leq x} \mu(n),\ L(x) = \sum_{n \leq x} \lambda(n),\ \pi(x) = \sum_{p \leq x} 1.$$

Based on extensive tables, Mertens conjectured that

$$|M(x)| \leq \sqrt{x}\ ,\ x > 0,$$

Von Sterneck that

$$|M(x)| < \frac{1}{2}\sqrt{x}\ ,\ x > 200,$$

and Polya that

$$L(x) \leq 0,\quad x \geq 2.$$

To these, we might add on the basis of tables

$$\pi(x) < \ell i(x)$$

although this was already disproved by Littlewood in 1914.

Several people have developed theorems to attack these conjectures over the years. The most useful theorem in this direction is that of Ingham [14] in 1942. In this paper, we will show that Ingham's Theorem yields as corollaries most of the usual approaches to these problems. Most of these approaches require numerical computation of (and with) many zeros of $\zeta(s)$ or related functions. However, one approach, which one of us likes to call proof by example, uses Ingham's Theorem in conjunction with explicit formulae to show that one good counterexample to any of these conjectures implies that there are infinitely many. Our proof that Ingham's Theorem can be used in this way for $M(x)$ and $L(x)$ is based on a method just developed by Anderson [1].

Based on examples of Neubauer and Lehman, we show that

$$\limsup_{x \to \infty} \frac{M(x)}{\sqrt{x}} > .557, \qquad \limsup_{x \to \infty} \frac{L(x)}{\sqrt{x}} > .023.$$

These results require no computations of values of $\zeta(s)$ at all. For technical reasons, it is still easier to apply Ingham's Theorem to show that

$$\limsup_{x \to \infty} \frac{\pi(x) - \ell i(x)}{x^{\frac{1}{2}}/\log x} = +\infty. \tag{2}$$

We do this here via $\psi(x) - x$ but it is possible to use the logarithmic form of Ingham's Theorem given by Stark [19] to prove (2) directly.

Although Ingham's Theorem is the most powerful approach to conjectures such as that of Mertens, more frivolous conjectures that fail to take into account the existence of complex zeros of $\zeta(s)$ can be most easily demolished by a theorem of Landau. We will begin with some of these. When averages such as $A(x)$ behave badly, number theorists will even investigate iterated averages. In this way Miller [11] was led to investigate the average,

$$\Lambda(x) = \frac{1}{x} \sum_{n \le x-1} M(n).$$

He conjectured $\Lambda(x) < 0$ for $x > 3$. At the Tenth Scandinavian Congress in 1946, Brun [3] even suggested that for integral x

$$\Lambda(x) = -2 + 12x^{-1} - \beta x^{-2}$$

"on the average", whatever this meant. There was an argument about the value of β. Brun suggested $\beta = 18$ while Siegel said $\beta = \frac{2\pi^2}{\zeta(3)} = 16.421\ldots$.

It occurred to Gupta [11] in 1949 to test these conjectures numerically. He found that $\Lambda(x) < 0$ for $3 < x < 20000$ but that $\Lambda(x)$ oscillated. Based on the size of the observed oscillations, Gupta conjectured that $\Lambda(x) = O(\log x)$ as

$x \to \infty$. (This was recently disproved by Anderson and Murty in a *Monthly* problem.) Later, Gupta, Cheema and Gupta [12] tested the conjectures for the value of β by tabulating

$$T_0(x) = x^2(\Lambda(x) + 2 - 12x^{-1}),$$

as well as $T_1(x), T_2(x),\ldots,T_5(x)$ for $1 \le x \le 750$ where

$$T_{j+1}(x) = \frac{1}{x} \sum_{n \le x} T_j(x).$$

Over the last half of the table, the values of $T_5(x)$ descended from -14 to -15, thereby settling nothing. Indeed, none of these conjectures can be true if the zeta function has complex zeros.

For a Dirichlet series

$$f(s) = \sum_{n=1}^{\infty} a_n n^{-s},$$

presumed convergent in some half plane $\sigma > \sigma_a > 0$, we have the Mellin transform,

$$\frac{f(s)}{s} = \int_1^{\infty} A(x)x^{-s}\,\frac{dx}{x}$$

with $A(x)$ given by (1), which converges absolutely in the same half plane. In the inverse transform, we will obtain

$$A_0(x) = \frac{A(x^+) + A(x^-)}{2}.$$

(We will use $M_0(x)$, $L_0(x)$ etc. for the $A_0(x)$ corresponding to $A(x) = M(x), L(x)$, etc.) It is easy to obtain information on integral averages of $A(x)$ since

$$A_{j+1}(x) = \frac{1}{x}\int_1^x A_j(t)dt = \int_1^x \left(\frac{x}{t}\right)^{-1} A_j(t)\,\frac{dt}{t}$$

gives $A_{j+1}(x)$ as the convolution of $A_j(x)$ and x^{-1}. Thus for $j \ge 0$,

$$\frac{f(s)}{s(s+1)^j} = \int_1^{\infty} A_j(x)x^{-s}\,\frac{dx}{x}$$

which is also absolutely convergent for $\sigma > \sigma_a$. If $f(s)$ has an analytic continuation to the left of $\sigma > \sigma_a$, we expect to get information on the behavior of $A_j(x)$ from the singularities of $f(s)\,[s(s+1)^j]^{-1}$. Likewise we expect to get information on summatory averages of $A(x)$ although this will be messier to obtain. Certainly $A_1(x)$ should be very similar to the summatory average of $A(x)$.

For $a_n = \mu(n)$ we will let $M_j(x) = A_j(x)$ (and similarly $L_j(x) = A_j(x)$ for $a_n = \lambda_n$ and so on for other functions). We have the inverse Mellin transform

$$M_j(x) = \frac{1}{2\pi i} \int_{2-i\infty}^{2+i\infty} \frac{x^s}{s(s+1)^j \zeta(s)} \, ds$$

$$= \sum_\rho \frac{x^\rho}{\rho(\rho+1)^j \zeta'(\rho)} - 2 + R_j(x)$$

where the sum over ρ is over the complex zeros of $\zeta(s)$ (written as simple zeros for simplicity) and $R_j(x)$ is formally the sum of the residues at $s = -1, -2, -4, \ldots$. For instance, when $j = 1$,

$$R_1(x) = 12x^{-1} + \frac{x^{-2}}{2\zeta'(-2)} + \ldots$$

$$= 12x^{-1} - \frac{2\pi^2}{\zeta(3)} x^{-2} + \ldots \,.$$

But the first complex zero, $\frac{1}{2} + i\gamma_1 \approx \frac{1}{2} + 14i$ and its conjugate already contribute an oscillatory cosine term of amplitude approximately $\frac{2x^{\frac{1}{2}}}{200}$. For small x, this term will be invisible compared to $-2 + R_1(x)$. But for $x \approx 40000$ it will be roughly equal to $-2 + R_1(x)$ and if the other zero terms are favorable we might expect the complex zeros to dominate $-2 + R_1(x)$ even earlier.

This it should be no surprise that Neubauer [17] in 1963 found

$$\Lambda(21068) > 0$$

(just barely past Gupta's table) or that for $10^6 < x < 10^8$, $\Lambda(x)/\log x$ oscillates wildly. Neubauer was primarily interested in the conjecture of Mertens and Von Sterneck. He found that for $x = 7760000000$, $M(x) = 47465$ giving

$$\frac{M(x)}{\sqrt{x}} = .557$$

and thus providing a large counterexample to Von Sterneck's conjecture.

2. Landau's Theorem. We have seen that morally, each $M_j(x)$ deserves to have oscillations of amplitude on the order of $x^{1/2}$. But how do we actually prove this? The first key theorem in this direction is due to Landau. It has many variants; the following version is suitable for us.

Landau's Theorem. Suppose $g(x)$ *is a piecewise continuous function bounded on finite intervals such that* $g(x)$ *has no sign changes past* x_0 *and*

$$G(s) = \int_1^\infty g(x)x^{-s}\,\frac{dx}{x} \tag{3}$$

converges absolutely in some half plane $\sigma > \sigma_a$. *Suppose further that* $G(s)$ *may be analytically continued leftward along the real axis to* $\sigma > \sigma_0$. *Then the integral* (3) *for* $G(s)$ *already converges absolutely for* $\sigma > \sigma_0$.

For example, we see that no $M_j(x)$ can possibly be eventually of one sign. Likewise, it will be easy to show that $\Lambda(x)$ or even the $T_j(x)$ of the last section cannot possibly be eventually of one sign. However, Landau's theorem allows us to do even more than this. Indeed, we have the following

Weak Corollary. Suppose $g(x)$ *is piecewise continuous, bounded on finite intervals,* $G(s)$ *is given by* (3) *for* $\sigma > \sigma_a$ *and that* $G(s)$ *analytically continues into a region including real* $s \geq \sigma_0$ *(with no singularity at* σ_0*) while* $G(s)$ *has a singularity at* $\sigma_0 + i\gamma$ *with* $\gamma \neq 0$. *Then for any* $\varepsilon > 0$

$$\limsup_{x \to \infty} \frac{g(x)}{x^{\sigma_0 - \varepsilon}} = \infty, \quad \liminf_{x \to \infty} \frac{g(x)}{x^{\sigma_0 - \varepsilon}} = -\infty.$$

The proof consists of replacing $g(x)$ by $g(x) + cx^{\sigma_0 - \varepsilon}$ in Landau's theorem. However, there is a better corollary, as noted for example in Grosswald [9]. If $g(x)$ has no sign changes past $x > x_0$, then $s = \sigma_0$ must be the "largest singularity" of $G(s)$ on the line $\sigma = \sigma_0$. We give here the case where the singularities are poles.

Strong Corollary. Under the hypotheses of the weak corollary, suppose that the singularity of G(s) *at* $\sigma_0 + i\gamma$ *is a first order pole with residue* r. *Then*

$$\limsup_{x\to\infty} \frac{g(x)}{x^{\sigma_0}} \geq |r|, \quad \liminf_{x\to\infty} \frac{g(x)}{x^{\sigma_0}} \leq -|r|.$$

<u>Proof</u>. Let

$$g_1(x) = \begin{cases} g(x) + cx^{-\sigma_0} & \text{for } x > x_0 \\ 0 & \text{for } x \leq x_0 \end{cases}$$

where x_0 and c are supposed to be given such that $g_1(x)$ is positive for all x. We let

$$G_1(s) = \int_1^\infty g_1(x)x^{-s}\frac{dx}{x}$$

which differs from G(s) by $c(s-\sigma_0)^{-1}$ plus an entire function. In particular, $G_1(s)$ has first order poles with residues c at σ_0 and r at $\sigma_0 + i\gamma$. By Landau's theorem,

$$|r| = \left|\lim_{\sigma\to\sigma_0^+} [(\sigma+i\gamma) - (\sigma_0+i\gamma)] \int_1^\infty g_1(x)x^{-(\sigma+i\gamma)}\frac{dx}{x}\right|$$

$$\leq \lim_{\sigma\to\sigma_0^+} (\sigma-\sigma_0)\int_1^\infty g_1(x)x^{-\sigma}\frac{dx}{x}$$

$$\leq c.$$

This takes care of the lim inf part and lim sup part follows from replacing g by -g.

For example, by Landau's theorem, M(x) has oscillations of $\pm\left|\frac{1}{\rho\zeta'(\rho)}\right|\sqrt{x}$ for zeros ρ of $\zeta(s)$. The biggest known value comes at the first zero which tells us that M(x) oscillates at least as much as $\pm\sqrt{x}/14$ about 0. Likewise $M_1(x)$ has oscillations of at least about $\pm\frac{\sqrt{x}}{200}$ and so can hardly approach -2 as $x\to\infty$. Even averages of $M_1(x)$ will have $\sqrt{x}$ oscillations and also can hardly approach -2. As another example, Alladi in a talk at the 1979 Western Number Theory conference asked whether

$$M(x) + x \sum_{n > x} \frac{\mu(n)}{n} = M(x) - x \sum_{n \le x} \frac{\mu(n)}{n}$$

can have infinitely many sign changes. We see that

$$\int_1^\infty x^{-s}[M(x)-x \sum_{n \le x} \frac{\mu(n)}{n}] \frac{dx}{x} = \frac{1}{s\zeta(s)} - \frac{1}{(s-1)\zeta(s)} = \frac{1}{s(s-1)\zeta(s)}$$

and so $M(x) + x \sum_{n \geq x} \frac{\mu(n)}{n}$ has oscillations of least approximately $\pm \frac{\sqrt{x}}{200}$ as $x \to \infty$. Again, such oscillations will be detected only for large x; for small x, the residue at s = 0 controls the situation.

3. Ingham's Theorem. For problems where large oscillations are required, Landau's theorem is usually not sufficient. In such cases, Ingham's theorem [14] is extremely useful. It certainly contains as corollaries all the usual approaches to such problems. We will give the case (as did Ingham) of first order poles. As in the previous section, we deal with a Mellin transform

$$G(s) = \int_1^\infty g(x)x^{-s} \frac{dx}{x}$$

which is presumed absolutely convergent for $\sigma > \sigma_a$. We assume that g(x) is real valued and that G(s) has an analytic continuation back to $\sigma = \sigma_0$ such that

$$H(s) = G(s) - [\frac{r_0}{s-\sigma_0} + \sum_\gamma \frac{r_\gamma}{s-(\sigma_0+i\gamma)}]$$

is analytic for $s = \sigma_0 + it$, $|t| < T$. Here γ denotes an element of a finite set of numbers, $0 < |\gamma| < T$. Of course, H(s) is the transform of

$$h(x) = g(x) - x^{\sigma_0} S_T(x)$$

where

$$S_T(x) = r_0 + \sum_{\substack{\gamma \\ |\gamma| < T}} r_\gamma x^{i\gamma}.$$

We expect that g(x) oscillates somewhat like $S_T(x)x^{\sigma_0}$. Ingham gave a precise version of this.

Ingham's Theorem. *Let*

$$S_T^*(x) = \frac{1}{T}\int_0^T S_t(x)\,dt = r_0 + \sum_{\substack{\gamma \\ |\gamma| < T}} r_\gamma\left(1-\frac{|\gamma|}{T}\right)x^{i\gamma}.$$

For any $T > 0$ *and* x_0 *whatsoever,*

$$\liminf_{x\to\infty} \frac{g(x)}{x^{\sigma_0}} \le S_T^*(x_0) \le \limsup_{x\to\infty} \frac{g(x)}{x^{\sigma_0}}.$$

Ingham proved his theorem for $\sigma_0 = 0$ and actually with milder hypotheses than analyticity for $H(s)$ on the line $\sigma = \sigma_0$. Also, the remark that any T and x_0 will do in Inghan's theorem is due to Haselgrove [13]. It is also possible to prove variants of Ingham's theorem for $G(s)$ with other singularities. For example, Stark [19] has a version with logarithmic singularities and has announced a version for other singularities which will no doubt appear someday.

Haselgrove applied Ingham's theorem to disprove Polya's conjecture. The Dirichlet series corresponding to $L(x)$ is

$$\sum_n \lambda(n)n^{-s} = \frac{\zeta(2s)}{\zeta(s)}$$

and the corresponding $S_T(x)$ is

$$S_T(x) = \frac{1}{\zeta(\frac{1}{2})} + \sum_{\substack{\gamma \\ |\gamma| < T}} \frac{\zeta(2\rho)}{\rho\zeta'(\rho)}\,x^{i\gamma}.$$

As usual, $\rho = \frac{1}{2} + i\gamma$ denotes the complex zeros of $\zeta(s)$. Thanks to Landau's theorem, it doesn't hurt to assume they all have real part $\frac{1}{2}$ up to height T. Haselgrove found that $S_{1000}^*(\exp(831.847)) = .00495$ thereby disproving Polya's conjecture. Lehman [16] later found $S_{1000}(\exp(20.62))$ is almost positive; this led him to find $L(x) = 1$ for $x = 906\,180\,359$. As of now, this is still the smallest known counterexample to Polya's conjecture. The largest value of $L(x)$ listed in Lehman's paper is $L(x) = 708$ at $x = 906\,400\,000$.

For the Mertens conjecture, we have

$$S_T(x) = \sum_{\substack{\gamma \\ |\gamma| < T}} \frac{x^{i\gamma}}{\rho\zeta'(\rho)} .$$

The best computational results at the moment are those of Te Riele [20]. Using 15000 zeros of $\zeta(s)$, he found that infinitely often,

$$\frac{M(x)}{\sqrt{x}} < -.84 \quad \text{and} \quad .86 < \frac{M(x)}{\sqrt{x}} .$$

Of course,

$$\sum_{\rho} \left| \frac{1}{\rho\zeta'(\rho)} \right| = + \infty$$

and so if the γ were linearly independent over $\mathbb{Z}$, Kronecker's theorem would imply that

$$\liminf_{x \to \infty} \frac{M(x)}{\sqrt{x}} = -\infty, \qquad \limsup_{x \to \infty} \frac{M(x)}{\sqrt{x}} = +\infty.$$

Although it is impossible to numerically verify the linear independence of a set of numbers over $\mathbb{Z}$, linear combinations with small coefficients can be checked. Ingham's theorem still allows us to say something. Let Γ denote the set of all positive γ under consideration and Γ' be a subset of Γ such that every γ in Γ' lies in the range $0 < \gamma < T$. Let $\{N_\gamma\}$ be a set of positive integers defined for $\gamma \in \Gamma'$. We say that the elements of Γ' are N_γ-independent in $\Gamma \cap [0,T]$ if for $\gamma_0 = 0$ or γ_0 in $\Gamma \cap [0,T]$,

$$\sum_{\gamma \in \Gamma'} n_\gamma \gamma = \gamma_0 \quad \text{with} \quad |n_\gamma| \le N_\gamma$$

implies every $n_\gamma = 0$ or γ_0 is in Γ', $n_{\gamma_0} = 1$ and all other $n_\gamma = 0$.

<u>*Corollary*</u>. *If the elements of* Γ' *are* N_γ*-independent in* $\Gamma \cap [0,T]$ *then*

(4) $$\liminf_{x \to \infty} \frac{g(x)}{x^{\sigma_0}} \le r_0 - \sum_{\gamma \in \Gamma'} \frac{2N_\gamma}{N_\gamma + 1} |r_\gamma| \left(1 - \frac{|\gamma|}{T}\right),$$

(5) $$\limsup_{x \to \infty} \frac{g(x)}{x^{\sigma_0}} \ge r_0 + \sum_{\gamma \in \Gamma'} \frac{2N_\gamma}{N_\gamma + 1} |r_\gamma| \left(1 - \frac{|\gamma|}{T}\right).$$

In particular, if the elements of Γ' *are* N_γ*-independent in* Γ*, then*

$$\liminf_{x\to\infty} \frac{g(x)}{x^{\sigma_0}} \le r_0 - \sum_{\gamma\in\Gamma'} \frac{2N_\gamma}{N_\gamma+1} |r_\gamma| ,$$

$$\limsup_{x\to\infty} \frac{g(x)}{x^{\sigma_0}} \ge r_0 + \sum_{\gamma\in\Gamma'} \frac{2N_\gamma}{N_\gamma+1} |r_\gamma| .$$

Proof. We write

$$A_T^*(u) = S_T^*(e^u) = r_0 + \sum_{\substack{\gamma \\ 0<\gamma<T}} a_\gamma\{\exp[i(\gamma u+\theta_\gamma)] + \exp[-i(\gamma u+\theta_\gamma)]\}$$

where

$$a_\gamma = |r_\gamma|(1 - \frac{|\gamma|}{T}), \quad \theta_\gamma = \arg(r_\gamma).$$

For any $U > 0$, we have

$$\frac{1}{2U}\int_{-U}^{U} A_T^*(u) \prod_{\gamma\in\Gamma'} \{\exp[\tfrac{i}{2}(\gamma u+\theta_\gamma)] + \exp[\tfrac{-i}{2}(\gamma u+\theta_\gamma)]\}^{2N_\gamma} du$$

$$\le \sup_u A_T^*(u) \cdot \frac{1}{2U}\int_{-U}^{U} \prod_{\gamma\in\Gamma'} \{\exp[\tfrac{i}{2}(\gamma u+\theta_\gamma)] + \exp[-\tfrac{i}{2}(\gamma u+\theta_\gamma)]\}^{2N_\gamma} du.$$

If we let $U \to \infty$ and use the N_γ-independence of Γ' in $\Gamma\cap[0,T]$, we get

$$r_0 \prod_{\gamma\in\Gamma'} \binom{2N_\gamma}{N_\gamma} + \sum_{\gamma\in\Gamma'} 2a_\gamma \binom{2N_\gamma}{N_\gamma+1} \prod_{\substack{\gamma'\in\Gamma \\ \gamma'\ne\gamma}} \binom{2N_{\gamma'}}{N_{\gamma'}} \le [\sup_u A_T^*(u)] \prod_{\gamma\in\Gamma'} \binom{2N_\gamma}{N_\gamma}.$$

This gives (5). Inequality (4) follows from (5) by replacing $g(x)$ by $-g(x)$; this replaces r_0 by $-r_0$ while leaving $|r_\gamma|$ unchanged and adding π to each θ_γ.

The special case of all the N_γ being equal yields a theorem of Diamond [4] (he did not use Ingham's theorem in his proof). Earlier, Bateman, et. al. [2] had proved the case of all $N_\gamma = 1$ as a corollary of Ingham's theorem essentially in the manner above. This method of proof is also essentially that used in one of the proofs of Kronecker's theorem. Indeed, Kronecker's theorem follows from the corollary by taking $g(x) = S_T(x)$ with all $r_\gamma = 1$ and all the N_γ sufficiently large. The strong corollary to Landau's theorem follows also by taking $\Gamma' = \{\gamma\}$ and $N_\gamma = 1$.

Saffari [18] had shown that Mertens' conjecture is false if the only relation,

$$\sum n_\gamma \gamma = 0,$$

among the first 28000 zeros of $\zeta(s)$ with $\sum |n_\gamma| < 28000$ is given by all $n_\gamma = 0$. This was vastly improved by Grosswald [10] who used the corollary with the first 75 zeros of $\zeta(s)$ and all $N_\gamma = 13$ to show that Merten's conjecture is false if the required 13-independence of the γ is true. Unfortunately, this still requires 27^{75} numerical verifications and is clearly not feasible on a computer. It is clear that using a finite T together with larger N_γ for small γ and smaller N_γ for larger γ would cut the required calculation still further but not to the point that it could really be done. As another example of what's involved here, Grosswald also found that the 1-independence of the first twenty zeros of $\zeta(s)$ in Γ would show that

$$\limsup_{x \to \infty} \frac{|M(x)|}{\sqrt{x}} > \frac{1}{3}.$$

Even this calculation would be expensive although Bateman, et. al. [2] have verified the N_γ-independence of Γ' in Γ' itself.

4. Proof by example. The method at the end of the previous section actually proceeds by finding an average value of $S_T^*(x)$ and so it is clear that hunting for good values of x should do better. But there is another way to hunt for good values of x which might be called proof by example. The method is based on the explicit formulas of number theory and first arose in 1951 in papers of Fawaz [6], [7]. It is clear from remarks at the end of Fawaz [7], that Ingham knew of the method but he does not seem to have published anything on it. It may be useful to first sketch how Ingham's theorem can be used in this situation.

An explicit formula in number theory begins with a Mellin transform,

$$G(s) = \int_0^\infty g(x) x^{-s} \frac{dx}{x} \quad (\sigma = \sigma_2)$$

absolutely convergent on some line $\sigma = \sigma_2$. The inverse transform is given by

(6) $$g(x) = \frac{1}{2\pi i} \int_{\sigma_2 - i\infty}^{\sigma_2 + i\infty} G(s) x^s ds.$$

Usually, there is a strip $\sigma_1 \leq \sigma \leq \sigma_2$ containing poles of G(s). We denote the poles of G(s) by $\rho = \beta + i\gamma$ and the residue of G(s) at $s = \rho$ by r_ρ. The line of integration in (6) is moved leftwards to $\sigma = \sigma_1$, picking up residues at the points $s = \rho$. Assuming for ease in writing, that G(s) has first order poles at each ρ, we get

(7) $$g(x) = \sum_\rho r_\rho x^\rho + \frac{1}{2\pi i} \int_{\sigma_1 - i\infty}^{\sigma_1 + i\infty} G(s) x^s \, ds.$$

Usually, there is a function h(x) such that the Mellin transform

$$\int_0^\infty h(x) x^{-s} \frac{dx}{x} = G(s) \quad (\sigma = \sigma_1)$$

is absolutely convergent on the line $\sigma = \sigma_1$. The explicit formula for g(x) then becomes

$$g(x) = \sum_\rho r_\rho x^\rho + h(x).$$

In applying Ingham's theorem, it usually may be assumed to be the case that the poles of G(s) in the strip $\sigma_1 \leq \sigma \leq \sigma_2$ occur on a particular line $\sigma = \sigma_0$ and are all of first order. Looking at the derivation of (7) more carefully, we find an assumption that

(8) $$G(s) \to 0 \text{ uniformly for } s = \sigma + iT_j, \quad \sigma_1 \leq \sigma \leq \sigma_2,$$

for some sequence $T_1, T_2, \ldots$ going to infinity. In this case, we find

(9) $$g(x) = \lim_{T_j \to \infty} x^{\sigma_0} S_{T_j}(x) + h(x).$$

When the limit as $T \to \infty$ of $S_T(x)$ exists, we also have

(10) $$\lim_{T \to \infty} S_T^*(x) = \lim_{T \to \infty} S_T(x)$$

and so Ingham's theorem implies that for any positive x_0,

$$\text{(11)} \qquad \liminf_{x \to \infty} \frac{g(x)}{x^{\sigma_0}} \le \frac{g(x_0)-h(x_0)}{x_0^{\sigma_0}} \le \limsup_{x \to \infty} \frac{g(x)}{x^{\sigma_0}} .$$

This is the method used by Stark [19] who used Ingham's theorem for logarithmic singularities to derive new oscillation results for arithmetic progression of primes. This was the first application of Ingham's theorem in the literature to the method of proof by example and was made possible by the fact that (10) holds.

A simple example illustrating this situation is the exact formula for $\psi(x) - x$ which is

$$-\sum_{\rho} \frac{x^\rho}{\rho} = \begin{cases} \psi_0(x) - x + \frac{\zeta'}{\zeta}(0) + \frac{1}{2}\log(1-x^{-2}) & x > 1 \\ \sum'_{n \le x^{-1}} \frac{\Lambda(n)}{n} - \gamma - \log x - x - \frac{1}{2}\log(\frac{1-x}{1+x}) & x < 1 \end{cases} .$$

By taking $x_0 = 1^-$, we find

$$\limsup_{x \to \infty} \frac{\psi(x)-x}{\sqrt{x}} = +\infty$$

which implies

$$\text{(12)} \qquad \limsup_{x \to \infty} \frac{\pi(x)-\ell i(x)}{x^{\frac{1}{2}}/\log x} = +\infty.$$

This is our version of Littlewood's theorem. Littlewood analyzed $x = 1$ in a more careful manner and got a better result. Ingham's theorem for logarithmic singularities leads to (12) directly (see Stark [19] who used it for $\pi(x;4,1)-\pi(x;4,3)$ and similar problems). Diamond [5] has proved (12) via Tauberian techniques and without even using the explicit formula for $\psi(x)$. He essentially reproves Ingham's theorem.

However, for many interesting oscillation problems, (9) holds only for a sequence $T = T_j \to \infty$. In this case it is not obvious that (10) holds for just sequences. Such is the case for Polya's conjecture which was treated by Fawaz and the Mertens conjecture treated by Jurkat [15]. It is perhaps for this reason that neither used Ingham's theorem although Fawaz at least relied heavily on Ingham's methods. The first paper to use Ingham's theorem in the method of proof by example for a problem where (9) holds only for a sequence $T = T_j$ is that of Anderson [1]. He proved a result of the form

$$\left|\frac{g(x_0)-h(x_0)}{x_0^{\sigma_0}}\right| \leq \limsup_{x \to \infty} \frac{|g(x)|}{x^{\sigma_0}}$$

to disprove an analogue of Mertens' conjecture for cusp forms proposed by Goldstein [8]. Our proof of Theorem 1 below which yields a result of the type (11) from Ingham's theorem is based upon Anderson's method.

Lemma 1. Suppose $f(u)$ *is a piecewise differentiable function such that for* $\mathrm{Re}(s) = \sigma_1$*, the extended Laplace transform*

$$F(s) = \int_{-\infty}^{\infty} f(u)e^{-su}du$$

exists as an absolutely convergent integral. Then for any σ_0 *and* u *at which* f *is continuous*

$$\lim_{\varepsilon \to 0^+} \frac{1}{2\pi i} \int_{\sigma_1-i\infty}^{\sigma_1+i\infty} F(s)e^{\varepsilon(s-\sigma_0)^2} e^{u(s-\sigma_0)} ds = e^{-\sigma_0 u} f(u).$$

Proof. The integral in question is

$$I = \frac{1}{2\pi i} \int_{\sigma_1-\sigma_0-i\infty}^{\sigma_1-\sigma_0+i\infty} F(s+\sigma_0)e^{\varepsilon s^2} e^{us} ds$$

and as such, is a convolution of two functions. In fact for $\mathrm{Re}(s) = \sigma_1-\sigma_0$

$$\int_{-\infty}^{\infty} [f(u)e^{-\sigma_0 u}]e^{-su}du = F(s+\sigma_0),$$

while

$$\int_{-\infty}^{\infty} \frac{1}{\sqrt{4\pi\varepsilon}} e^{-\frac{u^2}{4\varepsilon}} e^{-us}du = e^{\varepsilon s^2}$$

is the analytic continuation from $s = it$ of the usual Fourier transform formula for the Gauss kernel. Hence

$$I = \int_{-\infty}^{\infty} [f(t)e^{-\sigma_0 t}] \frac{1}{\sqrt{4\pi\varepsilon}} e^{-\frac{(u-t)^2}{4\varepsilon}} dt \to \frac{e^{-\sigma_0 u}}{2} [f(u^+)+f(u^-)]$$

as $\varepsilon \to 0^+$ (when the right side exists).

Theorem 1. *Suppose* $G(s)$ *is a meromorphic function in the strip* $\sigma_1 \le \sigma \le \sigma_2$ *with poles* $\rho = \beta+i\gamma$ *at least one of which is on the line* $\sigma = \sigma_0$ *with* $\sigma_1 < \sigma_0 < \sigma_2$ *where*

(13) $$\sigma_0 - \inf \beta \le \sup \beta-\sigma_0.$$

We suppose further that the number of poles of $G(s)$ *in this strip with* $|\gamma| < T$ *is* $O(\exp(T^{3/2}))$ *as* $T \to \infty$ *and that for some sequence* $T_1, T_2, \ldots$ *tending to infinity,* $G(s) = O(\exp(T^{3/2}))^*$ *uniformly for* $s = \sigma \pm iT_j$, $\sigma_1 \le \sigma \le \sigma_2$. *Suppose that* $g(x)$ *and* $h(x)$ *are piecewise differentiable real valued functions on* R^+ *such that for* $\sigma = \sigma_2$,

$$\int_0^\infty g(x)x^{-s} \frac{dx}{x} = G(s) \quad (\sigma = \sigma_2),$$

and for $\sigma = \sigma_1$

$$\int_0^\infty h(x)x^{-s} \frac{dx}{x} = G(s) \quad (\sigma = \sigma_1),$$

both integrals converging absolutely. Then for any x_0 *at which* g *and* h *are continuous,*

$$\liminf_{x\to\infty} \frac{g(x)}{x^{\sigma_0}} \le \frac{g(x_0)-h(x_0)}{x^{\sigma_0}} \le \limsup_{x\to\infty} \frac{g(x)}{x^{\sigma_0}}$$

Proof. Suppose $\limsup_{x\to\infty} \frac{g(x)}{x^{\sigma_0}} \ne +\infty$. By Landau's theorem, we may assume that there are no poles of $G(s)$ in the region $\sigma_2 \ge \sigma > \sigma_0$ and that any pole of $G(s)$ on the line $\sigma = \sigma_0$ is a first order pole. [Note that Landau's theorem and Ingham's theorem both extend to the case of functions $g(x)$ that are poorly behaved at $x = 0$. We simply apply these theorems to the function $g_1(x)$ which is $g(x)$ for $x > 1$ and 0 for $x \le 1$. The transform $G_1(s)$ differs from $G(s)$ by an

* or even $O(\exp(T^{2-\delta}))$.

analytic function of s in the strip $\sigma_1 < \sigma < \sigma_2$.] In the hypothesis (13), every pole of G(s) is on the line $\sigma = \sigma_0$; we denote them by $s = \rho = \sigma_0 + i\gamma$ and the residues by r_γ. Again by Landau's theorem, we may assume

$$r_\gamma = O(1) \quad \text{as } \gamma \to \infty.$$

Our hypothesis on the number of poles of G(s) in large rectangles implies that for any $\varepsilon > 0$, $\sum_\gamma r_\gamma e^{-\varepsilon\gamma^2}$ is absolutely convergent.

Let

$$A_{T,\varepsilon}(u) = \sum_{|\gamma| < T} r_\gamma \, e^{i\gamma u - \varepsilon\gamma^2}$$

and

$$A^*_{T,\varepsilon}(u) = \sum_{|\gamma| < T} r_\gamma \left(1 - \frac{|\gamma|}{T}\right) e^{i\gamma u - \varepsilon\gamma^2}.$$

By Ingham's theorem, for any u_0 and positive T and ε,

$$\limsup_{x \to \infty} \frac{g(x)}{x^{\sigma_0}} \geq \sup_u A^*_T(u) \int_{-\infty}^{\infty} \frac{1}{\sqrt{4\pi\varepsilon}} e^{-\frac{1}{4\varepsilon} t^2} dt$$

$$\geq \frac{1}{\sqrt{4\pi\varepsilon}} \int_{-\infty}^{\infty} A^*_T(u_0 + t) \, e^{-\frac{1}{4\varepsilon} t^2} dt$$

$$= \frac{1}{\sqrt{4\pi\varepsilon}} \int_{-\infty}^{\infty} \sum_{|\gamma| < T} r_\gamma \left(1 - \frac{|\gamma|}{T}\right) e^{i\gamma u_0 - \frac{1}{4\varepsilon} t^2 + i\gamma t} dt$$

$$= A^*_{T,\varepsilon}(u_0).$$

But now, the limit as $T \to \infty$ of $A_{T,\varepsilon}(u_0)$ exists and so

$$\limsup_{x \to \infty} \frac{g(x)}{x^{\sigma_0}} \geq \lim_{T \to \infty} A_{T,\varepsilon}(u_0)$$

$$= \lim_{j \to \infty} \left\{ \frac{1}{2\pi i} \int_{\sigma_2 - iT_j}^{\sigma_2 + iT_j} G(s) e^{\varepsilon(s-\sigma_0)^2} e^{u_0(s-\sigma_0)} ds \right.$$

$$- \frac{1}{2\pi i} \int_{\sigma_1 - iT_j}^{\sigma_1 + iT_j} G(s) e^{\varepsilon(s-\sigma_0)^2} e^{u_0(s-\sigma_0)} ds \Big\}$$

$$= \frac{1}{2\pi i} \int_{\sigma_2 - i\infty}^{\sigma_2 + i\infty} G(s) e^{\varepsilon(s-\sigma_0)^2} e^{u_0(s-\sigma_0)} ds - \frac{1}{2\pi i} \int_{\sigma_1 - i\infty}^{\sigma_1 + i\infty} G(s) e^{\varepsilon(s-\sigma_0)^2} e^{u_0(s-\sigma_0)} ds.$$

Finally, we may let $\varepsilon \to 0^+$ and use Lemma 1. With $u_0 = \log x_0$, the result is

$$\limsup_{x \to \infty} \frac{g(x)}{x^{\sigma_0}} \geq \frac{g(x_0) - h(x_0)}{x^{\sigma_0}} .$$

The lim inf part of the theorem again follows by simply replacing $g(x)$, $h(x)$ and $G(s)$ by $-g(x)$, $-h(x)$, $-G(s)$.

<u>*Corollary*</u>. *Suppose in addition to the hypotheses of Theorem 1 that equality holds in* (13) *and that*

$$g(x) = o(x^{\sigma_0}) \text{ as } x \to 0^+, \; h(x) = o(x^{\sigma_0}) \text{ as } x \to \infty.$$

Then

$$\limsup_{x \to \infty} \frac{g(x)}{x^{\sigma_0}} = \limsup_{x \to 0^+} \frac{-h(x)}{x^{\sigma_0}} , \quad \liminf_{x \to \infty} \frac{g(x)}{x^{\sigma_0}} = \liminf_{x \to 0^+} \frac{-h(x)}{x^{\sigma_0}} .$$

<u>Proof</u>. By hypothesis and Theorem 1,

(14) $$\limsup_{x \to \infty} \frac{g(x)}{x^{\sigma_0}} \to \limsup_{x \to 0^+} \frac{-h(x)}{x^{\sigma_0}}$$

Likewise,

(15) $$\liminf_{x \to 0^+} \frac{-h(x)}{x^{\sigma_0}} \geq \liminf_{x \to \infty} \frac{g(x)}{x^{\sigma_0}} .$$

If we replace x by x^{-1} in the integrals in Theorem 1 and s by $-s$, we will interchange g and h and so, we get from (15),

$$\liminf_{x \to 0^+} \frac{-g(1/x)}{x^{-\sigma_0}} \geq \liminf_{x \to \infty} \frac{h(1/x)}{x^{-\sigma_0}}$$

which is to say,

$$\liminf_{x \to \infty} \frac{-g(x)}{x^{\sigma_0}} \geq \liminf_{x \to 0^+} \frac{h(x)}{x^{\sigma_0}} .$$

This is indeed the opposite inequality to (14). As usual, the lim inf result follows from the lim sup result.

The example

$$g(x) = \begin{cases} x+1 & x \geq 1 \\ 0 & x < 1 \end{cases},$$

$$h(x) = \begin{cases} 0 & x > 1 \\ -x-1 & x \leq 1 \end{cases}$$

with $\sigma_1 = -1$, $\sigma_0 = 0$, $\sigma_2 = 2$ shows how necessary equality is in (13) for the corollary to hold. In applications, this usually comes from being able to assume a Riemann hypothesis.

Perhaps a good starting example is Polya's conjecture considered by Fawaz. In [6], he found the exact formula for $L(x)$, assuming the Riemann hypothesis and the simplicity of the zeros of $\zeta(s)$ (consequences of Polya's conjecture); for all $x > 0$,

$$L_0(x) = \frac{x^{1/2}}{\zeta(1/2)} + \sum_{\rho}{}' \frac{\zeta(2\rho)}{\rho\zeta'(\rho)} x^{\rho} + I(x)$$

where

(16) $$I(x) = \frac{1}{2\pi i} \int_{a-i\infty}^{a+i\infty} \frac{\zeta(2s)}{s\zeta(s)} x^s \, ds, \quad 0 < a < \frac{1}{2} .$$

Unlike the case of the Mertens conjecture, here there is only a pole at $s = 0$ (with a residue of 1) to the left of $\mathrm{Re}(s) = a$. Nevertheless, Fawaz was able to evaluate $I(x)$. For $M(x)$, one usually evaluates the residues, uses the functional equation to shift to the right half plane, replaces $1/\zeta(s)$ by its Dirichlet series and rearranges everything to get (20) below. Fawaz simply applied the functional equation first and used the Dirichlet series

$$\frac{\zeta(2s-1)}{\zeta(s)} = \sum_{n=1}^{\infty} c_n n^{-s}$$

to get

$$I(x) = 1 + 2 \sum_{n=1}^{\infty} \frac{c_n}{n} \left[C(\sqrt{nx}) + S(\sqrt{nx}) - 1\right] \tag{17}$$

where

$$C(y) = \int_0^y \cos\left(\frac{\pi v^2}{2}\right)dv = \frac{1}{2} - \int_y^{\infty} \cos\left(\frac{\pi v^2}{2}\right)dv$$

$$S(y) = \int_0^y \sin\left(\frac{\pi v^2}{2}\right)dv = \frac{1}{2} - \int_y^{\infty} \sin\left(\frac{\pi v^2}{2}\right)dv.$$

The same method for $M(x)$ leads to (20) directly.

The behavior of $I(x)$ for large x is easy to discuss. An integration by parts shows that

$$\left|\int_y^{\infty} e^{\frac{i\pi v^2}{2}} dv\right| = \left|\frac{i}{\pi y} e^{\frac{\pi i}{2} y^2} - \frac{i}{\pi} \int_y^{\infty} v^{-2} e^{\frac{\pi i}{2} v^2} dv\right| \leq \frac{2}{\pi y}$$

and hence,

$$|C(y) + S(y) - 1| \leq \frac{2\sqrt{2}}{\pi y}$$

(or, more obviously, $< \frac{4}{\pi y}$). It follows that

$$|I(x) - 1| \leq \frac{4\sqrt{2}}{\pi\sqrt{x}} \sum_{n=1}^{\infty} \frac{|c_n|}{n^{3/2}} = \frac{4\sqrt{2}}{\pi\sqrt{x}} \frac{\zeta(\frac{3}{2})\zeta(2)}{\zeta(3)} . \tag{18}$$

The behavior of $L(x)$ at ∞ is mirrored by the behavior of $I(x)$ at 0. Thus on the Riemann hypothesis, $I(x) = 0(x^{\frac{1}{2}-\varepsilon})$ as $x \to 0^+$ and this, together with (18), shows that the Mellin transform of $I(x)$,

$$\int_0^{\infty} x^{-s} I(x) \frac{dx}{x} \tag{19}$$

converges in the strip $0 < \mathrm{Re}(s) < \frac{1}{2}$. This accounts for the range on a in (16). If we use (17) as the definition of $I(x)$, and break the sum on n into two parts: $n \leq 1/x$ and $n > 1/x$, the crudest absolute value estimates give

$$I(x) = O(\log^2(\tfrac{1}{x})) \quad \text{as } x \to 0^+$$

without any hypothesis at all. This is not sufficient to show the convergence of (19) anywhere.

However, if we let

$$h(x) = \begin{cases} I(x) & x < 1 \\ I(x) - 1 & x > 1 \end{cases}$$

then

$$h(x) = O(x^{-\frac{1}{2}}) \quad \text{as} \quad x \to \infty$$

and so the Mellin transform

$$G(s) = \int_0^\infty x^{-s} h(x) \frac{dx}{x}$$

converges absolutely in the strip $-\frac{1}{2} < \mathrm{Re}(s) < 0$. Needless to say,

$$G(s) = \frac{1}{s} \left(\frac{\zeta(2s)}{\zeta(s)} - 1\right)$$

and so

$$h(x) = \frac{1}{2\pi i} \int_{a-i\infty}^{a+i\infty} \frac{1}{s} \left(\frac{\zeta(2s)}{\zeta(s)} - 1\right) x^2 ds, \quad -\frac{1}{2} < a < 0.$$

This shows that modulo the horizontal integrals, *Fawaz's exact formula for* $L(x)$ *with* $I(x)$ *given by* (17) *is valid without any hypothesis at all* (of course if $\zeta(s)$ has a multiple zero at $s = \rho$, the corresponding term in the sum over ρ must be replaced by an appropriate residue).

In Theorem 1, we take $g(x) = L(x) - 1$ for $x > 1$, $g(x) = 0$ for $0 < x < 1$. This gives $G(s)$ and $h(x)$ as above. The method of proof by example says that

$$\limsup_{x \to \infty} \frac{L(x)}{\sqrt{x}} \geq \frac{L(x_0) - I(x_0)}{\sqrt{x_0}}$$

for any x_0. Fawaz states that he was unable to find an x_0 such that $L(x_0) > I(x_0)$. He seems to have looked only between 0 and 2 but does not give any details. Thanks to Lehman, we have such an x_0. For x around $9 \cdot 10^8$, we see from (18) that $|I(x)| < 2$. Hence, with

$$x_0 = 906\ 400\ 000,$$

we get

$$\limsup_{x \to \infty} \frac{L(x)}{\sqrt{x}} > \frac{700}{\sqrt{x_0}} > .023.$$

The explicit formula for M(x) is

$$M_0(x) = \sum_{\rho} \frac{x^{\rho}}{\rho \zeta'(\rho)} + I(x) \tag{20}$$

where now

$$I(x) = -2 - 2 \sum_{n=1}^{\infty} \frac{\mu(n)}{n} \int_0^{\frac{2\pi x}{n}} \frac{\cos u - 1}{u}\, du.$$

We take

$$g(x) = \begin{cases} M(x) + 2 & x > 1 \\ 0 & x < 1 \end{cases}$$

$$G(s) = \frac{1}{s}\left(\frac{1}{\zeta(s)} + 2\right),$$

$$h(x) = \begin{cases} I(x) + 2 & x > 1 \\ I(x) & x < 1 \end{cases}$$

in Theorem 1. The Mellin transform of h(x) converges absolutely in the strip $-2 < \sigma < 0$ (on the Riemann hypothesis, this strip grows to $-2 < \sigma < \frac{1}{2}$). Thus

$$\liminf_{x \to \infty} \frac{M(x)}{\sqrt{x}} \le \frac{M(x_0) - I(x_0)}{\sqrt{x_0}} \le \limsup \frac{M(x)}{\sqrt{x}}.$$

Jurkat [15] noted that $M(x_0) - I(x_0)$ has a jump of 1 at $x_0 = 1$ and so, at the very least, Von Sterneck's conjecture is false. Unfortunately, hardly anything better comes out from the explicit values at $x = 1^-$ and 1^+. Indeed,

$$M(1^-) - I(1^-) = -.5054, \quad M(1^+) - I(1^+) = .495 .$$

In particular,

$$\limsup_{x \to \infty} \frac{|M(x)|}{\sqrt{x}} \geq .505.$$

This is Jurkat's result and is still the best result known that does not require a computer calculation. Jurkat did not use Ingham's theorem. Note, however, that Neubauer's example, once available, gives

$$\limsup_{x \to \infty} \frac{M(x)}{\sqrt{x}} > .557 .$$

As with Polya's conjecture, the trivial terms do not matter much for such an x. (The trivial terms do still matter around x = 200 where there is another counter-example to Von Sterneck's conjecture and we do not quite get .5 for $\limsup M(x)\, x^{-1/2}$ as a result.) Although he did not state it, it is clear that Jurkat [15, p. 152] noted this.

Goldstein [8] proposed an analogue of Merten's conjecture for cusp forms. Suppose

$$f(z) = \sum_{n=1}^{\infty} a_n \, e^{2\pi i n z}$$

is a cusp form of weight k on the whole modular group which is an eigenform for all Hecke operators and $a_1 = 1$. Let

$$F(s) = \sum a_n n^{-s}$$

and define $\mu(n;f)$ by

$$\frac{1}{F(s)} = \sum_{n=1}^{\infty} \mu(n;f) n^{-s} .$$

Goldstein defines

$$M(x;f) = \sum_{n \leq x} \mu(n;f)$$

and then asks if for x > 1,

$$|M(x;f)| < x^{k/2} .$$

Formally, things are very similar to the ordinary Mertens conjecture. For instance, using the functional equation for F(s),

$$(2\pi)^{-s}\Gamma(s)F(s) = (-1)^{\frac{k}{2}}(2\pi)^{-(k-s)}\Gamma(k-s)F(k-s),$$

we should get the exact formula for M(x;f) for x > 0:

$$M_0(x;f) = \sum_\rho{}' \frac{x^\rho}{\rho F'(\rho)} + \sum_{n=0}^{\infty} \underset{s=-n}{\text{res}} \left[\frac{x^s}{sF(x)}\right]$$

$$= \sum_\rho{}' \frac{x^\rho}{\rho F'(\rho)} + I(x).$$

(This would depend on verifying (8) for a sequence of $T_j \to \infty$. However, the hypothesis $G(s) = O(\exp(T_j^{3/2}))$ of Theorem 1 is trivial.) Again there is a jump of 1 across x = 1 but this time it is not centered at zero. Indeed, for large k Anderson [1] has found that

$$[M_0(x;f) - I(x)]\Big|_{x=1^+} = 1 + o(1)$$

and that the o(1) term is in fact positive for $k \equiv 2 \pmod 4$ (thanks to the $(-1)^{k/2}$ in the functional equation). Thus for large $k \equiv 2 \pmod 4$, the Mertens' conjecture for cusp forms if false.

5. The Gibb's phenomenon and Ingham's Theorem. Although we may replace the factor of $(1 - \frac{|y|}{T})$ in $S_T^*(x)$ by certain other integral kernels, and thereby improve the numerical results somewhat, it would be nice if Ingham's theorem held with $S_T(x)$ in place of $S_T^*(x)$. Unfortunately, there are counterexamples related to Gibb's phenomenon. A related question is why we didn't attempt to make use of Gibb's phenomenon applied to $S_T^*(x)$ for x near 1 and T large in the case of Mertens' conjecture. The answer is that there is no Gibb's phenomenon with $S_T^*(x)$. In the case of Fourier series, this says that the Gibb's phenomenon disappears with Cesaro sums. Interestingly enough, we can prove this directly from Ingham's theorem.

In this section, it will be convenient to deal with Laplace transforms rather than Mellin transforms. The variable change $x = e^u$ turns a Mellin transform into an extended Laplace transform,

$$F(s) = \int_{-\infty}^{\infty} f(u)e^{-us}du.$$

In using Ingham's theorem, $S_T(x)$ and $S_T^*(x)$ turns into

$$A_T(x) = \sum_{|\gamma| < T} r_\gamma \, e^{i\gamma u}$$

and

$$A_T^*(u) = \sum_{|\gamma| < T} r_\gamma(1- \frac{|\gamma|}{T}) \, e^{i\gamma u} .$$

The similarity of $A_\infty(u)$ to a Fourier series is readily apparent. Indeed, suppose we have a periodic function $f(u)$,

$$f(u+1) = f(u),$$

which generates the Fourier series,

$$f(u) \sim \sum_{n=-\infty}^{\infty} r_n e^{2\pi inu}.$$

As number theorists, we are obligated to investigate this relationship by examining the inverse Laplace transform of

$$\int_{-\infty}^{\infty} f(u)e^{-us}du.$$

Unfortunately, due to the periodic nature of f, this integral is divergent for all s. The closest we come to convergence is when s = it is purely imaginary and the Laplace transform turns into a Fourier transform.

This being the case, we turn to the two one sided Laplace transforms of f. Let

$$g(u) = \begin{cases} f(u) & u > 0 \\ 0 & u < 0 \end{cases},$$

$$h(u) = \begin{cases} 0 & u > 0 \\ -f(u) & u < 0 \end{cases}.$$

For $\sigma > 0$, we have the absolutely convergent transform

$$G(s) = \int_{-\infty}^{\infty} g(u)e^{-us}du.$$

Since $f(u+1) = f(u)$,

$$G(s) = \int_0^1 f(u)e^{-us}du + \int_0^{\infty} f(u+1)e^{-(u+1)s}du = \int_0^1 f(u)e^{-us}du + e^{-s}G(s),$$

which gives the well known expression,

$$G(s) = \frac{1}{1-e^{-s}} \int_0^1 f(u)e^{-us}du.$$

The original integral for $G(s)$ converges for $Re(s) > 0$ but this expression gives a meromorphic continuation of $G(s)$ to the entire plane with first order poles at $s = 2\pi in$, n in $\mathbb{Z}$.

Likewise, for $\sigma < 0$, we have the absolutely convergent transform

$$H(s) = \int_{-\infty}^{\infty} h(u)e^{-us}du$$

$$= \int_0^1 f(u)e^{-us}du - \int_{-\infty}^{0} f(u+1)e^{-(u+1)s}du$$

$$= \int_0^1 f(u)e^{-us}du + e^{-s}H(s).$$

Hence

$$H(s) = G(s).$$

Our explicit formula from number theory thus becomes

$$g_0(u) = \frac{1}{2\pi i} \int_{1-i\infty}^{1+i\infty} G(s)e^{us}ds$$

$$= \sum_{n=-\infty}^{\infty} \underset{s=2\pi in}{\text{res}} [G(s)e^{us}] + \frac{1}{2\pi i} \int_{-1-i\infty}^{-1+i\infty} G(s)e^{us}ds$$

$$= \sum_{n=-\infty}^{\infty} r_n e^{2\pi inu} + h_0(u)$$

where

$$r_n = \int_0^1 f(u)e^{-2\pi inu}\,du.$$

In other words, the explicit formula for g(u) is nothing more than the Fourier series for f(u). The rigorous justification of this uses the Riemann Lebesgue Lemma to move the line of integration to the left and, more importantly, requires starting with a function nice enough that the inverse Laplace transform gives the correct answer. This may seem like a silly way of deriving Fourier series, but there are good reasons for wanting to find ways of deriving the theory of Fourier series from the theory of Fourier integrals (see the early pages of Terras [21], for example).

Now, take for example the periodic function defined by

$$f(u) = \begin{cases} 1 & 0 < u < \frac{1}{2} \\ 0 & -\frac{1}{2} < u < 0. \end{cases}$$

Since here

$$A_T(u) = \sum_{|n| \leq \frac{T}{2\pi}} r_n\, e^{2\pi inu},$$

the Gibb's phenomenon shows that Ingham's theorem for this function would be false if $A_T^*(u)$ were replaced by $A_T(u)$. Further, with $T = (N+1^-)2\pi$,

$$A_T^*(u) = \sum_{|n| \leq N} \frac{N+1-|n|}{N+1}\, r_n e^{2\pi inu} = \frac{1}{N+1} \sum_{n=0}^{N} A_{2\pi n}(u)$$

deals with the Cesaro sum of the Fourier series for f. By Ingham's theorem,

$$1 = \limsup_{u \to \infty} f(u) \geq A_T^*(u)$$

for all T and u. Thus, the Gibb's phenomenon has disappeared for this f. As a result, it disappears in all situations. For example, the inverse transform for

$$\int_0^\infty [e^{-u/2} M(e^u) - f(u)]\, e^{us}\, du$$

converges uniformly near u = 0 and this shows that the Gibb's phenomenon disappears for Mertens also.

REFERENCES

1. Anderson, Robert J. On the Mertens conjecture for cusp forms, Mathematika 26 (1979), 236-249.

2. Bateman, P.T., Brown, J.W., Hall, R.S., Kloss, K.E., Stemmler, Rosemarie M., Linear relations connecting the imaginary parts of the zeros of the zeta function, in Computers in Number Theory, Academic Press, New York, 1971, 11-19.

3. Brun, Viggo, La somme des facteurs de Möbius, C.R. Dixième Congrès Math. Scandinaves 1946, 40-53.

4. Diamond, Harold G., Two oscillation theorems, Lecture Notes in Math. 251 (1972), 113-118.

5. ________________, Changes of sign of $\pi(x) - \ell i(x)$, L'Enseignement Mathematique 21 (1975), 1-14.

6. Fawaz, A.Y., The explicit formula for $L_0(x)$, P.L.M.S. (3) 1 (1951), 86-103.

7. __________, On an unsolved problem in the analytic theory of numbers, Quart. J. Math. Oxford Ser. (2) 3 (1952), 282-295.

8. Goldstein, L.J., A necessary and sufficient condition for the Riemann hypothesis for zeta functions attached to eigenfunctions of the Hecke operators, Acta Arith. 15 (1968/69), 205-215.

9. Grosswald, Emil, On some generalizations of theorems by Landau and Polya, Israel J. Math. 3 (1965), 211-220.

10. ______________, Oscillation theorems, Lecture Notes in Math. 251 (1972), 141-168.

11. Gupta, Hansraj, On a conjecture of Miller, J. Indian Math. Soc. (N.S.) 13 (1949), 85-90.

12. Gupta, H., Cheema, M.S., Gupta, O.P., On Möbius means, Res. Bull. Panjab Univ., no. 42 (1954).

13. Haselgrove, C.B., A disproof of a conjecture of Polya, Mathematika 5 (1958), 141-145.

14. Ingham, A.E., On two conjectures in the theory of numbers, Amer. J. Math. 64 (1942), 313-319.

15. Jurkat, W., On the Mertens conjecture and related general Ω-theorems, Proc. Symp. Pure Math. 24, A.M.S., Providence, 1973, 147-158.

16. Lehman, R. Sherman, On Liouville's function, Math. Comp. 14 (1960), 311-320.

17. Neubauer, Gerhard, Eine empirische Untersuchung zur Mertensschen Funktion, Numer. Math. 5 (1963), 1-13.

18. Saffari, Bahman, Sur la fausseté de la conjecture de Mertens, C.R. Acad. Sci. Paris Ser A-B 271 (1970), A1097-A1101.

19. Stark, H.M., A problem in comparative prime number theory, Acta Arith. 18 (1971), 311-320.

20. te Riele, H.J.J., Computations concerning the conjecture of Mertens, J. Reine Angew. Math., 311/312 (1979), 356-360.

21. Terras, Audrey, Fourier Analysis on symmetric spaces and applications to number theory, to be published.

Robert Anderson: Tufts University

H.M. Stark: M.I.T. and U.C.S.D.

ON THE GENERALIZED DENSITY HYPOTHESIS, I

by

Larry Joel Goldstein*

Dedicated to my Friend and Teacher Emil Grosswald

Department of Mathematics

University of Maryland

College Park, Maryland

MD80-10-LG

TR80-10

February 1980

1. Introduction

Let a be a rational integer and let $A = A(a) = \{p|p$ is a rational prime and a is a primitive root mod $p\}$. In 1927, Artin conjectured that $A(a)$ is infinite, provided that a is neither -1 nor a perfect square. The special case $a = 10$ was conjectured by Gauss in Art. 303 of his Disquisitiones Arithmeticae. To date, the conjecture has not been settled even for a single value of a. Artin put the conjecture in an even more precise form: For each rational prime q, let ζ_q be a primitive qth root of 1 and let

$$L_q = Q(\zeta_q, a^{1/q}).$$

Throughout, let k denote a square-free number (1 included), $k = q_1 \cdots q_r, q_i$ distinct primes (1 = the vacuous product). To each k, set

$$L_k = L_{q_1} \cdots L_{q_r}; L_1 = Q$$

$$n(k) = \deg(L_k/Q).$$

* Research supported by a University of Maryland Distinguished Teacher-Scholar Grant and by National Science Foundation Grant MCS-7902107 A01.

Then Artin made the following

Conjecture A ([1, p. viii]). *The set A has a natural density d(A), whose value is*

$$\sum_k \frac{\mu(k)}{n(k)},$$

where $\mu(k)$ denotes the Möbius function and the sum is over all square-free k.

Artin came upon this conjecture in the following way: Let Spl_k denote the set of rational primes which split completely in the number field L_k. If q is a rational prime, then, using elementary algebraic number theory, it is easy to show that

$$p \in Spl_q \iff p \equiv 1 \pmod q \text{ and } a^{(p-1)/q} \equiv 1 \pmod p.$$

Therefore, elementary reasoning shows that

$$p \in A \iff p \in \bigcap_q (Spl_1 - Spl_q),$$

where the intersection is over all rational primes q and Spl_1 denotes the set of all rational primes. If we formally expand the intersection, we arrive at the formula

$$A = \sum_k \mu(k) Spl_k,$$

where each prime is counted with multiplicity 1 or 0. The classical result which asserts that Spl_k has a natural density $1/n(k)$ then suggests Artin's conjecture.

In 1967, Hooley [2] showed that Artin's conjecture is true modulo the truth of the Riemann hypotheses for the Dedekind zeta functions of the fields L_k (GRH). Furthermore, Hooley showed, via an unconditional argument, that

$$a \neq -1, m^2 \Rightarrow \sum_k \frac{\mu(k)}{n(k)} > 0.$$

In particular, this shows that the GRH implies the infinitude of A(a) if a is neither -1 nor a perfect square.

The Riemann hypotheses for the Dedekind zeta functions of L_k may be replaced by the corresponding Riemann hypotheses for the fields $L_k' = \mathbb{Q}(\sqrt[k]{a})$ (Hooley [3, p. 59]). Moreover, if

$$\zeta_{L_k}(s) \neq 0 \quad \text{for } \mathrm{Re}(s) > 1 - \frac{1}{2e} - \delta(\delta > 0),$$

then A(a) is infinite provided $a \neq -1, m^2$ (Hooley [3, p. 61]).

The purpose of this paper is to prove a generalization of Hooley's main result [2]. We shall make two assumptions concerning the distribution of the zeros of the Dedekind zeta function. Each of these assumptions is far weaker than the generalized Riemann hypotheses, and, in fact, may be susceptible to proof in the forseeable future. Under the two assumptions, we show that Artin's conjecture is true. Furthermore, it is a direct consequence of our result that Artin's conjecture is consistent with the existence of zeros of the Dedekind zeta function arbitrarily close to the line Re(s) = 1.

Let L be an algebraic number field of degree n_L and let the absolute value of its discriminant be d_L. Throughout, let $c_1, c_2, \ldots$ denote absolute constants. Our first assumption is a zero-free region for $\zeta_L(s)$:

HYPOTHESIS 1: <u>There exists an absolute constant</u> c_1 <u>such that</u> $\zeta_L(\sigma+it) \neq 0$ <u>in the region</u> $\sigma \geq 1 - \frac{c_1}{\log(|t|+2)}$.

Note that Hypothesis 1 implies, among other things, the non-existence of Siegel zeros for c_1 sufficiently large. Indeed, Hypothesis 1 implies the non-existence of real zeros σ for which $\sigma \geq 1-c_1'$, c_1' absolute. Our second hypothesis is a zero-density theorem for $\zeta_L(s)$. Let $\sigma > 0$, $T > 0$ and let $N_L(\sigma,T)$ denote the number of zeros $\rho = \beta+i\gamma$ of $\zeta_L(s)$ in the rectangle $\sigma \leq \beta \leq 1$, $0 \leq \gamma \leq T$.

HYPOTHESIS 2: $N_L(\sigma,T) \ll (n_L T)^{2(1-\sigma)}(\log d_L^{1/n_L} T)^A$, <u>where</u> $A \geq 1$ <u>and</u> A <u>and the implied constant are both absolute and</u> $T \geq 2$.

In the next section, we shall provide motivation for the last assumption. For now, we note that each of our assumptions is consistent with the existence of zeros of $\zeta_L(s)$ which are arbitrarily close to the line Re(s) = 1. Furthermore, both assumptions are implied by GRH. (As to Hypothesis 2, see the next section.)

Let $x \geq 1$ and set

$$\pi(x,A) = \sum_{\substack{p \leq x \\ p \in A}} 1.$$

Then our main result is the following:

THEOREM 1.1: *Suppose that Hypotheses* 1 *and* 2 *are valid for all the fields* $L = L_k$. *Then*

$$\pi(x,A) = \Big(\sum_k \frac{\mu(k)}{n(k)}\Big)\frac{x}{\log x} + O\Big(\frac{x \log\log x}{\log^2 x}\Big)(x \to \infty),$$

where the O-term constant depends only on a *and* A. *That is, Conjecture* A *is true modulo Hypotheses* 1 *and* 2.

2. The Generalized Density Hypothesis

Let L be an algebraic number field of degree n_L and discriminant of absolute value d_L and let $\zeta_L(s)$ denote its Dedekind zeta function. Let $\rho = \beta + i\gamma$ be a typical zero of $\zeta_L(s)$ and let $\frac{1}{2} \leq \sigma \leq 1$, $T > 0$. Set

$$N_L(T) = \#\{\rho \mid 0 \leq \gamma \leq T\}$$

$$N_L(\sigma,T) = \#\{\rho \mid \sigma \leq \beta \leq 1,\ 0 \leq \sigma \leq T\}.$$

We shall be concerned with upper bounds for $N_L(\sigma,T)$ of the form

$$N_L(\sigma,T) \ll n_L^{\phi(\sigma)} T^{\theta(\sigma)} (\log d_L^{1/n_L} T)^A$$

where the implied constant does not depend on L,T or σ. There exist many such bounds valid under varied hypotheses. We provide a few examples.

EXAMPLE 1 (Trivial bound). By [4, Lemma 5.4], we have for $t \geq 0$ that

$$N_L(t+1)-N_L(t) \ll \log d_L + n_L \log(t+2),$$

so that for $T \geq 0$

$$N_L(T) = \sum_{t=0}^{T} (N_L(t+1)-N_L(t))$$

$$\ll T \log d_L + n_L T \log T$$

$$= n_L T \log d_L^{1/n_L} T.$$

EXAMPLE 2 (GRH). Assume the GRH for $\zeta_L(s)$. Then from Example 1, we see that for $T \geq 2$, we have

$$N_L(\sigma,T) = 0 \text{ for } \sigma > \frac{1}{2}.$$

$$N_L(\tfrac{1}{2},T) \ll n_L T \log d_L^{1/n_L} T.$$

EXAMPLE 3 (Cyclotomic Fields). Assume $L = Q(\zeta_f)$, ζ_f a primitive fth root of unity. Then the large sieve results of Montgomery [5, p. 98] give a bound for $N_L(\sigma,T)$. Indeed, we have

$$\zeta_L(s) = \prod_{\chi} L(s,\chi),$$

where χ runs over all Dirichlet characters modulo f and $L(s,\chi)$ denotes the Dirichlet L-series. Let

$$N(\sigma,T,\chi) = \#\{\rho = \beta+i\gamma \mid \rho \text{ is a zero of } L(s,\chi),\ \sigma \leq \beta \leq 1,\ 0 \leq \gamma \leq T\}.$$

Then

$$N_L(\sigma,T) = \sum_{\chi} N(\sigma,T,\chi),$$

so that by a well-known result of Montgomery, for $T \geq 2$, we have

$$N_L(\sigma,T) \ll (fT)^{\frac{3(1-\sigma)}{2-\sigma}} (\log fT)^9 \quad (\tfrac{1}{2} \leq \sigma \leq \tfrac{4}{5})$$

$$N_L(\sigma,T) \ll (fT)^{\frac{2(1-\sigma)}{\sigma}} (\log fT)^{14} \quad (\tfrac{4}{5} \le \sigma \le 1).$$

Moreover, in this case, $n_L = \phi(f)$, where $\phi(f)$ denotes the Euler function and

$$d_L = f^{\phi(f)} / \prod_{p|f} p^{\phi(f)/(p-1)},$$

so that

$$d_L^{1/n_L} \gg \sqrt{f}$$

and

$$n_L \log \log n_L \gg f.$$

Thus, we see that for $\varepsilon > 0$, we have

$$N_L(\sigma,T) \ll_\varepsilon (n_L T)^{\frac{3(1-\sigma)}{2-\sigma}} (\log d_L^{1/n_L} T)^{9+\varepsilon} (\tfrac{1}{2} \le \sigma < \tfrac{4}{5})$$

$$N_L(\sigma,T) \ll_\varepsilon (n_L T)^{\frac{2(1-\sigma)}{\sigma}} (\log d_L^{1/n_L} T)^{14+\varepsilon} (\tfrac{4}{5} \le \sigma \le 1).$$

EXAMPLE 4 (The Density Hypothesis). Let $L = \mathbb{Q}$. The estimate

$$N_{\mathbb{Q}}(\sigma,T) \ll T^{2(1-\sigma)}(\log T)^A, \; A \ge 1$$

is known as the density hypothesis and has been studied by Turan and others. The result [5, p. 101] of Montgomery contains the density hypothesis in the restricted range $\frac{9}{10} \le \sigma < 1$. This estimate asserts that there are few zeros near the line $\text{Re}(s) = 1$, yet is best possible for $\sigma = \frac{1}{2}$. In analogy to the density hypothesis, we may consider the estimate

$$N_L(\sigma,T) \ll (n_L T)^{2(1-\sigma)}(\log d_L^{1/n_L} T)^A, \; A \ge 1, \; T \ge 2. \qquad (2\text{-}2)$$

This also asserts that few zeros (of $\zeta_L(s)$) are near $\text{Re}(s) = 1$; yet, for $\sigma = \frac{1}{2}$, the estimate is essentially best possible for $A = 1$. Indeed, by a classical result of Landau, coupled with the GRH, we would have

$$N_L(\tfrac{1}{2},T) \sim \frac{n_L}{2\pi} T \log T \quad (T \to \infty).$$

We shall call the estimate (2-2) the generalized density hypothesis.

3. Three Preliminary Results

In what follows, $\sum_\rho$ shall mean a sum over the non-trivial zeros $\rho = \beta+i\gamma$ of $\zeta_L(s)$. Let $T \geq 2$, $x \geq 2$.

PROPOSITION 3.1: *Suppose that Hypothesis 1 holds for $\zeta_L(s)$. Then,*

$$\sum_{\substack{\rho \\ 0<\beta<\frac{1}{2} \\ |\gamma|\leq T}} \left|\frac{x^\rho}{\rho}\right| \ll n_L x^{\frac{1}{2}} \log T \log d_L^{1/n_L} T, \quad T \geq 2$$

where the implied constant is absolute.

PROOF. By Hypothesis 1 and the functional equation for $\zeta_L(s)$, we see that there exists an absolute constant c_2 such that $\zeta_L(s)$ has no zero in the region

$$0 \leq \operatorname{Re}(s) \leq c_2, \quad 0 \leq \operatorname{Im}(s) \leq 1.$$

Therefore, $1/|\rho| \ll 1/(|\gamma|+1)$, with an absolute implied constant. Thus,

$$\sum_{\substack{0<\beta<\frac{1}{2} \\ |\gamma|\leq T}} \left|\frac{x^\rho}{\rho}\right| \ll x^{\frac{1}{2}} \sum_{|\gamma|\leq T} \frac{1}{|\gamma|+1}$$

$$= x^{\frac{1}{2}} \sum_{t=1}^{T} \frac{N_L(t)-N_L(t-1)}{t}$$

$$\ll x^{\frac{1}{2}} \left(\frac{N_L(T)}{T} + \sum_{t=1}^{T} \frac{N_L(t)}{t^2} \right.$$

$$\ll n_L x^{\frac{1}{2}} \log T \log d_L^{1/n_L} T + n_L x^{\frac{1}{2}} \log d_L^{1/n_L} T \sum_{t=1}^{T} \frac{1}{t}$$

$$\ll n_L x^{\frac{1}{2}} \log T \log d_L^{1/n_L} T.$$

PROPOSITION 3.2: *Assume that Hypotheses 1 and 2 hold for* $\zeta_L(s)$. *Then*

$$\sum_{\substack{\rho \\ \beta > \frac{1}{2} \\ |\gamma| \leq T}} \left|\frac{x^\rho}{\rho}\right| \ll x \log x (\log d_L^{1/n_L} T)^A + \frac{x(\log d_L^{1/n_L} T)^A}{T}$$

$$+ n_L x^{\frac{1}{2}} \log x \log T (\log d_L^{1/n_L} T)^A.$$

PROOF. Note that

$$\sum_{\substack{\rho \\ \beta \geq \frac{1}{2} \\ |\gamma| \leq T}} \left|\frac{x^\rho}{\rho}\right| \ll \int_{\frac{1}{2}-}^{1} x^\sigma \, d_\sigma \int_0^T \frac{1}{1+t} \, d_t N_L(\sigma,t).$$

Moreover, by applying integration by parts to the Stieltjes integrals on the right, we obtain

$$\sum_{\substack{\rho \\ \beta \geq \frac{1}{2} \\ |\gamma| \leq T}} \left|\frac{x^\rho}{\rho}\right| \ll \int_{\frac{1}{2}-}^{1} x^\sigma \, d_\sigma \left[\frac{N_L(\sigma,T)}{T+1} + \int_0^T \frac{N_L(\sigma,T)}{(1+t)^2} \, dt\right]$$

$$= \frac{2}{T+1}\left[- x^{\frac{1}{2}} N_L(\tfrac{1}{2},T) - \log x \int_{\frac{1}{2}}^{1} x^\sigma N_L(\sigma,T) d\sigma\right]$$

$$-x^{\frac{1}{2}} \int_0^T \frac{N_L(\frac{1}{2},t)}{(1+t)^2}\, dt$$

$$- \log x \int_{\frac{1}{2}}^1 x^\sigma \int_0^T \frac{N_L(\sigma,t)}{(1+t)^2}\, dt\, d\sigma. \qquad (3.1)$$

The above integrals may be estimated using Hypothesis 2:

$$\int_{\frac{1}{2}}^1 x^\sigma N_L(\sigma,T)d\sigma \ll \int_{\frac{1}{2}}^1 x^\sigma (n_L T)^{2(1-\sigma)} (\log d_L^{1/n_L} T)^A\, d\sigma$$

$$= x(\log d_L^{1/n_L} T)^A \int_{\frac{1}{2}}^1 \left(\frac{n_L T}{\sqrt{x}}\right)^{2(1-\sigma)} d\sigma$$

$$\ll x(\log d_L^{1/n_L} T)^A \left[1 + \frac{n_L T}{\sqrt{x}}\right] \qquad (3.2)$$

$$\int_0^T \frac{N_L(\frac{1}{2},t)}{(1+t)^2}\, dt \ll \int_0^2 \frac{n_L(\log d^{1/n_L})^A}{1+t^2}\, dt + \int_2^T \frac{n_L t(\log d_L^{1/n_L} t)^A}{1+t^2}\, dt$$

$$\ll n_L \log T(\log d_L^{1/n_L} T)^A. \qquad (3.3)$$

$$\int_{\frac{1}{2}}^1 x^\sigma \int_0^T \frac{N_L(\sigma,t)}{(1+t)^2}\, dt\, d\sigma$$

$$\ll x(\log d_L^{1/n_L})^A \int_{\frac{1}{2}}^1 \left(\frac{n_L}{\sqrt{x}}\right)^{2(1-\sigma)} d\sigma + x(\log d_L^{1/n_L} T)^A \int_{\frac{1}{2}}^1 \left(\frac{n_L}{\sqrt{x}}\right)^{2(1-\sigma)} \int_2^T \frac{t^{2(1-\sigma)}}{1+t^2}\, dt\, d\sigma$$

$$\ll x(\log d_L^{1/n_L})^A \left(1+\frac{n_L}{\sqrt{x}}\right) + x(\log d_L^{1/n_L} T)^A \log T \int_{\frac{1}{2}}^1 \left(\frac{n_L}{\sqrt{x}}\right)^{2(1-\sigma)} d\sigma$$

$$\ll x(\log d_L^{1/n_L}T)^A \left(1+\frac{n_L}{\sqrt{x}}\right) + x(\log d_L^{1/n_L}T)^A \log T \left(1+\frac{n_L}{\sqrt{x}}\right)$$

$$\ll n_L x^{\frac{1}{2}} \log T(\log d_L^{1/n_L}T)^A + x \log T(\log d_L^{1/n_L}T)^A \qquad (3.4)$$

Combining equation (3.1) with the estimates (3.2) - (3.4) yields the desired result.

PROPOSITION 3.3: *Assume that Hypothesis 1 holds for $\zeta_L(s)$. Then there exists an absolute constant c_3 such that*

$$\sum_{\substack{\rho \\ \beta \geq \frac{1}{2} \\ |\gamma| \leq T}} \left|\frac{x^\rho}{\rho}\right| \ll n_L x \exp\left(-c_3 \frac{\log x}{\log T}\right) \log T \log d_L^{1/n_L}T.$$

PROOF. If $\rho = \beta+i\gamma$ is a zero of $\zeta_L(s)$ and $|\gamma| \leq T$, then $\beta \leq 1-c_1/\log(T+2)$. Therefore

$$\sum_{\substack{\rho \\ \beta \geq \frac{1}{2} \\ |\gamma| \leq T}} \left|\frac{x^\rho}{\rho}\right| \ll x \exp\left(-c_4 \frac{\log x}{\log T}\right) \sum_{\substack{\rho \\ \beta \geq \frac{1}{2} \\ |\gamma| \leq T}} \frac{1}{|\rho|}$$

$$\ll x \exp\left(-c_4 \frac{\log x}{\log T}\right) \sum_{\substack{\rho \\ \beta \geq \frac{1}{2} \\ |\gamma| \leq T}} \frac{1}{|\gamma|+1}$$

$$\ll x \exp\left(-c_4 \frac{\log x}{\log T}\right) \sum_{t=1}^{T} \frac{N_L(t)-N_L(t-1)}{t}$$

$$\ll n_L x \exp\left(-c_4 \frac{\log x}{\log T}\right) \log T \log d_L^{1/n_L}T.$$

4. Uniform Versions of the Tchebotarev Density Theorem

Let L/K be a Galois extension of algebraic number fields and let $G = \mathrm{Gal}(L/K)$. Let P be a prime of L which does not ramify in L/K and let $(\frac{L/K}{P})$ denote its Artin symbol. Let C be a conjugacy class of G and set

$$\pi_{L/K}(x,C) = \sum_{\substack{NP \leq x \\ (\frac{L/K}{P}) \subseteq C}} 1$$

$$\theta_{L/K}(x,C) = \sum_{\substack{NP \leq x \\ (\frac{L/K}{P}) \subseteq C}} \log NP$$

$$\psi_{L/K}(x,C) = \sum_{\substack{NP^m \leq x \\ m \geq 1 \\ (\frac{L/K}{P}) \subseteq C}} \log NP$$

In this section, we shall prove two sets of (conditional) estimates for these functions, which may be regarded as a uniform version of the Tchebotarev density theorem [4]. If we only assume Hypothesis 1, then we may derive an asymptotic formula for each function, albeit with a rather poor error term, namely

THEOREM 4.1: *Assume that Hypothesis 1 holds for $\zeta_L(s)$. Then there exist an absolute constant c_5 such that*

$$\pi_{L/K}(x,C) - \frac{|C|}{|G|}\,x \ll \frac{|C|}{|G|}\,(\log d_L)\,x\,\exp(-c_5\sqrt{\log x}) + b(L/K)\log x$$

$$\theta_{L/K}(x,C) - \frac{|C|}{|G|}\,x \ll \frac{|C|}{|G|}\,(\log d_L)\,x\,\exp(-c_5\sqrt{\log x}) + b(L/K)\log x$$

$$\psi_{L/K}(x,C) - \frac{|C|}{|G|}\,x \ll \frac{|C|}{|G|}\,(\log d_L)\,x\,\exp(-c_5\sqrt{\log x}) + b(L/K)\log x$$

for all $x \geq 2$, *where all implied constants are independent of* x, L, K *and* C *and where* $b(L/K) = \sum_{p \text{ ramifies in } L/K} \log p$

By assuming both Hypotheses 1 and 2, we may derive upper bounds for the above functions. As functions of x, these upper bounds are larger than predicted by the asymptotic formulae of Theorem 4.1. However, the dependence of the upper bound on the parameters of L and K is much more satisfactory than in the estimates of Theorem 4.1. Namely, we have the following result.

THEOREM 4.2: *Assume that Hypotheses* 1 *and* 2 *hold for* $\zeta_L(s)$. *Then*

$$\pi_{L/K}(x,C) \ll \frac{|C|}{|G|}\{x(\log d_L^{1/n_L}x)^A + n_L x^{\frac{1}{2}} \log x(\log d_L^{1/n_L}x)^A\} + b(L/K) + n_K x^{\frac{1}{2}} \log x$$

$$\theta_{L/K}(x,C) \ll \frac{|C|}{|G|}\{x \log x(\log d_L^{1/n_L}x)^A + n_L x^{\frac{1}{2}} (\log x)^2(\log d_L^{1/n_L}x)^A\} + b(L/K) \log x + n_K x^{\frac{1}{2}} \log^2 x$$

$$\psi_{L/K}(x,C) \ll \frac{|C|}{|G|}\{x \log x(\log d_L^{1/n_L}x)^A + n_L x^{\frac{1}{2}}(\log x)^2(\log d_L^{1/n_L}x)^A\} + b(L/K) \log x + n_K x^{\frac{1}{2}} \log^2 x$$

for all $x \geq 2$, *where all implied constants are independent of* x, L, K *and* C.

It is instructive to compare Theorems 4.1 and 4.2 with the following result:

THEOREM 4.3 (*Lagarias-Odlyzko [4, Theorem 1.1]*) *Assume that the GRH holds for* $\zeta_L(s)$. *Then*

$$\pi_{L/K}(x,C) - \frac{|C|}{|G|} Li(x) \ll \frac{|C|}{|G|} n_L x^{\frac{1}{2}} (\log d_L^{1/n_L}x) + \log d_L$$

for all $x \geq 2$, *where the implied constant is independent of* x, L, K *and* C. *Similar estimates hold for* $\theta_{L/K}(x,C)$ *and* $\psi_{L/K}(x,C)$.

To deduce Theorems 4.1 and 4.2 requires the following result.

PROPOSITION 4.4 (Lagarias Odlyzko [4, Theorem 7.1]): *Let* $x \geq 2$, $T \geq 2$. *Then*

$$\psi_{L/K}(x,C) - \frac{|C|}{|G|} x \ll$$

$$\frac{|C|}{|G|} \left\{\frac{x \log x + T}{T} \log d_L + n_L \log x + \frac{n_L x \log x \log T}{T}\right\}$$

$$+ b(L/K) \log x + \frac{n_K x \log^2 x}{T} + R(x,T),$$

where

$$b(L/K) = \sum_{P \text{ ramifies in } L/K} \log NP$$

$$R(x,T) = \frac{|C|}{|G|} \sum_{\substack{\rho \\ |\gamma| < T}} \left|\frac{x^\rho}{\rho}\right|$$

and where this last sum is over non-trivial zeros ρ *of* $\zeta_L(s)$.

Remarks: 1. Note that Lagarias and Odlyzko prove this result with $b(L/K)$ replaced by $\log d_L$. However, a careful examination of their proof yields the slightly stronger statement above.

2. In Theorem 7.1 of [4], $R(x,T)$ is replaced by a more complicated expression $S(x,T)$, but it is easy to see that $R(x,T)$ dominates $S(x,T)$.

PROOF OF THEOREM 4.1: Set $T = \exp(\cdot \sqrt{\log x})$ in Proposition 4.4 to obtain

$$\psi_{L/K}(x,C) - \frac{|C|}{|G|} x \ll \frac{|C|}{|G|} (\log d_L)\, x \exp(-c_6\sqrt{\log x}) + R(x,T) + b(L/K)\log x,$$

where we have used the estimate $n_L \ll \log d_L$ which follows from the Minkowski bound. Moreover, from Propositions 3.1 and 3.3, we have

$$\sum_{\substack{\rho \\ |\gamma| < T}} \left|\frac{x^\rho}{\rho}\right| \ll (\log d_L)\, x \exp(-c_6\sqrt{\log x}),$$

from which follows the desired estimate for $\psi_{L/K}(x,C)$. The corresponding estimates for $\theta_{L/K}(x,C)$ and $\pi_{L/K}(x,C)$ now follow using familiar arguments.

PROOF OF THEOREM 4.2: Set $T = x^{\frac{1}{2}} + 1$ in Proposition 4.4 to obtain

$$\psi_{L/K}(x,C) - \frac{|C|}{|G|}\, x \ll \frac{|C|}{|G|}\, n_L x^{\frac{1}{2}} \log x \log d_L^{1/n_L} x + b(L/K)\log x + n_K x^{\frac{1}{2}} \log^2 x$$

$$+ R(x,T).$$

Now apply Propositions 3.1 and 3.2 to obtain

$$\sum_{\substack{\rho \\ |\gamma| \le T}} \left|\frac{x^\rho}{\rho}\right| \ll n_L x^{\frac{1}{2}} (\log x)^2 (\log d_L^{1/n_L} x)^A$$

$$+ x \log x (\log d_L^{1/n_L} x)^A.$$

Combining the last two estimates yields the desired estimate for $\psi_{L/K}(x,C)$. The corresponding estimates for $\theta_{L/K}(x,C)$ and $\pi_{L/K}(x,C)$ then follow in the usual fashion.

Let

$$\mathrm{Spl}(L/K) = \{P \mid (\tfrac{L/K}{P}) = \text{identity}\}.$$

Then Spl(L/K) consists precisely of the primes of K which split completely in L. Furthermore, let C_0 = the identity conjugacy class of G and set

$$\pi_{L/K}(x,\mathrm{Spl}) = \pi_{L/K}(x,C_0)$$

$$\theta_{L/K}(x,\mathrm{Spl}) = \theta_{L/K}(x,C_0)$$

$$\psi_{L/K}(x,\mathrm{Spl}) = \psi_{L/K}(x,C_0).$$

Then we have the following results:

COROLLARY 4.5: Let L/Q be a Galois extension. Assume that Hypothesis 1 holds for $\zeta_L(s)$. Then for $x \ge 2$, we have

$$\pi_{L/Q}(x,\mathrm{Spl}) - \frac{1}{n_L}\,\mathrm{Li}(x) \ll (\log d_L^{1/n_L})\, x \exp(-c_8\sqrt{\log x}) + b(L/K)\log x$$

$$\theta_{L/Q}(x,Spl)-\frac{1}{n_L}x \ll (\log d_L^{1/n_L})x\exp(-c_8\sqrt{\log x})+b(L/K)\log x$$

$$\psi_{L/Q}(x,Spl)-\frac{1}{n_L}x \ll (\log d_L^{1/n_L})x\exp(-c_8\sqrt{\log x})+b(L/K)\log x.$$

COROLLARY 4.6: *Assume that Hypotheses* 1 *and* 2 *hold for* $\zeta_L(s)$. *Let* L/Q *be a Galois extension. Then for* $x \geq 2$ *we have*

$$\pi_{L/Q}(x,Spl) \ll \frac{1}{n_L}x(\log d_L^{1/n_L}x)^A + x^{\frac{1}{2}}\log x(\log d_L^{1/n_L}x)^A + \log d_L + n_K x^{\frac{1}{2}}\log x$$

$$\theta_{L/Q}(x,Spl) \ll \frac{1}{n_L}x\log x(\log d_L^{1/n_L}x)^A + x^{\frac{1}{2}}\log^2 x(\log d_L^{1/n_L}x)^A + \log d_L\log x+n_K x^{\frac{1}{2}}\log^2 x$$

$$\psi_{L/Q}(x,Spl) \ll \frac{1}{n_L}x\log x(\log d_L^{1/n_L}x)^A + x^{\frac{1}{2}}\log^2 x\,(\log d_L^{1/n_L}x)^A + \log d_L\log x+n_K x^{\frac{1}{2}}\log^2 x.$$

5. Conditional Proof of Artin's Conjecture

In this section, we prove Theorem 1. Throughout we assume Hypotheses 1 and 2. We begin with some notations. Set

$$\pi(x,A) = \sum_{\substack{p \leq x \\ p\in A}} 1,$$

$$A(\xi,\eta) = \bigcap_{\xi \leq q \leq \eta} (Spl_1 - Spl_q) \qquad (1 \leq \xi \leq \eta),$$

$$\pi(x;\xi,\eta) = \sum_{\substack{p \le x \\ p \in A(\xi,\eta)}} 1,$$

$$A^*(\xi,\eta) = \bigcup_{\xi \le q \le \eta} Spl_q. \quad (1 \le \xi \le \eta)$$

$$\pi^*(x;\xi,\eta) = \sum_{\substack{p \le x \\ p \in A^*(\xi,\eta)}}$$

Note that

$$A \subseteq A(\xi,\eta)$$

for all $\xi \le \eta$, so that for $x \ge 1$, we have

$$\pi(x,A) \le \pi(x;\xi,\eta) \qquad (1 \le \xi \le \eta).$$

In particular, we see that

$$\pi(x,A) \le \pi(x;1,\eta) \qquad (\eta \ge 1).$$

Moreover, if p and q are primes, $p \le x$, $q > x$, then $p \not\equiv 1 \pmod q \Rightarrow$ p does not split completely in the cyclotomic field $\mathbb{Q}(\zeta_q) \Rightarrow p \in Spl_1 - Spl_q$. Therefore,

$$\pi(x,A) = \pi(x;1,x).$$

If $1 \le \xi < x$, then

$$\pi(x;1,x) \ge \pi(x\,;1,\xi) - \pi^*(x;\xi,x),$$

so that for all ξ, $1 \le \xi \le x$, we have

$$\pi(x;1,\xi) - \pi^*(x,\xi,x) \le \pi(x,A) \le \pi(x;1,\xi). \qquad (5\text{-}1)$$

Let us set $\xi = \log^B x$, where B is an absolute constant, to be determined later.

<u>PROPOSITION 5.1</u>: <u>If</u> $\xi = \log^B x$, <u>then</u>

$$\pi(x;1,\xi) = \left(\sum_k \frac{\mu(k)}{n(k)}\right)\frac{x}{\log x} + O\left(\frac{x \log\log x}{\log^{B+1} x}\right) \quad (x \to \infty)$$

<u>where the O-term constant depends only on</u> a <u>and</u> B.

<u>PROOF</u>. For the sake of brevity, set

$$\pi_k(x,\mathrm{Spl}) = \pi_{L_k/\mathbb{Q}}(x,\mathrm{Spl}),$$

$$d_k = d_{L_k}.$$

Furthermore, for $\xi \geq 1$, set

$$S(\xi) = \{k \mid k = q_1\cdots q_r, q_i \text{ distinct primes, } q_i \leq \xi \ (1 \leq i \leq r)\}.$$

Note that $1 \in S(\xi)$, corresponding to the vacuous product of q_i's. It is easy to see that for any $\xi \geq 1$, we have for $x \geq 2$,

$$\pi(x;1,\xi) = \sum_{k \in S(\xi)} \mu(k)\pi_k(x,\mathrm{Spl}).$$

Therefore, by Corollary 4.5, we see that

$$\pi(x;1,\xi) = \sum_{k \in S(\xi)} \mu(k)\left\{\frac{1}{n(k)}\,\mathrm{Li}(x) + O(b(L_k/\mathbb{Q})\log x) + O(\log d_k^{1/n(k)} \cdot x\exp(-c_6\sqrt{\log x}))\right\}, \tag{5-2}$$

where c_6 is an absolute constant. Furthermore by [2, p. 215], we have that

$$d_k \ll n_k^{c_9 n_k}, \tag{5-3}$$

where c_9 is an absolute constant. Also, it is clear that $n_k \ll k^2$, so that

$$\log d_k^{1/n(k)} \ll \log n_k \ll \log k, \tag{5-4}$$

with an absolute implied constant. Thus, if $k \in S(\xi)$ and $\xi = \log^B x$, then

$$\log d_k^{1/n(k)} \ll \log k \ll (\log x)^B,$$

where the implied constant depends only on B. Next, note that since the only primes which can possibly ramify in $L_k/\mathbb{Q}$ are the primes dividing ak, we see that if $k \in S(\xi)$, then

$$b(L_k/\mathbb{Q}) = \sum_{p \text{ ramifies in } L_k/\mathbb{Q}} \log p$$

$$\le \sum_{p|ak} \log p \le \log ak.$$

Therefore, we have

$$\pi(x;1,\xi) = \Big(\sum_{k \in S(\xi)} \frac{\mu(k)}{n(k)}\Big) \operatorname{Li}(x)+O(x \exp(-c_{10}\sqrt{\log x})) \qquad (5\text{-}5)$$

where c_{10} and the O-term constant depend only on B. To complete the proof, we use an explicit formula for n(k). Let h be the largest integer such that a is a perfect hth power and let $k_1 = k/\gcd(h,k)$. Furthermore, let $a = a_1a_2^2$, a_1 square-free,

$$\varepsilon(k) = \begin{cases} 2 & \text{if } 2a_1|k \text{ and } a_1 \equiv 1 \pmod 4 \\ 1 & \text{otherwise.} \end{cases}$$

Then [2, p. 214, eq. (13)]

$$n(k) = \frac{k_1\phi(k)}{\varepsilon(k)} .$$

In particular, there exist constants c_{11} and c_{12}, depending only on a, such that

$$c_{11}k\phi(k) \le n(k) \le c_{12}k\phi(k).$$

Suppose that $k = q_1 \ldots q_r$ a product of distinct primes, $q_i \le \xi$. Then

$$\phi(k) = (q_1-1)\ldots(q_r-1)$$

$$= k(1-\frac{1}{q_1})\ldots(1-\frac{1}{q_r})$$

$$\ge k \prod_{q \in \xi} (1-\frac{1}{q})$$

$$\geq \frac{c_{13}k}{\log \xi}$$

$$\geq \frac{c_{14}k}{\log \log x},$$

where c_{14} depends only on B. Thus, if $k = q_1 \ldots q_r$, $q_i \leq \xi$, we have

$$n(k) \geq \frac{c_{15}k^2}{\log \log x}, \; c_{15} = c_{15}(B). \tag{5-6}$$

Thus,

$$\sum_{k > \xi} \frac{1}{n(k)} \ll \log \log x \sum_{k > \xi} \frac{1}{k^2}$$

$$\ll \frac{\log \log x}{\xi} = \frac{\log \log x}{(\log x)^B},$$

where the implied constants depend only on a and B. Since

$$S(\xi) \supseteq \{k | k \text{ square-free}, k \leq \xi\},$$

we see that

$$\sum_{k \in S(\xi)} \frac{\mu(k)}{n(k)} = \sum_{k \leq \xi} \frac{\mu(k)}{n(k)} + O(\sum_{k > \xi} \frac{1}{n(k)})$$

$$= \sum_{k} \frac{\mu(k)}{n(k)} + O(\sum_{k > \xi} \frac{1}{n(k)})$$

$$= \sum_{k} \frac{\mu(k)}{n(k)} + O(\frac{\log \log x}{\log^B x}),$$

where the O-term constants depend at most on a and B. Thus, by (5-5),

$$\pi(x;1,\xi) = (\sum_{k} \frac{\mu(k)}{n(k)}) \frac{x}{\log x} + O(\frac{x \log \log x}{\log^{B+1} x})$$

where the O-term constant depends only on a and B. This completes the proof of Proposition 5.1.

<u>PROPOSITION 5.2</u>: <u>Let</u> $x \geq 2$, $\xi = \log^B x$. <u>Then</u>

$$\pi^*(x;\xi,x) \ll \frac{x \log\log x}{\log^2 x},$$

<u>where the implied constant depends only on</u> a <u>and</u> B.

<u>PROOF</u>: Set

$$\xi_1 = x^{\frac{1}{2}} \log^{-B} x, \quad \xi_2 = x^{\frac{1}{2}} \log x.$$

Then clearly, we have

$$\pi^*(x,\xi,x) \leq \pi^*(x,\xi,\xi_1) + \pi^*(x;\xi_1,\xi_2) + \pi^*(x;\xi_2,x).$$

Hooley [2, pp. 211-12] has proved that

$$\pi^*(x;\xi_1,\xi_2) = O(\frac{x \log\log x}{\log^2 x})$$

$$\pi^*(x;\xi_2,x) = O(\frac{x}{\log^2 x}),$$

where the O-term constants depend only on a. Therefore, we have the estimate

$$\pi^*(x;\xi,x) \ll \pi^*(x;\xi,\xi_1) + \frac{x \log\log x}{\log^2 x} \tag{5-7}$$

Next, note that

$$\pi^*(x;\xi,\xi_1) \leq \sum_{\xi \leq q \leq \xi_1} \pi_q(x,\mathrm{Spl}). \tag{5-8}$$

Moreover, by Corollary 4.6, we have

$$\pi_k(x,\mathrm{Spl}) \ll \frac{1}{n(k)} x(\log d_k^{1/n(k)} x)^A + x^{\frac{1}{2}} \log x(\log d_k^{1/n(k)} x)^A$$
$$+ b(L_k/Q) + n_k x^{\frac{1}{2}} \log x. \tag{5-9}$$

Since the only primes which can possibly ramify in L_k/Q are the prime divisors of ak, we see that

$$b(L_k/Q) = \sum_{p \text{ ramifies in } L_k/Q} \log p \ll \log k,$$

with an implied constant depending only on a. Moreover, by (5-4), we have

$$(\log d_k^{1/n(k)}x)^A \ll \log^A kx,$$

so that by (5-9), we have

$$\pi_k(x,\mathrm{Spl}) \ll \frac{1}{n(k)}\, x \log^A kx + x^{\frac{1}{2}} \log x \log^A kx.$$

Therefore, by (5-9), we have

$$\pi^*(x;\xi,\xi_1) \ll x \log^A x \sum_{\xi \le q \le \xi_1} \frac{1}{n(q)} + x^{\frac{1}{2}} \log^{A+1} x \sum_{\xi \le q \le \xi_1} 1$$

$$\ll x \log^{A+1} x \sum_{\xi \le q \le \xi_1} \frac{1}{q^2} + \xi_1\, x^{\frac{1}{2}} \log^{A+1} x$$

$$\ll \frac{x \log^{A+1} x}{\xi} + \xi_1\, x^{\frac{1}{2}} \log^{A+1} x$$

$$\ll x \log^{A+1-B} x.$$

Set B = A+3 to obtain

$$\pi^*(x;\xi,\xi_1) \ll \frac{x}{\log^2 x},$$

so that by (5-7), we derive

$$\pi^*(x;\xi,x) \ll \frac{x \log\log x}{\log^2 x}.$$

PROOF OF THEOREM 1.1: By (5-1) and Proposition 5.2, we deduce that

$$|\pi(x,A)-\pi(x;1,\xi)| \ll \frac{x \log\log x}{\log^2 x}.$$

Now by Proposition 5.1 and the fact that $B \ge 2$, we deduce that

$$\left|\pi(x,A)-\left(\sum_k \frac{\mu(k)}{n(k)}\right)\frac{x}{\log x}\right| \ll \frac{x \log\log x}{\log^{B+1} x} + \frac{x \log\log x}{\log^2 x}$$

$$\ll \frac{x \log\log x}{\log^2 x},$$

which completes the proof.

BIBLIOGRAPHY

1. Artin, E., The Collected Papers of Emil Artin, S. Lang and J. Tate, eds. Addison-Wesley, 1965.

2. Hooley, C., "On Artin's Conjecture," J. Reine Angew. Math. 225 (1967), 209-220.

3. Hooley, C., Applications of Sieve Methods to the Theory of Numbers, Cambridge University Press, 1976.

4. Lagarias, J. and Odlyzko, A., "Effective Versions of the Chebotarev Density Theorem" in Algebraic Number Fields, A. Frohlich, ed., Academic Press, 1977.

5. Montgomery, H., Topics in Multiplicative Number Theory, Springer-Verlag Lecture Notes in Mathematics, vol. 227, 1971.

THE ZEROS OF HURWITZ'S ZETA-FUNCTION ON $\sigma = 1/2$

Steven M. Gonek

Dedicated to Professor Emil Grosswald

1. Introduction.

Let $s = \sigma + it$ be a complex variable. For a fixed α, $0 < \alpha \le 1$, Hurwitz's zeta-function is defined in the half-plane $\sigma > 1$ by

$$\zeta(s,\alpha) = \sum_{n=0}^{\infty} (n+\alpha)^{-s},$$

and except for a simple pole at $s = 1$, may be analytically continued throughout the complex plane. The resemblance of $\zeta(s,\alpha)$ to Riemann's zeta-function, $\zeta(s)$, is in certain ways superficial. For besides the two cases $\zeta(s,1/2) = (2^s-1)\zeta(s)$ and $\zeta(s,1) = \zeta(s)$, $\zeta(s,\alpha)$ possesses neither a functional equation nor an Euler product. It is therefore not surprising that the zeros of these functions are distributed differently. For instance, we note the following:

1. While $\zeta(s)$ has no zeros in $\sigma > 1$, $\zeta(s,\alpha)$ has infinitely many (provided $\alpha \ne 1/2$ or 1). In particular the analogue of the Riemann hypothesis for $\zeta(s,\alpha)$ is false. This was proved by Davenport and Heilbronn [3] when α is rational ($\ne 1/2$ or 1) or transcendental, and by Cassels [1] when α is an algebraic irrational. One may also prove a quantitative version of this result [2; p. 1780]. Namely, for any $\delta > 0$, the number of zeros of $\zeta(s,\alpha)$ ($\alpha \ne 1/2$ or 1) in the rectangle $1 < \sigma < 1+\delta$, $0 < t < T$ is $\asymp T$ for sufficiently large T.

2. Let σ_1,σ_2 be fixed with $1/2 < \sigma_1 < \sigma_2 < 1$. Then $\zeta(s,\alpha)$ has infinitely many zeros in the strip $\sigma_1 < \sigma < \sigma_2$ when α is rational ($\ne \frac{1}{2}$ or 1) or transcendental. The rational case is due to S.M. Voronin [8] (see also S.M. Gonek [5]), the transcendental case to S.M. Gonek [5]. Here too one can show that the number of zeros up to height T is $\asymp T$ for all large T. On the other hand, well-known zero-density estimates imply that $\zeta(s)$ has at most $o(T)$ zeros in such a rectangle.

Pursuing these contrasts further, one might naturally ask whether the line $\sigma = 1/2$ is special to $\zeta(s,\alpha)$ as it is to $\zeta(s)$. We know that as T tends to infinity, the number of zeros of either function in the strip $0 < t < T$ is $\sim \frac{T}{2\pi} \log T$. For $\zeta(s)$, N. Levinson [7] showed that more than 1/3 of these zeros lie on $\sigma = 1/2$; it is widely held that the correct proportion is 1. In this paper, our purpose is to show that for certain values of α the proportion of zeros of $\zeta(s,\alpha)$ on $\sigma = 1/2$ is definitely less than 1. Specifically, we shall prove the following result.

THEOREM. Let $\alpha = \frac{1}{3}, \frac{2}{3}, \frac{1}{4}, \frac{3}{4}, \frac{1}{6}$ or $\frac{5}{6}$. There is a positive constant $c < 1$ such that the number of zeros of $\zeta(s,\alpha)$ (counted according to their multiplicities) on the segment $[1/2, 1/2 + iT]$ is $\leq (c+o(1)) \frac{T}{2\pi} \log T$ as T tends to infinity.

The author would like to take this opportunity to thank Professor Hugh L. Montgomery for bringing this problem to his attention and Professor Patrick X. Gallagher for pointing out an error in the original manuscript.

2. An Auxiliary Lemma.

To prove our theorem we require information about the number of zeros common to two L-functions. This is provided by the lemma below which is essentially due to A. Fujii [4; Theorem 1].

Recall that two Dirichlet characters not induced by the same primitive character are called *inequivalent*. We denote by $L(s,\chi)$ the Dirichlet L-function with character χ.

LEMMA. Suppose χ_1 and χ_2 are inequivalent characters. Let $\rho_1 = \beta_1 + i\gamma_1$ denote a zero of $L(s,\chi_1)$ with $0 < \beta_1 < 1$, and write $m_i(\rho_1)$ for the multiplicity of ρ_1 as a zero of $L(s,\chi_i)$ $(i = 1,2)$. Then there exists a positive constant $c < 1$ such that

(1)
$$\sum_{0 \leq \gamma_1 \leq T}{}' \min_{i=1,2} m_i(\rho_1) \leq (c+o(1)) \frac{T}{2\pi} \log T$$

as T tends to infinity, where $\sum'$ means the sum is over distinct zeros ρ_1.

PROOF. We see from the proof of Theorem 1 in Fujii [4; §3,2] that for distinct primitive characters χ_1, χ_2 there exists a positive constant $c_1 < 1$ such that as T tends to infinity

$$\sum_{\substack{0 \le \gamma_1 \le T \\ m_1(\rho_1) > m_2(\rho_1)}}' 1 \ge (c_1+o(1)) \frac{T}{2\pi} \log T. \tag{2}$$

Indeed, (2) holds even when χ_1, χ_2, or both χ_1 and χ_2 are imprimitive as long as they are inequivalent. To see this, note that if χ_i^* induces χ_i $(i = 1,2)$ and χ_1, χ_2 are inequivalent, then χ_1^*, χ_2^* are distinct primitive characters. (Of course if χ_i is primitive $\chi_i = \chi_i^*$.) Therefore (2) is true for the pair $L(s,\chi_1^*)$, $L(s,\chi_2^*)$. But $L(s,\chi_i)$ and $L(s,\chi_i^*)$ have the same zeros in $0 < \sigma < 1$. Hence (2) is valid for the pair $L(s,\chi_1)$, $L(s,\chi_2)$ as well. (In the statement of his theorem, Fujii assumes χ_1 and χ_2 have the same modulus. However, he later points out (in §4) that this assumption is unnecessary.) Now

$$\sum_{0 \le \gamma_1 \le T}' \min_{i=1,2} m_i(\rho_1) = \sum_{\substack{0 \le \gamma_1 \le T \\ m_1(\rho_1) \le m_2(\rho_1)}}' m_1(\rho_1) + \sum_{\substack{0 \le \gamma_1 \le T \\ m_1(\rho_1) > m_2(\rho_1)}}' m_2(\rho_1)$$

$$\le \sum_{\substack{0 \le \gamma_1 \le T \\ m_1(\rho_1) \le m_2(\rho_1)}}' m_1(\rho_1) + \sum_{\substack{0 \le \gamma_1 \le T \\ m_1(\rho_1) > m_2(\rho_1)}}' (m_1(\rho_1)-1)$$

$$= \sum_{0 \le \gamma_1 \le T}' m_1(\rho_1) - \sum_{\substack{0 \le \gamma_1 \le T \\ m_1(\rho_1) > m_2(\rho_1)}}' 1.$$

The first sum on the last line is the total number of zeros of $L(s,\chi_1)$ in $0 < \sigma < 1$, $0 < t < T$, and is therefore equal to $(1+o(1)) \frac{T}{2\pi} \log T$ as T tends to infinity. Using this and (2) we conclude that

$$\sum_{0 \le \gamma_1 \le T}' \min_{i=1,2} m_i(\rho_1) \le (1-c_1+O(1)) \frac{T}{2\pi} \log T.$$

This establishes (1) with $c = 1-c_1$.

3. Proof of the Theorem.

For the sake of convenience, we carry out the proof of the Theorem only for $\alpha = 1/3$ and $2/3$. The modifications required to prove the other cases are minor and will be discussed at the end of this section. Throughout we write $e(x)$ for $e^{2\pi i x}$.

We begin with the identity (see Davenport and Heilbronn [3; p. 181])

$$\zeta(s, \tfrac{a}{q}) = \frac{q^s}{\phi(q)} \sum_{\chi} \bar{\chi}(a)L(s,\chi), \tag{3}$$

where $1 \le a < q$, $(a,q) = 1$, and the sum is over all $\phi(q)$ characters mod q. Take $q = 3$ and assume that a is either 1 or 2. We are then summing over $\phi(3) = 2$ characters in (3), both of which are real. Thus

$$\frac{2}{3^s} \zeta(s, \tfrac{a}{3}) = L(s,\chi_0) + \chi(a)L(s,\chi),$$

where χ_0 and χ are the principal and nonprincipal characters, respectively, mod 3. Since $L(s,\chi_0) = (1-3^{-s})\zeta(s)$, the last equation becomes

$$\frac{2}{3^s} \zeta(s, \tfrac{a}{3}) = (1-3^{-s})\zeta(s) + \chi(a)L(s,\chi). \tag{4}$$

REMARK. As will become apparent, it is essential to our proof that the sum in (3) reduce to two terms. This is why the reduced fraction α in the Theorem must have denominator 3,4 or 6.

Now write

$$\xi(s) = \frac{1}{2} s(s-1)\pi^{-s/2}\Gamma(\tfrac{s}{2})\zeta(s) \tag{5}$$

and

$$\xi(s,\chi) = (\tfrac{\pi}{3})^{-\frac{s+1}{2}} \Gamma(\tfrac{s+1}{2})L(s,\chi). \tag{6}$$

Using (5) and (6) to replace $\zeta(s)$ and $L(s,\chi)$ in (4) by $\xi(s)$ and $\xi(s,\chi)$, and then multiplying both sides of (4) by $(\frac{\pi}{3})^{-\frac{s+1}{2}}\Gamma(\frac{s+1}{2})$, we find (after simplifying) that

$$(7)\qquad \sqrt{\frac{12}{\pi}}\,(3\pi)^{-s/2}\Gamma(\tfrac{s+1}{2})\zeta(s,\tfrac{a}{3}) = \sqrt{\frac{12}{\pi}}\,\frac{(3^{s/2}-3^{-s/2})}{s(s-1)}\,\frac{\Gamma(\frac{s+1}{2})}{\Gamma(\frac{s}{2})}\,\xi(s)+\chi(a)\xi(s,\chi).$$

We write this more briefly as

$$(8)\qquad A(s)\zeta(s,\tfrac{a}{3}) = B(s)\xi(s) + \chi(a)\xi(s,\chi),$$

where

$$(9)\qquad A(s) = \sqrt{\frac{12}{\pi}}\,(3\pi)^{-s/2}\Gamma(\tfrac{s+1}{2})$$

and

$$(10)\qquad B(s) = \sqrt{\frac{12}{\pi}}\,\frac{(3^{s/2}-3^{-s/2})}{s(s-1)}\,\frac{\Gamma(\frac{s+1}{2})}{\Gamma(\frac{s}{2})}\,.$$

Since $A(s)$ never vanishes, the zeros of the right-hand side of (8) are precisely those of $\zeta(s,\frac{a}{3})$. Thus, $\zeta(1/2+it_0,\frac{a}{3}) = 0$ if and only if the terms on the right-hand side of (8) cancel or vanish for $s = 1/2+it_0$. Since $B(s) \neq 0$ on $\sigma = 1/2$ we see that $1/2+it_0$ is a zero of $\zeta(s,\frac{a}{3})$ if and only if:

I. $\xi(1/2+it_0) \neq 0$, $\xi(1/2+it_0,\chi) \neq 0$, *and* $B(1/2+it_0) = -\chi(a)\,\dfrac{\xi(1/2+it_0,\chi)}{\xi(1/2+it_0)}$,

or

II. $\xi(1/2+it_0) = \xi(1/2+it_0,\chi) = 0$.

Writing $N(T)$ for the number of zeros (counting multiplicities) of $\zeta(s,\frac{a}{3})$ on $[1/2, 1/2+iT]$ $(T > 0)$, $N_I(T)$ for the number of these zeros arising from condition I, and $N_{II}(T)$ for the number arising from II, we see that

$$(11)\qquad N(T) = N_I(T) + N_{II}(T).$$

We estimate $N(T)$ by combining estimates for $N_I(T)$ and $N_{II}(T)$.

First consider $N_I(T)$. From the relation $\overline{\xi(s,\chi)} = \xi(\bar{s},\chi)$ (χ is real) and the functional equation

$$\xi(1-s,\chi) = \frac{i\sqrt{3}}{\tau(\chi)}\ \xi(s,\chi),$$

where $\tau(\chi) = \sum_{n=1}^{3} \chi(n)e(\frac{n}{3})$, one easily finds that $\xi(1/2 + it,\chi)$ is real.

Similarly $\xi(1/2 + it)$ is real. Thus if t_0 satisfies I, $B(1/2 + it_0)$ is real. If $T \geq T_0 > 0$ and if $N_I'(T_0,T)$ denotes the number of solutions of

$$\arg B(1/2 + it) \equiv 0 \pmod{\pi}$$

with $t \in [T_0,T]$, it follows that $N_I'(T_0,T)$ is an upper bound for the number of distinct $t_0 \in [T_0,T]$ that satisfy I. We now prove that there exists a T_0 such that $N_I'(T_0,T) \ll T$ for all $T \geq T_0$, and that $1/2 + it_0$ is a simple zero of $\zeta(s, \frac{a}{3})$ if t_0 satisfies I and $t_0 \geq T_0$. These two assertions and the fact that $\zeta(s, \frac{a}{3})$ has only finitely many zeros on $[1/2, 1/2 + iT_0]$ clearly imply that

(12) $$N_I(T) \ll T \qquad (T \geq T_0).$$

To estimate $N_I'(T_0,T)$ we examine $\frac{d}{dt} \arg B(1/2 + it)$. (The derivative exists for all t since $B(s)$ is analytic and nonzero in $0 < \sigma < 1$.) By (10)

$$\arg B(1/2 + it) = \arg\left(\frac{-1}{t^2 + 1/4}\right) + \arg e\left(\frac{t \log 3}{4\pi}\right)$$

$$+ \arg\left(1 - \frac{1}{\sqrt{3}}\, e\left(\frac{-t \log 3}{2\pi}\right)\right)$$

$$+ \arg\left(\Gamma\left(\tfrac{3}{4} + i\,\tfrac{t}{2}\right)/\Gamma\left(1/4 + i\,\tfrac{t}{2}\right)\right)$$

or

(13) $$\arg B(1/2 + it) = \pi + \frac{t \log 3}{2} + \arctan\left(\frac{\sin(t \log 3)}{\sqrt{3} - \cos(t \log 3)}\right)$$

$$+ \arg\left(\Gamma\left(\tfrac{3}{4} + i\,\tfrac{t}{2}\right)/\Gamma\left(1/4 + i\,\tfrac{t}{2}\right)\right),$$

where the choice of arguments is immaterial. The sum of the derivatives of the first three terms on the right-hand side of (13) is equal to

$$\frac{\log 3}{4-2\sqrt{3}\cos(t\log 3)} .$$

Observing that

$$\frac{d}{dt}\arg\Gamma(\sigma+ it) = \mathrm{Re}\,\frac{\Gamma'}{\Gamma}(\sigma+ it)$$

and using the formula (see Ingham [6; p. 57])

$$\frac{\Gamma'}{\Gamma}(s) = \log s + O(\frac{1}{|s|})$$

which is valid in $|\arg s| < \pi-\delta$ for any $\delta > 0$, we find that

$$\frac{d}{dt}\arg(\Gamma(\tfrac{3}{4} + i\tfrac{t}{2})/\Gamma(1/4 + i\tfrac{t}{2})) \ll \frac{1}{t+1}$$

for $t \geq 0$. Thus

$$\frac{d}{dt}\arg B(1/2 + it) = \frac{\log 3}{4-2\sqrt{3}\cos(t\log 3)} + O(\frac{1}{t+1}) \qquad (t \geq 0).$$

From this we see that there exists a $T_0 > 0$ such that $\frac{d}{dt}\arg B(1/2 + it)$ is bounded and greater than zero for $t \geq T_0$. That is, $\arg B(1/2 + it)$ is an increasing function with bounded derivative on $[T_0,\infty)$. Clearly this implies that

$$N_I'(T_0,T) \ll T \qquad (T \geq T_0).$$

Now suppose that $1/2 + it_0$ is a zero of $\zeta(s, \frac{a}{3})$ arising from condition I and that $t_0 \geq T_0$ (T_0 as above). Differentiating the right-hand side of (8) with respect to t and evaluating at $s = 1/2 + it_0$, we obtain

$$(14)\quad \xi(1/2 + it_0)(\tfrac{d}{dt})_{t_0}B(1/2 + it)$$

$$+ B(1/2 + it_0)(\tfrac{d}{dt})_{t_0}\xi(1/2 + it)+\chi(a)(\tfrac{d}{dt})_{t_0}\xi(1/2 + it,\chi).$$

The second and third terms are real since $\chi(a)$, $\frac{d}{dt}\,\xi(1/2 + it)$, $\frac{d}{dt}\,\xi(1/2 + it,\chi)$ and $B(1/2 + it_0)$ are. (Recall that $B(1/2 + it_0)$ is real whenever t_0 satisfies I.)

If we write $B(1/2 + it) = |B(1/2 + it)|e(\frac{\arg B(1/2 + it)}{2\pi})$, the first term in (14) becomes

$$(15)\quad \xi(1/2 + it_0)e(\frac{\arg B(1/2 + it_0)}{2\pi})\{(\frac{d}{dt})_{t_0}|B(1/2 + it)|+i(\frac{d}{dt})_{t_0}\ \arg B(1/2 + it)\}.$$

Since t_0 satisfies I, $e(\frac{\arg B(1/2 + it_0)}{2\pi}) = \pm 1$ and $\xi(1/2 + it_0)$, which is real, does not equal zero. Also $(\frac{d}{dt})_{t_0}$ arg $B(1/2 + it) > 0$ for $t_0 \geq T_0$ (this is how T_0 was chosen), and $\frac{d}{dt}\,|B(1/2 + it)|$ is real for all t. It follows that (15) and therefore (14) have nonvanishing imaginary parts. Thus $1/2 + it_0$ is a simple zero of the right-hand side of (8) or, what is the same thing, of $\zeta(s, \frac{a}{3})$. This finally establishes (12).

We now turn to $N_{II}(T)$. Let $m(z)$, $m_1(z)$, and $m_2(z)$ be the multiplicities of the point z as a zero of $\zeta(s, \frac{a}{3})$, $\zeta(s)$, and $L(s,\chi)$ repsectively. By (5), $\zeta(s)$ and $\xi(s)$ have the same zeros in $0 < \sigma < 1$; the same is true for $L(s,\chi)$ and $\xi(s,\chi)$ in light of (6). Thus t_0 satisfies II if and only if $1/2 + it_0$ is a common zero of $\zeta(s)$ and $L(s,\chi)$. In particular, $\frac{1}{2} + it_0$ is a zero of $\zeta(s)$ on $\sigma = 1/2$. Letting $\rho = \beta + i\gamma$ denote a typical zero of $\zeta(s)$, we then have

$$(16)\qquad N_{II}(T) = \sum_{\substack{0 \leq \gamma \leq T \\ \beta = 1/2}}{}' \ m(\rho),$$

where as usual $\sum'$ means the sum is over distinct zeros ρ. In order to estimate this we need to consider the numbers $m(\rho)$. From (8) and the fact that $B(s) \neq 0$ on $\sigma = 1/2$, it immediately follows that

$$m(1/2 + i\gamma)\begin{cases} = \min\limits_{i=1,2} m_i(1/2 + i\gamma) \text{ if } m_1(1/2 + i\gamma) \neq m_2(1/2 + i\gamma) \\ \geq m_1(1/2 + i\gamma) \text{ if } m_1(1/2 + i\gamma) = m_2(1/2 + i\gamma). \end{cases}$$

However, the lower bound this provides for $m(1/2 + i\gamma)$ in the case $m_1(1/2 + i\gamma) = m_2(1/2 + i\gamma)$ is of no use to us since we seek an upper bound for $N_{II}(T)$. We remedy this by proving that, except for finitely many γ, if $m_1(1/2 + i\gamma) = m_2(1/2 + i\gamma)$ then $m(1/2 + i\gamma) = m_1(1/2 + i\gamma)$ or $m_1(1/2 + i\gamma) + 1$, with the latter holding at most $O(T)$ times for $\gamma \in [0,T]$.

To show this set $m_1(1/2 + i\gamma) = m_2(1/2 + i\gamma) = k \geq 1$. Then the kth derivative of the right-hand side of (8) with respect to t evaluated at $s = 1/2 + i\gamma$ is

$$(17) \qquad B(1/2 + i\gamma)\left(\frac{d}{dt}\right)_\gamma^k \xi(1/2 + it) + \chi(a)\left(\frac{d}{dt}\right)_\gamma^k \xi(1/2 + it,\chi).$$

Since the zeros of $B(s)\xi(s) + \chi(a)\xi(s,\chi)$ are those of $\zeta(s, \frac{a}{3})$, we see that $m(1/2 + i\gamma) > k$ if and only if (17) vanishes. By the definition of k, the kth derivatives of the two ξ-functions are nonzero at $1/2 + i\gamma$. Hence (17) vanishes only if its terms cancel. Since $\chi(a)$, $(\frac{d}{dt})^k\ \xi(1/2 + it)$, and $(\frac{d}{dt})^k \xi(1/2 + it,\chi)$ are real, this occurs only if $B(1/2 + i\gamma)$ is real. But we have already seen that $B(1/2 + it)$ is real at most $O(T)$ times on $[0,T]$. Thus $m_1(1/2 + i\gamma) = m_2(1/2 + i\gamma)$ implies that $m(1/2 + i\gamma) = m_1(1/2 + i\gamma)$ ($= k$) except for possibly $O(T)$ values of $\gamma \in [0,T]$. Suppose now that (17) does vanish at $1/2 + i\gamma$ (so that $B(1/2 + i\gamma)$ is real). Taking the $k+1^{st}$ derivative of the right-hand side of (8) with respect to t and evaluating at $s = 1/2 + i\gamma$, we obtain

$$(18) \quad (k+1)\left[\left(\frac{d}{dt}\right)_\gamma^k \xi(1/2 + it)\right]\left[\left(\frac{d}{dt}\right)_\gamma B(1/2 + it)\right] + B(1/2 + i\gamma)\left(\frac{d}{dt}\right)_\gamma^{k+1} \xi(1/2 + it)$$
$$+ \chi(a)\left(\frac{d}{dt}\right)_\gamma^{k+1} \xi(1/2 + it,\chi).$$

As in our analysis of (14), we find that the second and third terms are real and that the first has nonvanishing imaginary part when γ is large. Thus (18) is nonzero and $m(1/2 + i\gamma) = k+1 = m_1(1/2 + i\gamma) + 1$ (for large γ).

To summarize: there exists a $T_0 > 0$ such that if $1/2 + i\gamma$ is a zero of $\zeta(s)$ with $\gamma \geq T_0$, then

$$m(1/2 + i\gamma) = \min_{i=1,2} m_i(1/2 + i\gamma) \text{ or } \min_{i=1,2} m_i(1/2 + i\gamma) + 1;$$

the second case occurs at most $O(T)$ times on $[T_0, T]$.

We can now bound $N_{II}(T)$. Writing (16) as

$$N_{II}(T) = \sum_{\substack{T_0 \le \gamma \le T \\ \beta = 1/2}}' m(\rho) + O(1)$$

and using the previous result, we have

$$N_{II}(T) = \sum_{\substack{T_0 \le \gamma \le T \\ \beta = 1/2}}' \min_{i = 1,2} m_i(\rho) + O(T)$$

$$= \sum_{\substack{0 \le \gamma \le T \\ \beta = 1/2}}' \min_{i = 1,2} m_i(\rho) + O(T)$$

$$\le \sum_{0 \le \gamma \le T}' \min_{i = 1,2} m_i(\rho) + O(T),$$

where the final sum is over the distinct zeros ρ of $\zeta(s)$ with $0 < \beta < 1$, $0 \le \gamma \le T$. Applying the Lemma to the last sum (note that $\zeta(s)$ is an L-function) we see that as T tends to infinity

(19) $$N_{II}(T) \le (c+o(1)) \frac{T}{2\pi} \log T,$$

where c is a positive constant < 1.

The proof of the Theorem for $\alpha = 1/3$ and $2/3$ now follows from (11), (12), and (19).

Our proof carries over to the cases $\alpha = 1/4$, $3/4$. $1/6$, and $5/6$ with only slight changes in the formulae. For instance, if $\alpha = a/4$, $a = 1$ or 3, then corresponding to (8), (9), and (10) we have

$$A(s)\zeta(s, \tfrac{a}{4}) = B(s)\xi(s) + \chi(a)\xi(s,\chi),$$

$$A(s) = \frac{4}{\sqrt{\pi}} (4\pi)^{-s/2}\Gamma(\tfrac{s+1}{2}),$$

and

$$B(s) = \frac{4}{\sqrt{\pi}} \frac{(2^s-1)}{s(s-1)} \frac{\Gamma(\frac{s+1}{2})}{\Gamma(\frac{s}{2})},$$

where χ is the nonprincipal character mod 4.

When $\alpha = \frac{a}{6}$, $a = 1$ or 5, the situation is only slightly more complicated. The nonprincipal character χ mod 6 is induced by the primitive character χ^* mod 3. Also, for the principal character χ_0 mod 6 we have $L(s,\chi_0) = (1-2^{-s})(1-3^{-s})\zeta(s)$. Thus, in place of (4) we obtain

$$\frac{2}{6^s}\zeta(s, \tfrac{a}{6}) = (1-2^{-s})(1-3^{-s})\zeta(s)+\chi^*(a)(1+2^{-s})L(s,\chi^*),$$

and instead of (8), (9), (10) we have

$$A(s)\zeta(s, \tfrac{a}{6}) = B(s)\xi(s) + \chi^*(a)\xi(s,\chi^*),$$

$$A(s) = \sqrt{\frac{12}{\pi}} \frac{(12\pi)^{-s/2}}{(1+2^{-s})} \Gamma(\tfrac{s+1}{2}),$$

and

$$B(s) = \sqrt{\frac{12}{\pi}} \frac{(3^{s/2}-3^{-s/2})(1-2^{-s})}{s(s-1)(1+2^{-s})} \frac{\Gamma(\frac{s+1}{2})}{\Gamma(\frac{s}{2})}.$$

In either case $A(s) \neq 0$ for $0 < \sigma < 1$ and $\frac{d}{dt} \arg B(\frac{1}{2} + it)$ is bounded and > 0 for all large t.

4. A Conjecture.

We expect the Lemma, and therefore the Theorem, to be far from best possible. Indeed, it is generally held that no two L-functions with inequivalent characters have common zeros in $0 < \sigma < 1$. On this assumption we would have $N_{II}(T) \ll T$ instead of (19) and this along with (11) and (12) implies that $N(T) \ll T$. It is plausible to suppose that these bounds are valid for other rational values of α so we make the following

CONJECTURE. If α is rational, $0 < \alpha < 1$, and $\alpha \neq 1/2$, then $\zeta(s,\alpha)$ has $\ll T$ zeros on $[1/2, 1/2 + iT]$.

REFERENCES

1. J.W.S. Cassels, Footnote to a note of Davenport and Heilbronn, J. London Math. Soc. 36 (1961), 177-184.
2. H. Davenport, The collected works of Harold Davenport, vol. 4, Academic Press, New York, 1977.
3. H. Davenport and H. Heilbronn, On the zeros of certain Dirichlet series, I. J. London Math. Soc. 11 (1936), 181-185.
4. A. Fujii, On the zeros of Dirichlet L-functions (V), Acta Arith. 28 (1976), 395-403.
5. S.M. Gonek, Analytic properties of zeta and L-functions, Thesis, University of Michigan, 1979.
6. A.E. Ingham, The distribution of prime numbers, Cambridge University Press, London, 1932.
7. N. Levinson, More than one-third of the zeros of Riemann's zeta-function are on $\sigma = 1/2$, Advances in Math. 13 (1974), 383-436.
8. S.M. Voronin, On the zeros of zeta-functions of quadratic forms, Trudy Mat. Inst. Steklov 142 (1976), 135-147. See also: Proc. Steklov Inst. Math. 3 (1979), 143-155.

Department of Mathematics
University of Rochester
Rochester, NY 14627

GAPS BETWEEN CONSECUTIVE ZETA ZEROS

Julia Mueller

Dedicated to Emil Grosswald

Denote by γ_n the imaginary part of the zero $\rho_n = \frac{1}{2}+i\gamma_n$ of the Riemann zeta function. Let

$$\gamma = \overline{\lim}_n (\gamma_n-\gamma_{n-1})\frac{\log \gamma_n}{2\pi}$$

and

$$\mu = \underline{\lim}_n (\gamma_n-\gamma_{n-1})\frac{\log \gamma_n}{2\pi} .$$

A. Selberg has shown that assuming the Riemann hypothesis, there exists an absolute positive constant c such that for all positive integers r,

$$\overline{\lim}_n (\gamma_{n+r}-\gamma_n)\frac{\log \gamma_n}{2\pi r} > 1 + \frac{c}{\sqrt{r}}$$

and

$$\underline{\lim}_n (\gamma_{n+r}-\gamma_n)\frac{\log \gamma_n}{2\pi r} < 1 - \frac{c}{\sqrt{r}} .$$

It follows from this that $\lambda > 1$ and $\mu < 1$. He has also shown, on the same hypothesis, that for each integer $k \geq 1$, we have

$$(1) \qquad \sum_{0 < \gamma_n \leq T} (\gamma_n-\gamma_{n-1})^k = O_k(T(\log T)^{1-k}).$$

The true orders of magnitude for λ and μ are not known. However, one suspects that $\lambda = \infty$ and $\mu = 0$. In this note we will show that $\lambda > 1.9$ as well as some other new results on the distribution of prime numbers and the zeros of the Riemann zeta function in short intervals.

For each positive integer k, let

$$I(T,\alpha,k) = \int_T^{2T} (S(t+\tfrac{\alpha}{L})-S(t))^{2k}dt$$

where $L = (2\pi)^{-1} \log T$ and S is the remainder term in the Riemann-von Mangoldt formula, when the formula is written as

$$N(T) = \int_0^T \frac{1}{2\pi} \log \frac{t}{2\pi}\, dt + S(T) = M(T) + S(T).$$

Selberg has remarked that (1) is a consequence of the following formulae for $I(T,\alpha,k)$:

(2a) $$I(T,\alpha,k) << c_k\, T(\log(2+\alpha))^k$$

for $1 << \alpha << \log T$, where the c_k's are absolute constants for all $k \geq 1$. In particular, for each function α of T for which $\alpha \to \infty$ as $T \to \infty$ we have

(2b) $$I(T,\alpha,k) \sim c_k T(2 \log \alpha)^k.$$

Derivations of (2a) and (2b) may be found in [1].

Denote by $R(x)$ the remainder term in the prime number theorem

$$\psi(x) = x+R(x) \ ,$$

where $\psi(x)$ is von Mangoldt's function. Then analogous to $I(T,\alpha,1)$ we may consider the integral

$$J(T,\beta) = \int_1^{T^\beta} (R(x+\tfrac{x}{T}) - R(x))^2 \frac{dx}{x^2} \ ,$$

for each positive β. This integral was first considered by Selberg [6]. He showed that the Riemann hypothesis implies that

(3) $$J(T,\beta) << T^{-1} \log^2 T$$

for each positive β; Montgomery [5] has shown that the implicit constant here can be taken $<< \beta$. From (3), Selberg has shown that

(4) $$\sum_{P_n \leq x} \frac{(P_n - P_{n-1})^2}{P_n} << \log^3 x.$$

A comparison between (1) and (4) is both interesting and illuminating. We remark that Erdös has conjectured that

$$\sum_{P_n \leq x} (P_n - P_{n-1})^2 << x \log x \ .$$

Since the function $I(T,\alpha,k)$ has an asymptotic formula (see (2b)), a natural question arises whether this might also be the case for $J(T,\beta)$. The following theorem provides an answer to this question. We remark that from now on the symbols β and α will have the following meaning: $\beta = \beta(T)$ for which $\beta \to \infty$ as $T \to \infty$ and $\alpha = \alpha(T)$ for which $\alpha \to 0$ as $T \to \infty$.

THEOREM 1. *Assume the Riemann hypothesis. The following two hypotheses are equivalent:*

(A) $$J(T,\beta) \sim \beta T^{-1} \log^2 T \qquad (T \to \infty)$$

(B) $$I(T,\beta) \sim \alpha T \qquad (T \to \infty) .$$

Here $I(T,\alpha)$ *is defined to be* $I(T,\alpha,1)$.

We will first show that hypothesis (B) may be reformulated as a hypothesis of "essential simplicity" of the zeros. To be more precise, let

$$N^*(T) = \sum_{0 < \gamma \leq T}{}' \; m^2(\gamma).$$

Here and later the dash indicates that the sum is over distinct values of γ. Denote by $N(T,U)$ the number of pairs of zeros $\frac{1}{2} + i\gamma$, $\frac{1}{2} + i\gamma$, with $0 < \gamma, \gamma' \leq T$ and $0 < \gamma'-\gamma \leq U$. The "essential simplicity" hypothesis states

(C_1) $$N^*(T) \sim TL. \quad (T \to \infty)$$

and

(C_2) $$N(T,U) = o(TL) \qquad (UL \to 0 \text{ as } T \to \infty).$$

We remark that $N^*(T) \gtrsim^{(+)} TL$ and $N(T,U) \geq 0$.

THEOREM 2. *Hypothesis* (B) *is equivalent to hypotheses* (C_1) *and* (C_2).

PROOF. We will first prove formula (5) from which the equivalence of (B) and (C_1) and (C_2) will follow immediately. Let $\alpha = UL$; then we have

---

(+) By $f \gtrsim g$ we mean $\overline{\lim}\, f/g \geq 1$.

$$UN^*(T) + 2\int_0^U N(T,u)du = I(T,\alpha) + o(T\alpha) \tag{5}$$

with $\alpha \to 0$ as $T \to \infty$.

Putting

$$F(t+U) - F(t) = \Delta_u F(t).$$

we have

$$\int_T^{2T} (\Delta_u N)^2 dt = \int_T^{2T} (\Delta_u M)^2 dt + 2\int_T^{2T} (\Delta_u M)(\Delta_u S)dt + \int_T^{2T} (\Delta_u S)^2 dt. \tag{6}$$

For $T \to \infty$ and α sufficiently small, we have

$$\Delta_u M(t) \sim \alpha$$

and hence

$$\int_T^{2T} (\Delta_u M)^2 dt \sim \alpha^2 T. \tag{7}$$

To estimate the second integral on the right-hand side of (6) we use the bound $S(T) << L$ to get

$$\int_T^{2T} (\Delta_u S)(t)dt = \int_{2T}^{2T+U} S(t)dt - \int_T^{T+U} S(t)dt << \alpha.$$

Since $\Delta_u M$ is monotonic and $<< \alpha$, it follows that for sufficiently small α, we have

$$\int_T^{2T} (\Delta_u M)(\Delta_u S)dt << \alpha^2. \tag{8}$$

Combining (6), (7), and (8) we get, for $T \to \infty$ and $\alpha \to 0$.

$$\int_T^{2T} (\Delta_u N)^2 dt = I(T,\alpha) + o(\alpha T). \tag{9}$$

On the other hand, it can be shown that

$$\int_T^{2T} (\Delta_u N)^2 dt = UN^*(T) + 2\int_0^U N(T,u)du + O(L^2). \tag{10}$$

It is now immediate that (5) follows from (9) and (10).

Our next aim is to prove Theorem 1, but first we will state the following lemmas. For the proofs of the lemmas, see [4].

Let

$$A^*(T) = \sum_{\gamma > 0}{}' \, a(\gamma)\, m^2(\gamma)$$

and

$$A(T,\beta) = \sum_{\substack{\gamma > 0 \\ \gamma' - \gamma > 0}} a(\gamma)\, S((\gamma'-\gamma)L\beta)$$

where

$$a(t) = a(t,T) = \left(\frac{\sin t/2T}{t/2T}\right)^2 \quad \text{and} \quad S(b) = \left(\frac{\sin \pi b}{\pi b}\right)^2 .$$

We remark that $A^*(T) \gtrsim \frac{1}{2} T \log T$ and $A(T,\beta) \geq 0$.

LEMMA 1. *Hypothesis* (C_1) *is equivalent to*

(11) $$A^*(T) \sim \frac{1}{2} T \log T \quad (T \to \infty).$$

LEMMA 2. *Hypothesis* (C_2) *is equivalent to*

(12) $$A(T,\beta) = o(TL) \quad (T \to \infty).$$

LEMMA 3. *We have*

(13) $$\int_0^\beta J(T,b)\,db = \frac{\beta^2 \log T}{T^2} \{A^*(T) + 2A(T,\beta) + o(TL)\} \quad (T \to \infty).$$

PROOF OF THEOREM 1. From hypothesis (A) and the bound in (3) we get

(14) $$\int_0^\beta J(T,b)\,db \sim \frac{1}{2}\,\beta^2\, \frac{\log^2 T}{T} \quad (T \to \infty).$$

Combining (13) and (14) we get

(15) $$A^*(T) + 2A(T,\beta) \sim \frac{1}{2} T \log T \quad (T \to \infty).$$

Since $A(T,\beta) \geq 0$ and $A^*(T) \gtrsim \frac{1}{2} T \log T$, it is clear that (15) implies (11) and (12) and hence (by lemmas 1 and 2) $(C_1$ and (C_2). Conversely, from assumptions (C_1) and (C_2) we get (14). Since $J(T,\beta)$ is a positive increasing function of β, we have

$$(\beta\varepsilon)^{-1} \int_{\beta(1-\varepsilon)}^{\beta} J(T,b)\,db \lesssim J(T,\beta) \lesssim (\beta\varepsilon)^{-1} \int_{\beta}^{\beta(1+\varepsilon)} J(T,b)\,db$$

for each $\epsilon > 0$. But the integrals here are asymptotic to $\beta^2\epsilon\ (1+\frac{1}{2}\epsilon)\ T^{-1}\log^2 T$. Letting $\epsilon \to 0$ we get (A).

Our last result is also our main result, that is

THEOREM 3. *Assume the Riemann hypothesis. We have* $\lambda > 1.9$.

PROOF. In 1917 Hardy and Littlewood [3] showed that

(16) $$\int_1^T |\zeta(\tfrac{1}{2}+it)|^2 dt \sim T\log T \qquad (T \to \infty).$$

Recently, S. Gonek [2] showed that assuming the Riemann hypothesis,

(17) $$\sum_{0<\gamma\le T} |\zeta(\tfrac{1}{2}+i(\gamma+cL^{-1}))|^2 = \frac{T}{2\pi}\log^2 T\left(1-\left(\frac{\sin \pi c}{\pi c}\right)^2\right) + O(T\log T)$$

as $T \to \infty$, uniformly for $|c| \le \frac{1}{2}L$.

For each positive α, we may integrate (17) with respect to c over the interval $[-\frac{\alpha}{2}, \frac{\alpha}{2}]$. This gives

(18) $$\sum_{0<\gamma\le T} \int_{\gamma-\frac{\alpha}{2}L^{-1}}^{\gamma+\frac{\alpha}{2}L^{-1}} |\zeta(\tfrac{1}{2}+it)|^2 dt \sim T\log T\cdot A(\alpha) \qquad (T\to\infty),$$

where

$$A(\alpha) = \int_{-\frac{1}{2}\alpha}^{\frac{1}{2}\alpha} \left(1-\left(\frac{\sin \pi c}{\pi c}\right)^2\right)dc.$$

Since $A(\alpha)$ has the power series expansion

$$A(\alpha) = \frac{2}{\pi}\sum_{k=1}^{\infty}(-1)^{k+1}\frac{(\pi\alpha)^{2k+1}}{(2k+2)!\,(2k+1)},$$

its value can be easily determined. We have

α	1.8	1.9	2.0
$A(\alpha)$	0.8979	0.9973	1.0972

.

It is clear that for $\alpha = 1.9$, the right hand side of (18) is $< T \log T$. Comparing this with (16), we get $\lambda > 1.9$.

The author would like to thank Professor Selberg for several valuable discussions.

Department of Mathematics
Fordham University
Bronx, N.Y. 10458

School of Mathematics
The Institute for Advanced Study
Princeton, N.J. 08540

References

1. Fujii, A., On the zeros of Dirichlet L-functions, I, Trans. Amer. Math. Soc. 196 (1974), 225-235.
2. Gonek, S.M., Analytic properties of zeta and L-functions, Dissertation, University of Michigan, 1979.
3. Hardy, G.H. and Littlewood, J.E., Contributions to the theory of the Riemann zeta function and the theory of the distribution of primes, Acta Math., 41 (1918), 119-196.
4. Mueller, J., Arithmetic equivalent of essential simplicity of zeta zeros, to appear.
5. Montgomery, H.L., Gaps between primes (unpublished).
6. Selberg, A., On the normal density of primes in short intervals and the difference between consecutive primes, Archiv. J. Math. og Naturvid. B47 (1943), No. 6.

ON THE REPRESENTATION OF THE SUMMATORY FUNCTIONS OF A CLASS OF ARITHMETICAL FUNCTIONS

James Lee Hafner

This paper is dedicated to Professor Emil Grosswald for his incalculable contribution to the body of mathematical knowledge and to the community of mathematical scientists.

1. Introduction

In this paper we establish some identities for the summatory functions (the Riesz sums) of a class of arithmetical functions associated with certain Dirichlet series. These identities involve a "main term" which is a residual function, and an "error term" which is a series of generalized Bessel functions. We first give two classical examples.

EXAMPLE 1. Let $r(n)$ be the number of representations of the integer n as a sum of two integer squares, counting order. Define the Dirichlet series $L(s)$ by

$$L(s) = \sum_{n=1}^{\infty} r(n)n^{-s} .$$

Hardy [9,11] proved the following identity. For $\rho \geq 0$

$$\Gamma(\rho+1)^{-1} \sum_{n\leq x}{}' r(n)(x-n)^{\rho} = \frac{\pi x^{1+\rho}}{\Gamma(\rho+2)} - \frac{x^{\rho}}{\Gamma(\rho+1)} + \frac{1}{\pi^{\rho}} \sum_{n=1}^{\infty} r(n)\left(\frac{x}{n}\right)^{(1+\rho)/2} J_{1+\rho}(2\pi\sqrt{nx}) ,$$

where J_ν is the ordinary Bessel function of order ν and the prime on the summation sign indicates the last term is to be multiplied by 1/2 if $\rho = 0$ and x is an integer. Chandrasekharan and Narasimhan [5] later showed that this identity is valid also for ρ in the range $-1/2 < \rho < 0$, if x is not an integer. In fact, the series of Bessel functions converges absolutely if $\rho > 1/2$ but only conditionally if $-1/2 < \rho \leq 1/2$.

EXAMPLE 2. Let $d(n)$ be the classical divisor function of Dirichlet, i.e., the number of ways of writing n as a product of two positive integers, counting

[1] Some of the results in this paper appeared in the author's Ph.D. dissertation written under the direction of Professor B.C. Berndt at the University of Illinois at Urbana-Champaign in 1980. Also partially supported by NSF Grant MCS 77-18723 A03 at the Institute for Advanced Study.

order. Associated with this arithmetical function is the Dirichlet series $\zeta^2(s) = \sum_{n=1}^{\infty} d(n)n^{-s}$ where $\zeta(s)$ is the classical Riemann zeta function. Chandrasekharan [4, Chapter 8] gives the following identity. For $\rho > -1/2$,

$$\Gamma(\rho+1)^{-1} \sum_{n\leq x}{}' d(n)(x-n)^{\rho} = \frac{x^{\rho+1}}{\Gamma(\rho+2)} \{\log x + \gamma - \frac{\Gamma'(\rho+2)}{\Gamma(\rho+2)}\}$$

$$+ \frac{x^{\rho}}{4\Gamma(\rho+1)} + \frac{1}{\pi^{2\rho+1}} \sum_{n=1}^{\infty} \frac{d(n)}{n^{1+\rho}} f_{\rho}(x\pi^2 n) \qquad (1.1)$$

where

$$f_{\rho}(x) = \frac{1}{2\pi i} \int_C \frac{\Gamma(1-s)\Gamma^2(s/2)x^{1+\rho-s}}{\Gamma(\rho+2-s)\Gamma^2((1-s)/2)} ds$$

and C is the oriented polygonal path with vertices $-1 - i\infty$, $-1 - iR$, $3/2 - iR$, $3/2 + iR$, $-1 + iR$, and $-1 + i\infty$, in that order, with R any positive number. (The prime on the summation sign has the same meaning as in the previous example.) Of course if $\rho < 0$, we must assume x is not an integer.

If ρ is a non-negative integer, then in fact

$$f_{\rho}(x) = \frac{-x^{(1+\rho)/2}}{2^{\rho}} \{Y_{\rho+1}(4\sqrt{x}) + \frac{2}{\pi}(-1)^{\rho} K_{\rho+1}(4\sqrt{x})\}$$

where Y_{ν} is the ordinary Bessel function of the second kind and K_{ν} is the modified Bessel function. See Watson [15, pp.64,78].

Again the infinite series on the right of (1.1) converges absolutely if $\rho > 1/2$ and conditionally if $-1/2 < \rho \leq 1/2$. The special case when ρ is a non-negative integer was known to Voronoi [14] and was later re-established by Hardy [10].

In both of these examples, identities of the type described in the first paragraph are obtained for values ρ where the series of Bessel functions converges only conditionally. Since these series cannot be reduced to ordinary Fourier series directly, their convergence is not obvious. We also point out that in these examples, the special case $\rho = 0$, which is of particular importance in questions involving the average order of the arithmetical functions, falls in the range where the series converge only conditionally. This fact was exploited by Hardy

[10] when he gave his famous one-sided omega theorems on the average orders of $r(n)$ and $d(n)$, and was again exploited by the author [8] to improve on Hardy's results. It is also known that for $\rho < -1/2$, these series are summable by Cesaro means of order k (i.e., (C,k) summable) for $k > -\rho - 1/2$.

The main ingredient used in establishing identities of this type is the fact that the Dirichlet series $L(s)$ and $\zeta^2(s)$ satisfy functional equations with gamma factors. We establish these kinds of identities for arithmetical functions associated with Dirichlet series satisfying very general functional equations of this type. We make this more precise in the following definition, first given by Chandrasekharan and Narasimhan [6].

DEFINITION 1.1. Let $\{a(n)\}$ and $\{b(n)\}$ be two sequences of complex numbers, not identically zero. Let $\{\lambda_n\}$ and $\{\mu_n\}$ be two strictly increasing sequences of positive numbers tending to ∞. Suppose the series

$$\phi(s) = \sum_{n=1}^{\infty} a(n)\lambda_n^{-s}$$

and

$$\psi(s) = \sum_{n=1}^{\infty} b(n)\mu_n^{-s}$$

converge in some half-plane and have abscissas of absolute convergence σ_a^* and σ_a, respectively. For each $\nu = 1,2,\ldots,N$, suppose that $\alpha_\nu > 0$ and β_ν is complex. Let

$$\Delta(s) = \prod_{\nu=1}^{N} \Gamma(\alpha_\nu s + \beta_\nu) .$$

If r is a real number, we say that ϕ and ψ satisfy the functional equation

$$\Delta(s)\phi(s) = \Delta(r-s)\psi(r-s)$$

if there exists in the s-plane a domain D that is the exterior of a compact set S and on which there exists a holomorphic function χ such that

(i) $\lim_{|t|\to\infty} \chi(\sigma + it) = 0$

uniformly in every interval $-\infty < \sigma_1 \leq \sigma \leq \sigma_2 < \infty$, and

(ii) $$\chi(s) = \begin{cases} \Delta(s)\phi(s) & \text{for } \sigma > \sigma_a^* \\ \Delta(r-s)\psi(r-s) & \text{for } \sigma < r-\sigma_a \end{cases} .$$

In examples 1 and 2 above, $\phi(s) = \psi(s) = \pi^{-s}L(s)$, $\Delta(s) = \Gamma(s)$, $r = 1$ and $\phi(s) = \psi(s) = \pi^{-s}\zeta^2(s)$, $\Delta(s) = \Gamma^2(s/2)$, $r = 1$, respectively.

For $x > 0$ and ρ a real number, define the summatory function $A_\rho(x)$ of the arithmetical function $a(n)$ by

$$A_\rho(x) = \Gamma(\rho+1)^{-1} \sum_{\lambda_n \le x}{}' a(n)(x-\lambda_n)^\rho$$

where the prime on the summation sign indicates the last term is to be multiplied by 1/2 if $\rho = 0$ and $x = \lambda_n$. Of course when ρ is negative, $A_\rho(x)$ is defined only for those positive x not equal to any λ_n.

Before stating our theorems, we need to establish some notation. First let $\alpha = \sum_{\nu=1}^N \alpha_\nu$. Let $a = \min\{\sigma_a - 2/\alpha, \ r/2-1/(2\alpha)\}$, $s_0 = \sup\{|s| : s \in S\}$ where S is the "singularity set" in Definition 1.1, and $t_0 = \max\{|\beta_\nu/\alpha_\nu| : \nu = 1,2,\ldots,N\}$. Choose a constant $c > \max\{\sigma_a^*, \sigma_a, s_0, t_0\}$. Choose another constant b so that $r-b$ is not an integer and $b > \max\{c,r\}$. Let R be a real number larger than both s_0 and t_0.

Now let C be the rectangle with vertices $c \pm iR$ and $r - b \pm iR$, taken in the counter-clockwise direction. (See the figure.) Then for $x > 0$ we let $Q_\rho(x)$ denote the residual function

$$Q_\rho(x) = \frac{1}{2\pi i} \int_C \frac{\Gamma(s)\phi(s)x^{\rho+s}}{\Gamma(s+\rho+1)}\, ds \quad .$$

Note that our choices of b, c and R ensure that the path C encircles all of S. In general, $Q_\rho(x)$ can be calculated very readily. In example 1,

$$Q_\rho(x) = \frac{\pi x^{1+\rho}}{\Gamma(\rho+2)} - \frac{x^\rho}{\Gamma(\rho+1)} \quad .$$

For real numbers p and q, we shall denote by $C_{p,q}$ the oriented polygonal path with vertices $p - i\infty$, $p - iR$, $q - iR$, $q + iR$, $p + iR$, and $p + i\infty$ in that order. (See the figure.)

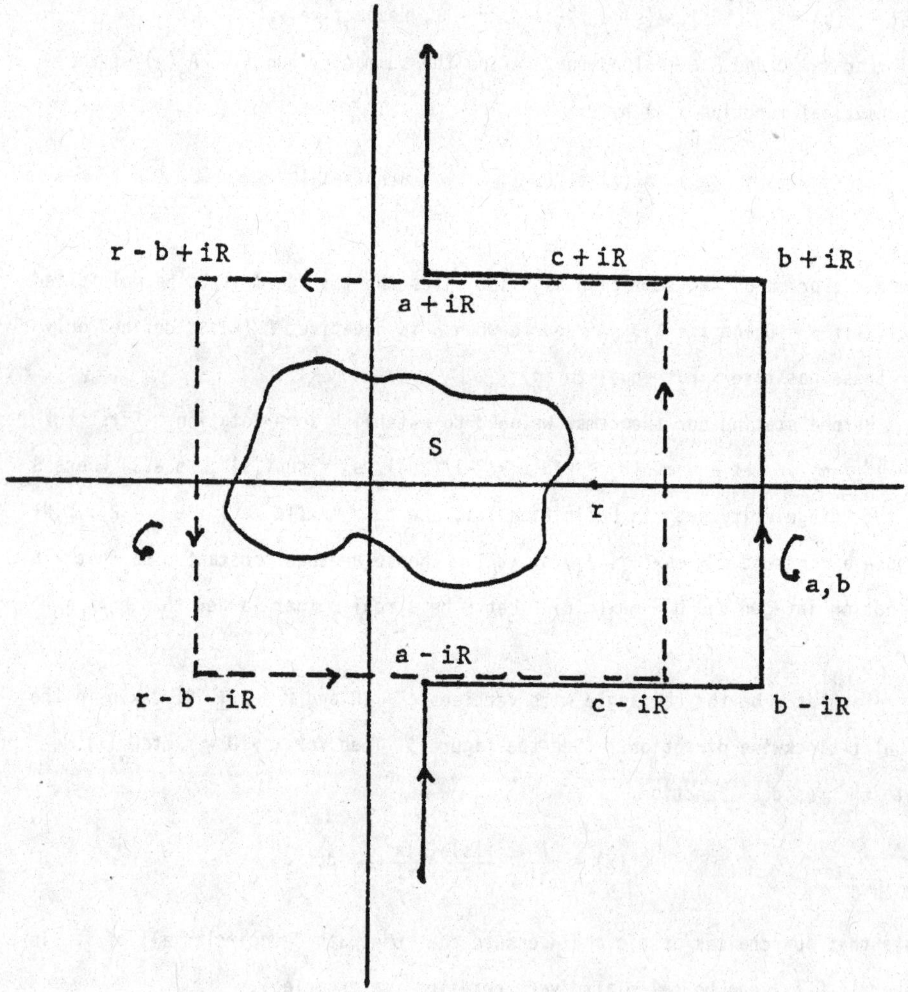
r - b + iR
c + iR
b + iR
a + iR
S
r
a - iR
r - b - iR
c - iR
b - iR
$C_{a,b}$

Next we define our so-called generalized Bessel functions. For $\rho > \min\{2\alpha\sigma_a - \alpha r - 4, -1\}$ and $x > 0$, let

$$f_\rho(x) = \frac{1}{2\pi i} \int_{C_{a,b}} G_\rho(s) x^{r+\rho-s}\, ds \tag{1.3}$$

where

$$G_\rho(s) = \frac{\Gamma(r-s)\Delta(s)}{\Gamma(r+\rho+1-s)\Delta(r-s)} \quad . \tag{1.4}$$

Applying Stirling's formula to the gamma factors in G_ρ shows that $f_\rho(x)$ is well-defined. We have referred to these functions as generalized Bessel functions because in the case $\Delta(s) = \Gamma(s)$, they are in fact certain multiples of the ordinary Bessel function J_ν. When $\Delta(s) = \Gamma^2(s/2)$, $\Gamma(s/2)\Gamma((s-p)/2)$ or $\Gamma^2((s+1)/2)$, they also involve Bessel functions (with certain restrictions on ρ, p and r). See, for example, Chandrasekharan and Narasimhan [6]. Furthermore, they have very similar asymptotic expansions as will be seen in Lemma 2.1.

We can now state our three theorems.

THEOREM A. *If* $x > 0$ *and* $\rho > 2\alpha\sigma_a - \alpha r - 1/2$, *then*

$$A_\rho(x) = Q_\rho(x) + \sum_{n=1}^{\infty} \frac{b(n)}{\mu_n^{r+\rho}} f_\rho(x\mu_n) \ . \tag{1.5}$$

THEOREM B. *Suppose that for* $\sigma > \sigma_a$,

$$\sup_{0\le t\le 1} \left| \sum_{k^{2\alpha} \le \mu_n \le (k+t)^{2\alpha}} b(n)\mu_n^{-\sigma+1/(2\alpha)} \right| = o(1) \quad , \tag{1.6}$$

as k *tends to* ∞. *Then the identity* (1.5) *is valid for* $\rho > 2\alpha\sigma_a - \alpha r - 3/2$ *and for those* $x > 0$ *where* $A_\rho(x)$ *is defined. Furthermore, the series converges uniformly on any finite closed interval in* $(0,\infty)$ *where* $A_\rho(x)$ *is continuous. The convergence is bounded on every finite interval when* $\rho = 0$.

THEOREM C. *Under the hypotheses of Theorem B, for* $x > 0$ *and* $\rho > -1$, *the series*

$$\sum_{n=1}^{\infty} \frac{b(n)}{\mu_n^{r+\rho}} f_\rho(x\mu_n)$$

is (C,k) summable for $k > \max\{2\alpha\sigma_a - \alpha r - 3/2 - \rho, 0\}$. The summability is uniform in every finite interval of continuity of $A_\rho(x)$ in $x > 0$ and the sum is $A_\rho(x) - Q_\rho(x)$.

We shall prove these three theorems in a series of lemmas. We should point out that the essence of a proof of Theorem A can be found in the work of Chandrasekharan and Narasimhan [6], but their purpose was not explicit representations of $A_\rho(x)$ but its order of magnitude. The proof, which we include here for completeness, is quite easy. The difficulty lies in the proof of Theorem B where the validity of (1.5) is established for a larger range for ρ, one in which the series does not converge absolutely. Chandrasekharan and Narasimhan [5] established Theorem B in the special case when $\Delta(s) = \Gamma(s)$. Berndt [1] generalized their method to include the cases where $\Delta(s) = \Gamma^m(s)$ and m is a positive integer. It is this method that we further generalize to arbitrary $\Delta(s)$. (Note that Berndt's theorem does not apply to example 2 as given. It does, however, yeild an identity for $\Gamma(\rho+1)^{-1} \sum'_{n\le x} d(n)(x^2-n^2)$. Our theorems can be applied to yield an identity for $\Gamma(\rho+1)^{-1} \sum'_{n\le x} d(n)(x^A-n^A)^\rho$ for any $A > 0$.) Our proof differs from Berndt's in some important technical details. These are indicated in the proof of Lemma 2.1. We also point out that Berndt [3] gave a second proof of Theorem B in the case $\Delta(s) = \Gamma(s)$, but with (1.6) replaced by another assumption involving the order of magnitude of $A_\rho(x) - Q_\rho(x)$.

Theorem C is a generalization to arbitrary $\Delta(s)$ of a theorem of Chandrasekharan and Narasimhan [5, Theorem IV] for the case $\Delta(s) = \Gamma(s)$. The extension of their proof of the general case is straight-forward and we omit it. (It depends on an equi-summability theorem of Zygmund [16, Theorem 9] and is of the same character as the proof of Theorem B.) We note that in [5], it is assumed that $\mu_n \ll n$ but this does not appear to be necessary. Furthermore, the range of ρ can be extended below -1 by choosing any smaller value for a.

2. The Generalized Bessel Functions

We begin the proofs of these theorems with a study of the functions $f_\rho(x)$ defined in (1.3).

LEMMA 2.1. For $x > 0$ and $\rho > \min\{2\alpha\sigma_a - \alpha r - 4, -1\}$,

$$\frac{d}{dx} f_{\rho+1}(x) = f_\rho(x) ; \tag{2.1}$$

and for any non-negative integer m,

$$f_\rho(x) = \sum_{\nu=0}^{m} e_\nu(\rho) x^{\theta_\rho - \nu/(2\alpha)} \cos(hx^{1/(2\alpha)} + k_\nu \pi) + O(x^{\theta_\rho - (m+1)/(2\alpha)}) + O(x^{r+\rho-b}), \tag{2.2}$$

where

$$\theta_\rho = r/2 - 1/(4\alpha) + \rho(1 - 1/(2\alpha)),$$
$$k_\nu = \beta_\rho + \nu/2 ,$$
$$\beta_\rho = -(\mu + \alpha r/2 + \rho/2 + 1/4) ,$$

μ and h are constants and $e_\nu(\rho)$, $\nu = 0,1,\ldots,m$, are constants depending on ρ. In particular,

$$e_0(\rho) = \left(\frac{2\alpha}{h}\right)^\rho \frac{1}{\sqrt{h\pi}} \tag{2.3}$$

and

$$e_1(\rho) = \left(\frac{2\alpha}{h}\right)^{\rho+1} \frac{1}{\sqrt{h\pi}} \left\{B_0 + \frac{\rho}{2}\left(r + 1 - \frac{1}{\alpha}\right) + \rho^2\left(1 - \frac{1}{2\alpha}\right)\right\} \tag{2.4}$$

where B_0 is an absolute constant.

PROOF. For the proof of (2.1), we note that the path of integration defining f_ρ is independent of ρ. Also since $f_{\rho+1}$ and f_ρ both converge absolutely and uniformly in $\rho > \min\{2\alpha\sigma_a - \alpha r - 4, -1\}$ and x in any bounded set, we may differentiate under the integral sign. The result (2.1) is then immediate from the identity $\Gamma(s + 1) = s\Gamma(s)$.

To prove (2.2) we deform the path $C_{a,b}$ into a path $C_{a,b'}$, where b' a certain constant depending on m. (In fact we take $b' > \max\{b, (m+1)/(2\alpha) - (\mu + \beta_\rho)/\alpha\}$.) By Cauchy's Theorem,

$$f_\rho(x) = \frac{1}{2\pi i} \int_{C_{a,b'}} G_\rho(s) x^{r+\rho-s}\, ds + \sum_{\substack{n \geq 0 \\ b<r+n<b'}} \operatorname*{res}_{s=r+n} G_\rho(s) x^{r+\rho-s} . \tag{2.5}$$

Since b and b' are constants,

$$\sum_{\substack{n \geq 0 \\ b<r+n<b'}} \operatorname*{res}_{s=r+n} G_\rho(s) x^{r+\rho-s} = O(x^{r+\rho-b}) . \tag{2.6}$$

Note that if ρ is an integer then $G_\rho(s)$ is analytic in $\sigma > b$ so that the sum in (2.6) is actually equal to zero. In this case (2.2) can be replaced by an asymptotic series.

In the remainder of the proof, the following asymptotic formula for the integral on the right hand side of (2.5) is developed:

$$\frac{1}{2\pi i} \int_{C_{a,b'}} G_\rho(s) x^{r+\rho-s}\, ds = \sum_{\nu=0}^{m} e_\nu(\rho) x^{\theta_\rho - \nu/(2\alpha)} \cos(hx^{1/(2\alpha)} + k_\nu \pi)$$

$$+ O(x^{\theta_\rho - (m+1)/(2\alpha)}) . \tag{2.7}$$

For a proof of (2.7) we refer the reader to Chandrasekharan and Narasimhan [7]. (The specific values of μ and h can also be found in [7].) We should point out that in their proof ρ is an integer. However, the computation in (2.5) and (2.6) allows the generalization of their proof to non-integral ρ. This difficulty was avoided by Berndt in [1] (the special case $\Delta(s) = \Gamma^m(s)$) because the singularities at $s = r + n$ arising from the factor $\Gamma(r-s)$ in the numerator of $G_\rho(s)$ are removed by one term in $\Delta(r-s) = \Gamma^m(r-s)$ appearing in the denominator. He therefore could produce a full asymptotic series. However, the full series is not required; by our choice of b, we can proceed with only a partial series. Inserting (2.6) and (2.7) in (2.5) completes the proof.

LEMMA 2.2. *For* $\rho > \min\{2\alpha\sigma_a - \alpha r - 4, -1\}$, *as* x *tends to* ∞,

$$f_\rho(x) = O(x^{\theta_\rho}) + O(x^{r+\rho-b}).$$

PROOF. This follows easily from (2.2) with $m = 0$.

LEMMA 2.3. *For each* $\rho > 2\alpha\sigma_a - \alpha r - 1/2$, *the series*

$$\sum_{n=1}^{\infty} \frac{b(n)}{\mu_n^{r+\rho}} f_\rho(x\mu_n) .$$

converges absolutely and uniformly in any finite interval in $x > 0$.

PROOF. This is an immediate consequence of Lemma 2.2 and our choice of $b > c > \sigma_a$.

3. PROOF OF THEOREM A

By the Perron formula,

$$A_\rho(x) = \frac{1}{2\pi i} \int_{c-i\infty}^{c+i\infty} \frac{\Gamma(s)\phi(s)x^{s+\rho}}{\Gamma(s+\rho+1)} ds$$

where c is the constant chosen in Section 1. Furthermore, by the usual procedure (see Chandrasekharan and Narasimhan [5]), we change the path of integration to $C_{r-c,r-b}$, and using the functional equation along with a change of variable from s to r-s, we have for $\rho > \max\{0, 2\alpha c - \alpha r\}$,

$$A_\rho(x) = Q_\rho(x) + \frac{1}{2\pi i} \int_{C_{c,b}} G_\rho(s)\psi(s)x^{r+\rho-s} ds .$$

Here $Q_\rho(x)$ and $G_\rho(s)$ are defined in (1.2) and (1.4), respectively. Now since $b > c > \sigma_a$, we can replace ψ by its Dirichlet series and invert the order of summation and integration. Thus we find that for $\rho > \max\{0, 2\alpha c - \alpha r\}$,

$$A_\rho(x) = Q_\rho(x) + \sum_{n=1}^{\infty} \frac{b(n)}{\mu_n^{r+\rho}} \frac{1}{2\pi i} \int_{C_{c,b}} G_\rho(s)(x\mu_n)^{r+\rho-s} ds . \tag{3.1}$$

Now if $\rho > \min\{2\alpha\sigma_a - \alpha r - 4, -1\}$, it follows from (1.3), the analyticity of the integrand for $|\mathrm{Im}s| > R$, Stirling's formula, and Cauchy's Theorem, that

$$\frac{1}{2\pi i} \int_{C_{c,b}} G_\rho(s)x^{r+\rho-s} ds = f_\rho(x).$$

Substituting this formula into (3.1) we find that (1.5) is valid provided $\rho > \max\{0, 2\alpha c - \alpha r, 2\alpha\sigma_a - \alpha r - 4\}$. But this identity remains unchanged in form by differentiation provided the differentiated series converges absolutely and uniformly for x in any bounded interval. But by Lemma 2.3 this occurs when $\rho > 2\alpha\sigma_a - \alpha r - 1/2$. This completes the proof of Theorem A.

4. Equiconvergence Theorems

We shall need some very deep results of A. Zygmund [16] involving trigonometric series. We refer the reader to Chandrasekharan [4, Chapter 8] for a careful exposition of the results needed and for the proofs of the first two lemmas that we quote below.

DEFINITION 4.1. Two series $\sum_{-\infty}^{\infty} a_j(x)$ and $\sum_{-\infty}^{\infty} b_j(x)$ are uniformly equiconvergent on an interval if

$$\sum_{j=-n}^{\infty} [a_j(x) - b_j(x)]$$

converges uniformly on that interval as n tends to ∞.

The following two lemmas are of central importance.

LEMMA 4.1. *Let $\{\alpha_n\}$ be a sequence of positive numbers tending to ∞. Suppose that $\alpha_{-n} = \alpha_n$. Let J be a closed interval contained in an interval I of length 2π. Let λ be a C^∞-function with compact support on I that equals one on J. Assume that*

$$\sum_{n=-\infty}^{\infty} |c(n)| < \infty .$$

Then if g is a function with period 2π that equals $\sum_{-\infty}^{\infty} c(n)\exp\{i\alpha_n x\}$ on I, then the Fourier series of g converges uniformly on J.

LEMMA 4.2. *With the same notation as Lemma 4.1, assume only that*

$$\sup_{0 \le t \le 1} \left| \sum_{k \le \alpha_n \le k+t} c(n) \right| = o(1),$$

as k tends to ∞, and

$$\sum_{n=-\infty}^{\infty} |c(n)|\alpha_n^{-1} < \infty .$$

Furthermore, let γ be a C^∞-function. Then the series

$$\gamma(x) \sum_{n=-\infty}^{\infty} c(n)\exp\{i\alpha_n x\}$$

is uniformly equiconvergent on J with the differentiated series of the Fourier series of a function with period 2π that on I equals

$$\lambda(x) \sum_{n=-\infty}^{\infty} c(n)W_n(x),$$

where $W_n(x)$ is an antiderivative of $\gamma(x)\exp\{i\alpha_n x\}$.

We now apply these lemmas to the particular series in question. Define for $y > 0$

$$S_\rho(y) = \sum_{n=1}^{\infty} \frac{b(n)}{\mu_n^{r+\rho}} f_\rho(y^{2\alpha}\mu_n) . \tag{4.1}$$

LEMMA 4.3. Suppose that

$$\rho > 2\alpha\sigma_a - \alpha r - 1/2 \tag{4.2}$$

and

$$\sup_{0\le t\le 1} \left| \sum_{k^{2\alpha}\le\mu_n\le(k+t)^{2\alpha}} b(n)\mu_n^{-r/2-\rho/(2\alpha)+1/(4\alpha)} \right| = o(1) \tag{4.3}$$

as k tends to ∞. Let I be any interval of length π/α, and let J be any subinterval of I of length less than π/α. Furthermore, let $g_K(y)$ be a function of period π/α which on I equals $\lambda(y)[S_\rho(y) - K]$, where K is a constant. Then for some constant K the series $2\alpha y^{2\alpha-1}S_{\rho-1}(y)$ is uniformly equiconvergent on J with the derived series of the Fourier series for $g_K(s)$.

PROOF. We examine the function

$$F_{\rho-1}(y) = 2\alpha y^{2\alpha-1} \sum_{n=1}^{\infty} \frac{b(n)}{\mu_n^{r+\rho-1}} \{f_{\rho-1}(y^{2\alpha}\mu_n)$$

$$- e_0(\rho-1)(y^{2\alpha}\mu_n)^{\theta_{\rho-1}} \cos(hy\mu_n^{1/(2\alpha)} + \beta_{\rho-1}\pi) \tag{4.4}$$

$$+ e_1(\rho-1)(y^{2\alpha}\mu_n)^{\theta_{\rho-1}-1/(2\alpha)} \sin(hy\mu_n^{1/(2\alpha)} + \beta_{\rho-1}\pi)\},$$

for $y > 0$. By Lemma 2.1, (2.2) with $m = 1$, (4.2) and (2.3) - (2.4),

$$|F'_{\rho-1}(y)| \leq Ay^B \sum_{n=1}^{\infty} |b(n)|\mu_n^{-r/2-\rho/(2\alpha)-1/(4\alpha)} < \infty ,$$

where A and B are constants. Hence $F_{\rho-1}$ is continuously differentiable for $y > 0$.

Let g be a function with period π/α that equals $F_{\rho-1}$ on I. Since $F_{\rho-1}$ is continuously differentiable, the Fourier series for g is uniformly convergent on J.

Now consider each of the terms in the series $F_{\rho-1}$. For the term

$$2\alpha y^{2\alpha-1} \sum_{n=1}^{\infty} \frac{b(n)}{\mu_n^{r+\rho-1}} (y^{2\alpha}\mu_n)^{\theta_{\rho-1}} \cos(hy\mu_n^{1/(2\alpha)} + \beta_{\rho-1}\pi), \tag{4.5}$$

in view of (4.2) and (4.3), we can apply Lemma 4.2 with

$$c(n) = \frac{b(n)}{\mu_n^{r+\rho-1-\theta_{\rho-1}}} = \frac{b(n)}{\mu_n^{r/2+\rho/(2\alpha)-1/(4\alpha)}}$$

and $\alpha_n = \mu_n^{1/(2\alpha)}$. Hence (4.5) is uniformly equiconvergent on J with the differentiated series of the Fourier series of a function with period π/α that on I equals

$$2\alpha\lambda(y) \sum_{n=1}^{\infty} \frac{b(n)}{\mu_n^{r+\rho-1}} \int_{y_0}^{y} t^{2\alpha-1}(t^{2\alpha}\mu_n)^{\theta_{\rho-1}} \cos(ht\mu_n^{1/(2\alpha)} + \beta_{\rho-1}\pi)dt,$$

where y_0 lies in J.

Next we examine the series

$$2\alpha y^{2\alpha-1} \sum_{n=1}^{\infty} \frac{b(n)}{\mu_n^{r+\rho-1}} (y^{2\alpha}\mu_n)^{\theta_{\rho-1}-1/(2\alpha)} \sin(hy\mu_n^{1/(2\alpha)} + \beta_{\rho-1}\pi). \tag{4.6}$$

By (4.2) and (4.3) and Lemma 4.1 with

$$c(n) = \frac{b(n)}{\mu_n^{r+\rho-1-\theta_{\rho-1}+1/(2\alpha)}} = \frac{b(n)}{\mu_n^{r/2+\rho/(2\alpha)+1/(4\alpha)}}$$

the Fourier series for the function of period π/α that equals (4.6) on I converges uniformly on J.

Thus it follows that $2\alpha y^{2\alpha-1} S_{\rho-1}(y)$ is uniformly equiconvergent on J with the differentiated series of the Fourier series of a function with period π/α that on I equals

$$\lambda(y) \sum_{n=1}^{\infty} \frac{b(n)}{\mu_n^{r+\rho-1}} \int_{y_0}^{y} 2\alpha t^{2\alpha-1} f_{\rho-1}(t^{2\alpha}\mu_n)dt$$

$$= \lambda(y) \sum_{n=1}^{\infty} \cdot \frac{b(n)}{\mu_n^{r+\rho-1}} \{f_\rho(y^{2\alpha}\mu_n) - f_\rho(y_0^{2\alpha}\mu_n)\}$$

$$= \lambda(y) [S_\rho(y) - S_\rho(y_0)]$$

$$= \lambda(y) [S_\rho(y) - K],$$

where K is a constant depending on y_0. We have used (2.1) and (4.2) so that by (2.8), $S_\rho(y_0)$ converges absolutely. This completes the proof.

5. Proof of Theorem B

We shall show that

$$2\alpha y^{2\alpha-1} \{A_\rho(y^{2\alpha}) - Q_\rho(y^{2\alpha})\} = 2\alpha y^{2\alpha-1} S_\rho(y), \tag{5.1}$$

where $S_\rho(y)$ is defined in (4.1) and y is in any interval J of length less than π/α. Under the hypotheses of the theorem, if $\rho > 2\alpha\sigma_a - \alpha r - 3/2$, then

(a) $\rho+1 > 2\alpha\sigma_a - \alpha r - 1/2$

and $r/2 + (\rho+1)/(2\alpha) + 1/(4\alpha) > \sigma_a$, so that

(b) $$\sup_{0\le t\le 1} \left| \sum_{k^{2\alpha}\le\mu_n\le(k+t)^{2\alpha}} b(n)\mu_n^{-r/2-\rho/(2\alpha)-1/(4\alpha)} \right| = o(1),$$

as k tends to ∞. We can thus apply Lemma 4.3 with ρ replaced by $\rho+1$. The right-hand side of (5.1) is then uniformly equiconvergent on J with the differentiated series of the Fourier series of a function of period π/α that on I equals $\lambda(y)\,[S_{\rho+1}(y) - K]$. But by (a) and (4.1), the definition of $A_\rho(x)$, and Theorem A, with ρ replaced by $\rho+1$ and x by $y^{2\alpha}$, we conclude that

$$\lambda(y)\,[S_{\rho+1}(y) - K] = \lambda(y)A_{\rho+1}(y^{2\alpha}) - \lambda(y)\,[Q_{\rho+1}(y^{2\alpha}) + K]$$

$$= \frac{\lambda(y)}{\Gamma(\rho+2)} \sum_{\lambda_n \le y^{2\alpha}}{}' a(n)\,(y^{2\alpha} - \lambda_n)^{\rho+1}$$

$$- \lambda(y)\,[Q_{\rho+1}(y^{2\alpha}) + K]$$

$$= \frac{\lambda(y)}{\Gamma(\rho+1)} \int_0^y \sum_{\lambda_n \le t^{2\alpha}}{}' a(n)\,(t^{2\alpha} - \lambda_n)^{\rho}\, 2\alpha t^{2\alpha-1}\,dt$$

$$- \lambda(y)\,[Q_{\rho+1}(y^{2\alpha}) + K]\,.$$

Now $\lambda(y) = 1$ on J, λ and $Q_{\rho+1}$ are C^∞-functions, and the derived series of the Fourier series of

$$\frac{1}{\Gamma(\rho+1)} \int_0^y \sum_{\lambda_n \le t^{2\alpha}}{}' a(n)\,(t^{2\alpha} - \lambda_n)\, 2\alpha t^{2\alpha-1}\,dt$$

is just the Fourier series of

$$\frac{2\alpha y^{2\alpha-1}}{\Gamma(\rho+1)} \sum_{\lambda_n \le y^{2\alpha}}{}' a(n)\,(y^{2\alpha} - \lambda_n)^{\rho}\,.$$

Thus the series $2\alpha y^{2\alpha-1}\,S_\rho(y)$ converges on J in the smae manner as the Fourier series for (5.2). Since J is arbitrary, this last remark is valid for any finite interval. Now this Fourier series for (5.2) converges exactly as described in Theorem B. This shows that the series $S_\rho(y)$, and so the series in (1.5), converges for $\rho > 2\alpha\sigma_a - \alpha r - 3/2$ as described, and completes the proof.

6. An Example

We generalize the two examples given in the introduction. Let K be an algebraic number field of degree $N = r_1 + 2r_2$ where r_1 is the number of real conjugates and $2r_2$ is the number of imaginary conjugates of K. Let m be a positive integer and let $d_m(n,K)$ be the coefficient of n^{-s} in the Dirichlet series for $\zeta_K^m(s)$, the mth power of the Dedekind zeta-function of K. Note that $4d_1(n, Q(\sqrt{-1})) = r(n)$ and $d_2(n, Q) = d(n)$ as given in the examples above. Put

$$A = 2^{mr_2} \pi^{mN/2} d^{-m/2}$$

where d is the absolute value of the discriminant of K. It is known (see Landau [12]) that $\phi(s) = A^{-s}\zeta_K^m(s)$ satisfies the functional equation

$$\Delta(s)\phi(s) = \Delta(1-s)\phi(1-s)$$

where

$$\Delta(s) = \Gamma(s/2)^{mr_1} \Gamma(s)^{mr_2} .$$

The above theorems then yield the following identity. For $\rho > (mN-3)/2$,

$$\Gamma(\rho+1)^{-1} \sum_{n\leq x}{}' d_m(n,K)(x-n)^\rho = x^{\rho+1} P_{m-1}(\log x) + \frac{\zeta_K^m(0)x^\rho}{\Gamma(\rho+1)}$$

$$+ \frac{1}{A^{2\rho+1}} \sum_{n=1}^{\infty} \frac{d_m(n,K)}{n^{1+\rho}} f_\rho(xA^2n)$$

where with $a = \min\{\frac{1}{2} - 1/(mN), 1 - 4/(mN)\}$ and $b = 3/2$,

$$f_\rho(x) = \frac{1}{2\pi i} \int_{C_{a,b}} \frac{\Gamma(1-s)\Delta(s)}{\Gamma(2+\rho-s)\Delta(1-s)} x^{1+\rho-s} ds$$

and where P_{m-1} is a polynomial of degree m - 1. The series converges absolutely if $\rho > (mN-1)/2$, and conditionally if $(mN-3)/2 < \rho \leq (mN-1)/2$.

For $-1 < \rho \leq (mN - 3)/2$, the series is (C,k) summable of order $k > (mN - 3 - 2\rho)/2$. In particular, for $\rho = 0$ the series is (C,k) summable $(k > (mN - 3)/2)$ with sum

$$\Delta(x) = \sum_{n\leq x}{}' d_m(n,K) - xP_{m-1}(\log x) - \zeta_K^m(0).$$

This fact can be applied to the study of the order of $\Delta(x)$, the so-called "error term in the Piltz divisor problem" as studied by Szegö and Walfisz [13] and

Berndt [2]. By applying methods similar to those used by the author in [8], their proofs can be significantly simplified.

This also generalizes some of the examples given by Berndt in [3]. There he had to assume that one of r_1 or r_2 be zero.

REFERENCES

1. B.C. Berndt, Identities involving the coefficients of a class of Dirichlet series I, _Trans. Amer. Math. Soc._, 137 (1969), 345-359.
2. _________, On the average order of a class of arithmetical functions, I, J. Number Theory 3 (1971), 184-203.
3. _________, Identities involving the coefficients of a class of Dirichlet series VII, _Trans. Amer. Math. Soc._, 201 (1975), 247-261.
4. K. Chandrasekharan, _Arithmetical Functions_, Springer-Verlag, New York, 1969.
5. K. Chandrasekharan and R. Narasimhan, Hecke's functional equation and arithmetical identities, _Ann. Math._ 74 (1961), 1-23.
6. _________, Functional equations with multiple gamma factors and the average order of arithmetical functions, _Ann. Math._, 76 (1962), 93-136.
7. _________, Approximate functional equations for a class of zeta-functions, _Math. Ann._ 152 (1963), 30-64.
8. J.L. Hafner, New omega theorems for two classical lattice point problems, Invent. Math. 63 (1981), 181-186.
9. G.H. Hardy, On the expression of a number as the sum of two squares, _Quart. J. Math._, 46 (1915), 263-283.
10. _________, On Dirichlet's divisor problem, _Proc. London Math. Soc._ (2) 15 (1916), 1-25.
11. _________, The average order of the arithmetical functions P(x) and $\Delta(x)$, _Proc. London Math. Soc._ (2) 15 (1916), 192-213.
12. E. Landau, _Einführung in die elementare und analytische Theorie der algebraischen Zahlen und der Ideale_, Chelsea, New York, 1949.
13. G. Szegö and A. Walfisz, Uber das Piltzsche Teilerproblem in algebraischen Zahlkörpern (Erste Abhandlung), Math. Zeit. 26 (1927), 138-156.

14. G. Voronoï, Sur une fonction transcendente et ses applications à la sommation de quelque series, Ann. Sci. Ecole Norm. Sup. (3) 21 (1904), 207 - 267, 459-553.

15. G.N. Watson, A Treatise on the Theory of Bessel Functions, 2nd ed., Cambridge Univ. Press, Cambridge, 1952.

16. A. Zygmund, On trigonometric integrals, Ann. Math., 48 (1947),393-440.

The University of Illinois at Urbana-Champaign
Urbana, Illinois 61801

The Institute for Advanced Study
Princeton, New Jersey 08540

P-ADIC L-FUNCTIONS AT s = 0 AND s = 1

Lawrence C. Washington*

Department of Mathematics
University of Maryland
College Park, Maryland
20742

To Emil Grosswald on the occasion of his retirement

Let χ be a nontrivial even Dirichlet character of conductor f and let

$$L(s,\chi) = \sum_{n=1}^{\infty} \frac{\chi(n)}{n^s}$$

be the associated Dirichlet L-series. It follows easily from the functional equation that $L(0,\chi) = 0$ and

$$\frac{L'(0,\chi)}{L(1,\bar{\chi})} = \frac{\tau(\chi)}{2} ,$$

where

$$\tau(\chi) = \sum_{a=1}^{f} \chi(a)e^{2\pi ia/f}.$$

In particular, $L'(0,\chi)/L(1,\bar{\chi})$ is algebraic. In this note, we show that the corresponding result is false for p-adic L-functions.

Define the generalized Bernoulli numbers by

$$\sum_{n=0}^{\infty} B_{n,\chi} \frac{t^n}{n!} = \sum_{a=1}^{f} \frac{\chi(a)t\, e^{at}}{e^{ft} - 1} .$$

It is well-known that for $n \geq 1$,

$$L(1-n,\chi) = -\frac{B_{n,\chi}}{n} .$$

The Teichmüller character ω is defined as follows: for $a \in \mathbb{Z}_p$, $p \nmid a$, $\omega(a)$ is the

* Research supported in part by the Alfred P. Sloan Foundation and the National Science Foundation.

unique (p-1)st root of unity in $\mathbb{Z}_p$ satisfying $\omega(a) \equiv a \pmod p$ (if $p = 2$, $\omega(a) = \pm 1 \equiv a \pmod 4$). The p-adic L-function $L_p(s,\chi)$ is the unique continuous p-adic valued function on $\mathbb{Z}_p$ such that for $n \geq 1$,

$$L_p(1-n,\chi) = -(1-\chi\omega^{-n}(p)p^{n-1}) \frac{B_{n,\chi\omega^{-n}}}{n} .$$

So $L_p(1-n,\chi)$ agrees with $L(1-n,\chi)$ when $n \equiv 0 \pmod{p-1}$, except for the Euler factor at p. If χ is an odd character then $L_p(s,\chi)$ vanishes identically. Therefore we restrict our attention to even χ.

$L_p(s,\chi)$ is similar in many ways to the classical L-function, but it appears that $L_p(s,\chi)$ does not satisfy a functional equation. However, Pierrette Cassou-Noguès, in a talk at Oberwolfach (March 1979), raised the question of whether or not there could be a relation between the values at $s = 0$ and $s = 1$. This would be nice, since $s = 1$ gives information about units and class numbers of real fields, while $s = 0$ gives information about relative class numbers of imaginary fields. Thus, one could hope for more precise results similar to Kummer's "$p|h^+(Q(\zeta_p)) \Rightarrow p|h^-(Q(\zeta_p))$." As a first test for such a relation, she suggested checking whether or not $L_p(0,\chi) = 0$ implies that $L_p'(0,\chi)/L_p(1,\bar{\chi})$ is algebraic.

We first note that if this result were true, it would be unlikely that the ratio would be a Gauss sum. Note that $\tau(\chi) = \sum \chi(a)\zeta_f^a$, where $\zeta_f = e^{2\pi i/f}$, so we have made a definite choice of a primitive f-th root of unity. In the p-adics, there is no canonical choice for ζ_f, and different choices of ζ_f give different values for $\tau(\chi)$, while $L_p'(0,\chi)/L_p(1,\bar{\chi})$ is of course unchanged.

THEOREM. Suppose χ is an even character and $L_p(0,\chi) = 0$. Then $L_p'(0,\chi)/L_p(1,\bar{\chi})$ is transcendental (over Q).

PROOF. We first note that $L_p(1,\bar{\chi}) \neq 0$ (Ax-Brumer; see [1]), and that if $L_p(0,\chi) = 0$ then $L_p'(0,\chi) \neq 0$ (Ferrero-Greenberg [2]). Both proofs rely on the p-adic version of Baker's work on linear forms in logarithms of algebraic numbers.

Since $B_{1,\chi\omega^{-1}} \neq 0$, it follows from the above that $L_p(0,\chi) = 0$ if and only if $\chi\omega^{-1}(p) = 1$. Let d be the conductor of $\chi\omega^{-1}$, so $(p,d) = 1$ and $f = pd$ is the conductor of χ ($f = 4d$ if $p = 2$). Let q be the smallest power of p such that $d|q-1$, let $\mathbb{F}_q$ be the field with q elements, and let Tr = trace to $\mathbb{Z}/p\mathbb{Z}$. Fix a primitive p-th root of unity ζ_p and let $\psi(a) = \zeta_p^{Tr(a)}$ be a character of the additive group of $\mathbb{F}_q$. Let θ be the d-th power residue character of $\mathbb{F}_q^\times$. Define the Gauss sum

$$\gamma_c = - \sum_{a \in \mathbb{F}_q^\times} \theta^{-c}(a)\psi(a).$$

Then $\gamma_c\bar{\gamma}_c = q$ = a power of p. Let $\{c_1,\dots,c_g\}$ be a set of representatives for $(\mathbb{Z}/d\mathbb{Z})^\times$ modulo the subgroup generated by p. Using work of Gross and Koblitz [3], Ferrero and Greenberg [2] showed that if $\chi\omega^{-1}(p) = 1$ then

$$L_p'(0,\chi) = \sum_{i=1}^{g} \chi\omega^{-1}(c_i)\log_p(\gamma_{c_i}),$$

where $\log_p$ is the p-adic logarithm.

The second result we need is due to Leopoldt [4]. Let ζ_f be a primitive f-th root of unity and let $\tau(\chi) = \sum_{a=1}^{f} \chi(a)\zeta_f^a$ be the Gauss sum (there is more than one type of Gauss sum, though an appropriate definition for a finite ring would include both γ_c and $\tau(\chi)$). Then

$$L_p(1,\bar{\chi}) = -\left(1-\frac{\bar{\chi}(p)}{p}\right)\frac{\tau(\bar{\chi})}{f}\sum_{a=1}^{f} \chi(a)\log_p(1-\zeta_f^a).$$

Since p divides f, $\bar{\chi}(p) = 0$, so the Euler factor disappears. Also, $f = pd$ must be composite (χ even $\Rightarrow$ $\chi\omega^{-1} \neq 1$ $\Rightarrow$ $d \neq 1$), so $1-\zeta_f^a$ is a global unit of $Q(\zeta_f)$ for $(a,f) = 1$. Since $\chi(a) = 0$ if $(a,f) \neq 1$, the sum only involves those a with $(a,f) = 1$.

Let $\{g_1,\dots,g_r\}$ be a maximal subset of $\{\gamma_{c_1},\dots,\gamma_{c_g}\}$ such that $\{p,g_1,\dots,g_r\}$ is multiplicatively independent. For convenience, let $g_0 = p$.

Let $\{\varepsilon_1,\ldots,\varepsilon_s\}$ be a maximal multiplicatively independent subset of $\{1-\zeta_f^a \mid (a,f) = 1\}$. We claim that the set

$$\{g_0,g_1,\ldots,g_r,\ \varepsilon_1,\ldots,\varepsilon_s\}$$

is multiplicatively independent.

Suppose

$$\prod_{i=0}^{r} g_i^{a_i} = \prod_{j=1}^{s} \varepsilon_j^{b_j}, \text{ with } a_i,b_j \in \mathbb{Z}.$$

Then, for any embedding into the complex numbers,

$$\left|\prod g_i^{a_i}\right|^2 = \prod (g_i\bar{g}_i)^{a_i}$$

is a unit and is also an integral power of p, therefore equals 1. So $\prod g_i^{a_i}$ is a unit of absolute value 1, hence a root of unity. Therefore

$$\prod g_i^{Na_i} = 1 \text{ for some } N > 0.$$

It follows that $a_i = 0$ for all i, consequently $b_j = 0$ for all j, too. This proves the claim.

Since the kernel of $\log_p$ is generated by p and roots of unity, it follows that

$$\{\log_p g_1,\ldots,\log_p g_r,\ \log_p \varepsilon_1,\ldots,\log_p \varepsilon_s\}$$

is linearly independent over Q.

By the choice of the g's and ε's, we may write

$$L_p'(0,\chi) = \sum_{i=0}^{r} A_i \log_p g_i = \sum_{i=1}^{r} A_i \log_p g_i$$

and

$$L_p(1,\bar{\chi}) = \sum_{j=1}^{s} B_j \log_p \varepsilon_j,$$

where A_i and B_j are algebraic over Q. Suppose now that $L_p'(0,\chi)/L_p(1,\bar{\chi}) = C$ is algebraic. Then

$$\sum A_i \log_p g_i - \sum CB_j \log_p \varepsilon_j = 0.$$

But Brumer's p-adic version of Baker's Theorem [1] states that if $\alpha_1, \ldots, \alpha_e$ are algebraic over Q and $\{\log_p \alpha_i\}$ is linearly independent over Q, then $\{\log_p \alpha_i\}$ is linearly independent over the algebraic closure of Q. Therefore $A_i = B_j = 0$ for all i,j. But then $L_p(1,\bar{\chi}) = 0$, which is not true. Therefore $L_p'(0,\chi)/L_p(1,\bar{\chi})$ is transcendental and the theorem is proved.

REFERENCES

1. A. Brumer, On the units of algebraic number fields, Mathematika 14 (1967), 121-124.
2. B. Ferrero and R. Greenberg, On the behavior of p-adic L-functions at s = 0, Inv. Math. 50 (1978), 91-102.
3. B. Gross and N. Koblitz, Gauss sums and the p-adic Γ-function, Ann. of Math. 109 (1979), 569-581.
4. K. Iwasawa, Lectures on p-adic L-functions, Ann. of Math. Studies 74, Princeton University Press, 1972.

Department of Mathematics
University of Maryland
College Park, Maryland 20742

SOME PROBLEMS AND RESULTS ON ADDITIVE AND MULTIPLICATIVE NUMBER THEORY

P. Erdös

To my old friend Emil Groswald, in friendship and admiration.

In this note I discuss some problems of a somewhat unconventional nature which recently occupied me and my collaborators. I will deal with divisors and prime factors of integers, some additive problems of a combinatorial nature and on differences of consecutive primes, squarefree numbers and more general sequences defined by divisibility properties.

1. Let $1 = d_1 < d_2 < \dots < d_{\tau(n)} = n$ be the sequence of consecutive divisors of n. Put

$$h_\alpha(n) = \sum_{i=1}^{\tau(n)-1} \left(\frac{d_{i+1}}{d_i} - 1\right)^\alpha. \tag{1.1}$$

Is it true that for every $\alpha > 1$ there is a constant C_α and infinitely many integers n for which $h_\alpha(n) < C_\alpha$? This question occurred to me a few weeks ago but I was unable to make any progress. In fact I could not prove the existence of C_α for any α. $n!$ or the least common multiple of the integers not exceeding n seem to be good candidates for integers with (1.1) bounded above.

I came to (1.1) by considering the sum $\sum_{i=1}^{\tau(n)-1} d_{i+1}/d_i$.

It is easy to see that

$$\sum_{i=1}^{\tau(n)-1} d_{i+1}/d_i > \tau(n) + \log n$$

and I asked myself the question whether it is true that

$$\liminf_{n \to \infty} \left(\sum d_{i+1}/d_i - \tau(n) - \log n\right) < \infty. \tag{1.2}$$

(1.2) would follow if (1.1) is bounded for an infinite set of n.

Srinivasan calls a number n practical if every $m \leq n$ is the sum of distinct divisors of n. It is well known and easy to see that the density of practical numbers is 0. Let S(n) be the smallest integer so that every $1 \le m \le n$ is the sum of S(n) or fewer distinct divisors of n(S(n) = 0 if n is not practical). In connection with problems on representation of the form

$$\frac{a}{b} = \frac{1}{x_1} + \dots + \frac{1}{x_k}, \quad 1 \le a < b, \ k \text{ minimal},$$

I needed integers n for which S(n) is small. I easily observed

$$S(n!) < n \text{ or } S(m) < \frac{\log m}{\log \log m}$$

for infinitely many m. I conjectured that for infinitely many n

(1.3) $$S(n) < (\log \log n)^C$$

but I could make no progress with (1.3), which is unsolved for more than 30 years. I offer 250 dollars for a proof or disproof of (1.3). In itself (1.3) is perhaps somewhat artificial and isolated but a proof or disproof of (1.3) might throw some light on more important problems.

I just notice that the investigation of $\max_{n \le x} S(n)$ might lead to nontrivial questions. At first I thought that S(n) < clog n holds for all n but this is easily seen to be false. Let m_k be the product of the first k primes and let q_k be the greatest prime less than $\sigma(m_k)$. It is easy to see that $n_k = q_k m_k$ is practical but q_k-1 needs for its representation $n_k^{c/\log \log n_k}$ divisors of n_k. Perhaps one could try to obtain an asymptotic formula for $\sum_{n=1}^{x} S(n)$.

My most interesting unsolved problem on divisors states that almost all integers have two consecutive divisors

(1.4) $$d_{i+1} < 2d_i$$

or in a sharper form: For almost all n, (and every $\varepsilon > 0$)

(1.5) $$\min_i d_{i+1}/d_i < 1 + c^{-\log \log n(\log 3 - 1 - \varepsilon)}$$

R.R. Hall and I proved that the exponent in (1.5) if true is best possible.

Denote by $\tau^+(n)$ the number of integers k for which n has a divisor in $(2^k,2^{k+1})$. I conjectured that for almost all n $\tau^+(n)/\tau(n) \to 0$, which of course would have implied (1.4). Tenenbaum and I recently disproved this, and we also proved a recent conjecture of Montgomery which stated that if $\tau^{(d)}(n)$ denotes the number of indices i for which $d_i | d_{i+1}$ then $\tau^{(d)}(n)/\tau(n) > \varepsilon$ holds for a sequence of positive density. Very likely $\tau^{(d)}(n)/\tau(n)$ has a distribution function, but this question we have not yet settled.

Denote by $\tau_r(n)$ the number of indices i for which $(d_i,d_{i+1}) = 1$. R.R. Hall and I studied $\tau_r(n)$ and we obtained various asymptotic inequalities for it, but we are very far from settling all the interesting questions which can be posed here. One of our questions stated: Let n be squarefree and $v(n) = k$ ($v(n)$ denotes the number of distinct prime factors of n). How large is $\max_{v(n)=k} \tau_r(n)$? Simonovits and I proved

$$(2^{1/2}+o(1))^k < \max_{v(n)=k} \tau_r(n) < (2-c)^k. \tag{1.6}$$

We proved (1.6) by the following lemma: Let $0 < x_1 < \ldots < x_k$, assume that the 2^k sums $\sum_{i=1}^{k} \varepsilon_i x_i$, $\varepsilon_i = 0$ or 1, are all distinct and order the sums $\sum_{i=1}^{k} \varepsilon_i x_i$ by size. Denote by g(k) the maximum number of consecutive sums $\sum_{i=1}^{k} \varepsilon_i x_i$, $\sum_{i=1}^{k} \varepsilon_i' x_i$, $\varepsilon_i \varepsilon_i' = 0$, for every $1 \le i \le k$. Clearly $g(k) = \max_{v(n)=k} \tau_r(n)$.

Simonovits and I proved that g(k) satisfies (1.6).Perhaps g(k) can be determined explicitly.

Let $p_1^{(n)} < \ldots < p_{v(n)}^{(n)}$ be the sequence of consecutive prime factors of n. Our knowledge of the properties of the prime factors of almost all integers is much more satisfactory than our knowledge of the divisors of n. Here I state only one result which can easily be obtained by the methods of probabilistic number theory: Put

$$\varepsilon_r = \frac{\log\log p_r^{(n)} - r}{r^{1/2}}.$$

The sequence

$$\frac{1}{\log\log\log n} \sum_{\varepsilon_r > c} \frac{1}{r}\ ;\ r = 1,2,\ldots,v(n),$$

has Gaussian distribution; $\varepsilon_r > 0$ does not have a distribution function. Also roughly speaking for almost all n

$$r^{1/2}\varepsilon_r = \log\log p_r^{(n)} - r$$

is dense in $(-C\, r^{1/2}, C\, r^{1/2})$. Here is a more exact special case. An old theorem of mine states that the r-th prime factor of n is for almost all n between $\exp\exp(r(1-\varepsilon))$ and $\exp\exp(r(1+\varepsilon))$. How close can in fact $p_r^{(n)}$ come to exp exp r for almost all n? It is easy to see that for almost all n the number of solutions of

$$|\log\log p_r^{(n)} - r| < \frac{1}{f(r) r^{1/2}}$$

tends to infinity if and only if $\sum_r \frac{1}{rf(r)} = \infty$. The proof is an easy consequence of sieve methods and elementary independence arguments.

It seems impossible to obtain similarly sharp estimates for the divisors of n; in fact such results are almost certainly not true mainly due to the lack of independence.

REFERENCES

1. P. Erdös and R.R. Hall, The propinquity of divisors, Bull. London Math. Soc. 11 (1979), 304-307.

2. P. Erdös and R.R. Hall, On some unconventional problems on the divisors of integers, J. Austral. Math. Soc., (Series A) 25 (1978), 479-485. See also P. Erdös, Some unconventional problems in number theory, Math. Magazine, 52 (1979), 67-70.

2. Let $1 \le a_1 < \ldots < a_k \le n$. Assume that the sums $a_i + a_j$ are all distinct. Denote $g(n) = \max k$. Turán and I conjectured

(2.1) $$g(n) = n^{1/2} + O(1),$$

but we are very far from being able to prove (2.1). The sharpest result known about g(n) states:

(2.2) $$n^{1/2}-n^{\frac{1}{2}-c} < g(n) < n^{1/2}+n^{1/4}+1.$$

Our original proof of the upper bound gives without much difficulty the following slightly sharper theorem: Let $1 \le a_1 < \ldots < a_k \le n$, $k = [(1+c)n^{1/2}]$. Then the number of distinct differences of the form a_i-a_j, $a_i > a_j$ is less than $(1-\varepsilon_c)\binom{k}{2}$. I do not know the best possible value of ε_c and probably the determination of the best possible value of ε_c will not be easy. This problem is perhaps of some interest but I have not investigated it carefully. A problem of Graham and Sloane in graph theory led me to conjecture that if $k > (1+c)n^{1/2}$, then the number of distinct sums a_i+a_j is also less than $(1-\varepsilon_c')\binom{k}{2}$. Unfortunately I noticed a few days ago that my conjecture is completely wrongheaded. To see this we define $k = [(1+o(1))\frac{2}{3^{1/2}}n^{1/2}]$ a's not exceeding n so that if $a_i+a_j = a_r+a_s$ then $a_i+a_j = n$. Let $1 \le a_1 < \ldots < a_\ell \le \frac{n}{3}$ be a maximal sequence for which all the sums a_i+a_j, $1 \le i \le j \le \ell$ are distinct. By our result with Turán we have $\ell = [(1+o(1))\,(\frac{n}{3})^{1/2}]$. Now put $a_{\ell+i} = n-a_{\ell-i+1}$. Our sequence has $(1+o(1))2(\frac{n}{3})^{1/2}$ terms and it is easy to see that all the sums a_i+a_j are distinct unless $a_i+a_j = n$.

The problem now remains: What is the largest value of c for which there is a sequence $1 \le a_1 < \ldots < a_k \le n$, $k = (1+o(1))cn^{1/2}$, so that the number of distinct sums a_i+a_j is $(1+o(1))\binom{k}{2}$? Trivially $c \le 2$ and it is not hard to show that $c < 2$. Perhaps $c < 2^{1/2}$ but at the moment I do not see how to show this. For the problem of Graham and Sloane it was more natural to assume that the number of distinct sums mod n should be $(1+o(1))\binom{k}{2}$. Here of course trivially $k < (2n)^{1/2}$ and probably $k < (1-c)(2n)^{1/2}$ but I have not yet been able to settle this problem.

REFERENCES

1. P. Erdös and P. Turán, On a problem of Sidon in additive number theory and on some related problems, J. London Math. Soc. 16 (1941), 212-216. For the sharpest result, see B. Lindström, An inequality for B_2-sequences, Journal

Comb. Theory, 6 (1969), 211-212. See also H. Halberstam and K.F. Roth, Sequences, Vol. I, Oxford Univ. Press. Oxford 1966.

3. Let $1 = q_1 < q_2 < \dots$ be the sequence of squarefree numbers. Many mathematicians investigated them from various points of view. Denote by $Q(x)$ the number of squarefree numbers not exceeding x. It is easy to see that $Q(x) = \frac{6}{\pi^2}x + O(x^{1/2})$; the prime number theorem gives $Q(x) - \frac{6}{\pi^2}x = o(x^{1/2})$. It is known that the error term cannot be $o(x^{1/4})$ and it was known for a long time that the Riemann hypothesis implies that $Q(x) - \frac{6}{\pi^2}x = o(x^{2/5})$ and this has been recently improved to $o(x^{1/3})$. We will not deal with these problems here.

The difference $q_{k+1}-q_k$ has been investigated a great deal. No doubt $q_{k+1}-q_k = o(q_k^{\varepsilon})$ for every $\varepsilon > 0$ if $k > k_0(\varepsilon)$, but we are very far from being able to prove this. The sharpest results are due to Richert, Rankin and Schmidt. They proved it for ε a little less than 2/9. I proved that for every $\alpha \leq 2$

$$(3.1) \qquad \sum_{q_k < x} (q_{k+1}-q_k)^{\alpha} = c_{\alpha}x + o(x).$$

Hooley proved that (3.1) holds for every $\alpha \leq 3$. There is no doubt that (3.1) holds for every $\alpha > 0$ but this seems hopeless at present. Put

$$f(x,c) = \sum_{q_k < x} \exp(c(q_{k+1}-q_k)).$$

I expect that

$$(3.2) \qquad f(x,c)/x \to \infty$$

for every $c > 0$ but cannot prove it for any c.

The reason for the difficulty of proving (3.2) is that I cannot give a uniform estimation for the density α_t of the indices k for which $q_{k+1}-q_k > t$. It is not difficult to show that $\alpha_t^{1/t} \to 0$, i.e. α_t tends to 0 faster than exponentially, but I have no uniform estimation for α_t and as far as I know there is no such estimation available in the literature; i.e. I have no good estimation for the number of indices k for which $q_{k+1}-q_k > t_n$, $q_k < n$, when t_n tends to infinity together with n.

I observed nearly 30 years ago that for infinitely many k,

$$q_{k+1}-q_k > (1+o(1))\frac{\pi^2}{12}\frac{\log k}{\log\log k}. \tag{3.3}$$

(3.3) follows easily from the Chinese remainder theorem, the prime number theorem and the sieve of Eratosthenes. I never was able to improve (3.3) and cannot exclude the unlikely possibility that (3.3) is best possible. More generally let $u_1 < u_2 < \dots$ be a sequence of integers satisfying

$$(u_i.u_j) = 1, \quad \sum_i \frac{1}{u_i} < \infty, \tag{3.4}$$

and denote by $a_1 < a_2 < \dots$ the set of integers not divisible by any of the u's. Put

$$u_1.u_2\dots u_{t_x} \le x < u_1\dots u_{t_x}u_{t_x+1}.$$

Analogously to (3.3) we obtain

$$\max_{a_k < x}(a_{k+1}-a_k) > (1+o(1))t_x\pi(1-\frac{1}{u_i})^{-1}. \tag{3.3'}$$

We will show that there are sequences satisfying (3.4) for which (3.3') is best possible. I stated this result in a previous paper. In fact we shall prove it in the following slightly stronger form: There is an infinite sequence of primes $p_1 < p_2 < \dots$, $\sum_i \frac{1}{p_i} < \infty$, so that for all $k > k_0(\varepsilon)$

$$a_{k+1}-a_k < (1+\varepsilon)t_x\prod_i(1-\frac{1}{p_i})^{-1}. \tag{3.5}$$

The proof of (3.5) is indeed easy. Let $p_1 < p_2 < \dots$ be an infinite sequence of primes which tend to infinity sufficiently fast. Let

$$p_k < x < x+L < p_{k+1}, \quad L = (1+\varepsilon)t_x\prod_i(1-\frac{1}{p_i})^{-1}. \tag{3.6}$$

To prove (3.5) we only have to show that there is at least one integer T,

$x < T < x+L$, which is not a multiple of any of the primes $p_1,\ldots,p_k$. If the p's increase sufficiently fast then $t_x = k-1$ or $t_x = k$. Let r be large but small compared to k. Then the number of integers in $(x,x+L)$ which are not multiples of any of the p's is by the sieve of Eratosthenes at least

$$(3.7) \qquad L \prod_{i=1}^{r} \left(1-\frac{1}{p_i}\right) - 2^r - k \sum_{i > r} \frac{1}{p_i} - k > 0,$$

by (3.6) and $t_x \geq k-1$, which completes the proof of (3.5).

The real problem here is: Is there a sequence $u_1 < u_2 < \ldots$, $(u_i,u_j) = 1$, $\sum_i \frac{1}{u_i} < \infty$, which satisfies (3.5) and the u_i do not tend to infinity very fast, say $u_i < i^C$ for some absolute constant C? I do not expect that such a sequence exists. I am fairly sure that there is a sequence satisfying (3.5) for which $u_i^{1/i} \to 1$.

I proved that every irreducible cubic polynomial represents infinitely many squarefree integers and Hooley that the set of integers n for which the cubic polynomial f(n) is squarefree has positive density. It seems hopeless at present to extend this result to quartic polynomials,; in fact there is no quartic polynomial about which we can prove that it represents infinitely many squarefree integers and of course it seems hopeless to prove that $2^n \pm 1$, $2^{2^n} \pm 1$ or $n! \pm 1$ represents infinitely many squarefree numbers. The sharpest results on the representation of power free numbers are due to Nair and to Huxley and Nair.

The analogue of the prime k tuple conjecture is true and was certainly known to L. Mirsky for a long time. It states: let $a_1,\ldots,a_k$ be a set of integers which does not contain a complete set of residues mod p^2 for every p;

then the density of integers n, for which the integers $n+a_i$, $i = 1,\ldots,k$, are all squarefree, is positive. There seems to be no possibility of extending this result for infinite sequences $A = \{a_1 < a_2 < \ldots\}$, where we assume that A does not contain a complete set of residues mod p^2. A is said to have property P if for every integer n, $n+a_i$ is squarefree for only a finite number of indices i. It is easy to see that there are sequences having property P. The simple proof is left to the reader. Probably a sequence having property P must increase fairly fast, but I have no results in this direction.

A is said to have property $\bar{P}$ (respectively $\bar{P}_\infty$) if there are infinitely many n for which $n+a_i$ is squarefree for all (respectively for all but finitely many) $a_i \in A$. A sequence having property $\bar{P}$ or $\bar{P}_\infty$ must no doubt also increase fast.

A is said to have property Q if for infinitely many n, $n+a_i$ is squarefree for all $a_i < n$. It is easy to see that if A increases sufficiently fast then it has property Q and in fact there is an n, $a_k < n < n_{7k+1}$ for which $n+a_i, i = 1,\ldots,k$, is always squarefree. I have no precise information about the rate of increase a sequence having property Q must have.

It would of course be interesting to investigate which special sequences (e.g. $2^n \pm 1$, $n! \pm 1$ etc.) have properties P, $\bar{P}$, $\bar{P}_\infty$ or Q, but as far as I know nothing is known here. These problems can of course be stated for other sequences than p^2, but we formulate only one such question: Is there an infinite sequence $a_1 < a_2 < \ldots$ so that there are infinitely many n for which for all $a_k < n$, $\{n+a_k\}$ always is a prime?

The prime k-tuple conjecture implies the existence of such a sequence. It would be of interest to obtain some estimates about the rate of growth of such a sequence.

It is easy to see that there is an infinite sequence A for which a_i+a_j, $1 \le i \le j$, is always squarefree. In fact one can find such a sequence which grows exponentially. Must such a sequence really increase so fast? I do not expect that there is such a sequence of polynomial growth.

Is there a sequence of integers $1 \le a_1 < a_2 < \ldots$ so that for every i,

$a_i \equiv t \pmod{p^2}$ implies $1 \leq t < p^2/2$? If such a sequence exists then clearly $a_i + a_j$ is always squarefree, but I am doubtful if such a sequence exists. I formulated this problem while writing these lines and must ask the indulgence of the reader if it turns out to be trivial.

Let $A(X)$ be the largest integer for which there is a sequence $1 \leq a_1 < \ldots < a_k \leq X$, $k = A(X)$, which does not form a complete set of residues mod p^2 (for every p). Trivially $A(X) = (1+o(1)) \frac{6}{\pi^2} X$ and Ruzsa pointed it out to me that for infinitely many X $A(X) > O(X)$. Probably this holds for all large X. It would be of some interest to estimate $A(X)$ as accurately as possible. This problem is of course of interest for other sequences than p^2. The sensational results of Hensley and Richards for the sequence of all primes are well known.

One final problem of this type: Let $(u_i, u_j) = 1$ and $a_1 < a_2 < \ldots$ an infinite sequence with the property R: for every u_i and $a_k > u_i$ there is an $a_j < u_i$ for which $a_k \equiv a_j \pmod{u_i}$. The set of all integers clearly always has property R, and if the u's are the set of all primes then no other set has property R. It is easy to see that if the u's are sufficiently thin then there are nontrivial sequences with property R. I am not sure if property R leads to interesting and fruitful questions.

Let $u_1 < u_2 < \ldots$ be a sequence of integers. I conjectured long ago that if $u_n/n \to \infty$ then $\sum_n u_n/2^n$ is irrational. Recently I proved this if we assume the slightly stronger hypothesis $u_{n+1} - u_n \to \infty$. I know of no example of a sequence $u_1 < \ldots$ for which $\limsup (u_{n+1} - u_n) = \infty$, and $\sum_n u_n/2^{u_n}$ is rational. I am sure that such sequences exist and perhaps I overlook an obvious idea. To my surprise and disappointment I could not prove that $\sum_n q_n/2^{q_n}$ is irrational where $q_1 < \ldots$ is the sequence of all squarefree numbers. In fact if $q_{i_1} < q_{i_2} < \ldots$ is any subsequence of the squarefree numbers then surely $\sum_n q_{i_m}/2^{q_{i_m}}$ is always irrational. Here again I perhaps overlook a trivial point.

In trying unsuccessfully to prove these conjectures I found a result which perhaps is of some interest:

THEOREM. <u>Let</u> $c > 0$ <u>be a sufficiently small absolute constant. Then for every</u> $x > x_0(c)$ <u>there are integers</u> $y_1 < y_2 < y_3 < y_4 < x$ <u>satisfying</u>

$$y_2-y_1 = y_4-y_3 = t > c(\log x)^2 \tag{3.8}$$

<u>for which the squarefree numbers in</u> (y_1,y_2) <u>and</u> (y_3,y_4) <u>are congruent by translation by</u> $y_3-y_1 = y_4-y_2$.

Denote by t_x the longest such interval. Unfortunately I have no good upper bound for t_x; surely $t_x = o(x^\varepsilon)$ and perhaps $t_x < (\log x)^c$. I. Ruzsa pointed out that it is unlikely that one can get a good result without some really new idea since we cannot exclude the existence of large gaps between the y's.

The proof of our Theorem will not be difficult. Denote by $f(n,t)$ the number of integers $m, n < m < n+t$, for which there is a $p > \frac{1}{100}\log x$, satisfying $m \equiv 0 \pmod{p^2}$. Clearly by the prime number theorem and (3.8)

$$\sum_{n=1}^{x} f(n,t) < tx \sum_{p > \frac{\log x}{100}} \frac{1}{p^2} < \frac{200\, t\, x}{\log x \log\log x} < \frac{200\, cx \log x}{\log\log x}. \tag{3.9}$$

Thus from (3.9) there are clearly at least $\frac{x}{2}$ values of n for which

$$f(n,t) < \frac{400\, c \log x}{\log\log x} = L. \tag{3.10}$$

Henceforth we will only consider these (at least) $\frac{x}{2}$ values of n which satisfy (3.10). We now give an upper bound for the number of patterns the integers $m \equiv 0 \pmod{q^2}$ can form in $(n,n+t)$ (q runs through all primes).

The number of these patterns is clearly less than

$$\binom{t}{L} \prod_{p \le \frac{\log x}{100}} p^2 < \left(\frac{t}{L}\right)^L c^L x^{1/10} = o(x^{1/2}) \tag{3.11}$$

for sufficiently small c. To prove (3.11) observe that the factor $\binom{t}{L}$ comes from the primes $p > \frac{\log x}{100}$ and the factor $\prod_{p \leq \frac{\log x}{100}} p^2$ from the primes $\leq \frac{\log x}{100}$.

Thus by (3.10) there are two intervals of length t, which by (3.11) can be assumed to be disjoint,in which the squarefree numbers are congruent.

It would be easy to get an explicit bound for c, but this is hardly worth the trouble since at the moment there is no reason to assume that the true order of magnitude of t_x is $(\log x)^2$.

REFERENCES

1. N. Nair, Power free values of polynomials I and II, Mathematika 23, (1976), 159-183 and Proc. London Math. Soc. (3) 38 (1979), 353-368.
2. H.M. Huxley and N. Nair, Power free values of polynomials III, Proc. London Math. Soc. (3) 41,(1980), 66-82. These papers have extensive references to the earlier literature.
3. C. Hooley, On the intervals between consecutive terms of sequences, Proc. Symp. Pure Math., vol. 24, Analytic number theory, Amer. Math. Soc., 1973, 129-140. This paper contains extensive references.
4. P. Erdös, On the difference of consecutive terms of sequences defined by divisibility properties, Acta Arithmetica 12,(1966-67), 175-182; E. Szemerédi, same title II, ibid 23 (1973), 359-361.

COMPUTATIONS AND GENERALIZATIONS ON A REMARK OF RAMANUJAN

by

Ronald Alter

Department of Computer Science
University of Kentucky

ABSTRACT

In a conversation with Ramanujan, G.H. Hardy mentioned that 1729 seemed to be a dull number. Ramanujan answered, "No, it is a very interesting number; it is the smallest number expressible as a sum of two cubes in two different ways." He also stated that the answer to the corresponding problem for fourth powers "... must be very large." In this paper, this problem and other generalizations to higher powers and larger sums is examined.

In particular, the computational results of Lander, Parkin and Selfridge ("...Equal Sums of Like Powers") are extended.

This paper is dedicated to my teacher, Professor Emil Grosswald, in honor of his sixty-eighth birthday.

I. Introduction.

The problem of representing a number as a sum of other numbers has been around for a long time. Among the most famous problems in this area are:

(I) Fermat's Last Theorem

(II) The Goldbach Conjecture

(III) The Euler Conjecture

(IV) The Waring Problem

One observes that each of the above problems is named after a particular individual who had something to do with the problem itself. The problems are listed above in chronological order. We will now discuss each of these problems, briefly, but not necessarily in the listed order. For further background information about these problems and their history, the reader is referred to Bell [1], Dickson [3], Hardy and Wright [5] and Newman [8].

The Goldbach Conjecture first appeared in a letter dated June 7, 1742 from Christian Goldbach to Leonard Euler. This letter, in fact, contained two related conjectures.

<u>CONJECTURE 1</u>. Every odd integer greater than 7 can be represented as a sum of three odd prime numbers.

<u>CONJECTURE 2</u>. Every even integer greater than 4 can be represented as a sum of two odd prime numbers.

Clearly, Conjecture 1 is true if the stronger Conjecture 2 is true. As a result, Conjecture 1 became known as the little Goldbach Conjecture while Conjecture 2 became known as the Goldbach Conjecture. Euler believed the conjectures to be true but confessed his inability to prove them. There was no significant progress on these conjectures until 1937 when the Russian mathematician, I.M. Vinogradov, proved that every odd number beyond a certain point is a sum of three odd prime numbers. A nice proof of this result can be found in Estermann [4]. Although much progress has been made, an equivalent result for Conjecture 2 has not yet been attained.

In 1637, Pierre Fermat claimed to discover a marvelous proof of his now famous "Last Theorem."

<u>CONJECTURE 3</u>. If x,y,z are integers and n is greater than 2 then the following equation has no integer solutions if $x \cdot y \cdot z \neq 0$.

$$x^n + y^n = z^n. \tag{1}$$

Fermat never published a proof of Conjecture 3. To this day, a complete proof still has not appeared. For many values of n, Conjecture 3 is known to be true. The conjecture states that no n^{th} power of an integer can be expressed as a sum of two other n^{th} powers. Euler generalized this conjecture by stating that no k^{th} power can be expressed as a sum of less than k, k^{th} powers of positive integers. This leads to Euler's Conjecture.

CONJECTURE 4. The following equation has no solutions in positive integers for $3 \leq s \leq k-1$

$$x_1^k + x_2^k + \ldots + x_s^k = y^k. \tag{2}$$

The Euler Conjecture was believed to be true until 1967 when Lander and Parkin [6], using a CDC 6600 computer, found via an extensive search, a counter-example. They were able to express a fifth power as a sum of four fifth powers. To be more precise:

$$27^5 + 84^5 + 110^5 + 133^5 = 144^5. \tag{3}$$

To this day no other counterexample has been found, for fifth or any other powers of k.

In 1770 the English algebraist E. Waring made the following conjecture without proving any single case of the problem nor offering any suggestion for its solution.

CONJECTURE 5. Every integer $n > 0$ is the sum of a fixed least number g(s) of sth powers of integers that are greater than or equal to zero.

In 1909 Hilbert settled Waring's Conjecture. He proved the existence of g(s) for every s but did not find its numerical value for any s. Many mathematicians have worked on the problem of determining g(s) for various values of s. Today, the exact value of g(s) is known for all s except s = 4 and 5. (I have been told that g(5) has recently been solved. However, I have not yet been able to verify this.)

Conjecture 2 (Goldbach) deals with the representation of integers as sums of prime numbers. Conjecture 1 (Fermat) and Conjecture 3 (Euler) are both concerned with the representation of an nth power of an integer as a sum of a fixed number of nth powers. But Conjecture 4 (Waring) deals with the representation of any positive integer as a sum of a fixed number of powers of integers. In this paper we will discuss a combinatorial variation of the sums of powers problem. We are interested in the number of different representations. To be more precise, we want to find $r(m,n,s)$, where

(4) $r(m,n,s)$ = the smallest integer than can be expressed as a sum of m positive n^{th} powers in s different ways.

It is easy to see that $r(2,2,2) = 50$, since

(5) $$50 = 5^2 + 5^2 = 1^2 + 7^2.$$

The answer to $r(2,3,2)$ lies in one of the interesting anecdotes in the history of mathematics. (See Newman [8, p. 375].) G.H. Hardy wrote that the Indian mathematician, Ramanujan, "could remember the idiosyncrasies of numbers in an almost uncanny way." He recalled visiting him when he was lying ill. Hardy had ridden in taxicab number 1729. He remarked to Ramanujan that 1729 seemed to him to be a rather dull number, and he hoped it was not an unfavorable omen. "No", Ramanujan replied, "it is a very interesting number, it is the smallest number expressible as a sum of two cubes in two different ways."

Returning to the Hardy-Ramanujan conversation, Hardy, upon hearing about 1729, asked Ramanujan if he knew the answer to the corresponding problem for fourth powers. Ramanujan replied, after a moment's thought, that he could see no obvious example and thought that the first such number must be very large. Thus, Ramanujan was able to mentally compute that

(6) $$r(2,3,2) = 1729 = 1^3 + 12^3 = 9^3 + 10^3.$$

He was unable to mentally compute that

(7) $$r(2,4,2) = 635{,}318{,}657 = 59^4 + 158^4 = 133^4 + 134^4.$$

It is interesting to note that (7) was known to Euler (See Dickson [3]).

II. Results on $r(m,n,s)$.

In this paper we are interested in generalizations of Results (6) and (7) to higher powers, larger sums and more representations. Some work has already been done on these generalizations. Most of the known results can be found in Lander, Parkin and Selfridge [7], where among other things they have a bibliography consisting of 40 different references. A lot of the earlier results can be found in Dickson [3]. In [7] the authors study the general diophantine equation,

$$(8) \qquad \sum_{i=1}^{k} X_i^n = \sum_{i=1}^{m} Y_i^n, \quad 1 \leq k \leq m.$$

They identify the parameters k,m and n for which solutions exist and they find the least solution for certain sets. They produce, among other results, tables which give for various values of k and m, the least n for which a solution to (8) is known. For the purpose of this paper we are primarily interested in the case when $k = m$. However we do not restrict ourselves to only looking at two representations as they do in [7]. We in fact find complete solutions for three and four representations as well as for the case $s = 2$. Current research deals with the problem of finding parametric solutions to Eq. (8) once $r(m,n,2)$ is found for some particular m and n. A good example of this type of research can be found in Brudno [2]. He deals with Eq. (8) when $k = m = 3$ and $n = 6$. He points out that parametric solutions for $n > 6$ are difficult and currently unknown. In our paper we first find $r(m,n,s)$ for some particular m_0,n_0,s_0 then $r(m,n,s)$ is found for all n_0,s_0 and $m \geq m_0$.

For simplicity we will use the notation that is standard in partition theory. That is, we will write x^r to represent the term x repeated r times in the expression in which it appears.

If one of the parameters in $r(m,n,s)$ is 1 we have some immediate results.

CASE (i). $m = 1$. Since an integer can only be represented as a positive n^{th} power in one way we have:

$$(9) \qquad r(1,n,s) = 0 \text{ if } s > 1.$$

CASE (ii). $s = 1$. Since, $m = \sum_{i=1}^{m} 1^n = (1^m)$, it follows that

(10) $$r(m,n,1) = m \text{ for all } m \text{ and } n.$$

CASE (iii). $n = 1$. This is basically a problem in partition theory which we will not discuss here. However there are some immediate results.

(11) $$r(m,1,2) = m + 2.$$

(12) $$r(m,1,3) = \begin{cases} 6 & \text{if } m = 2 \\ m+3 & \text{if } m \geq 3. \end{cases}$$

(13) $$r(2,1,s) = 2s.$$

The above follow since:

$$m+2 = (1^{m-1},3) = (1^{m-2},2^2).$$
$$r(2,1,3) = 6 = (1,5) = (2,4) = (3,3).$$
$$m+3 = (1^{m-1},4) = (1^{m-2},2,3) = (1^{m-3},2^3).$$
$$2s = (1,2s-1) = (2,2s-2) = \ldots = (s-1,s+1) = (s,s).$$

One can also show that

(14) $$r(m,1,s) \leq s+m \text{ if } s \geq 3 \text{ and } m \leq s.$$

Next we generalize Result (5) by finding the smallest integer that can be expressed as a sum of m squares in two different ways.

(15) $$r(m,2,2) = \begin{cases} 50 & \text{if } m = 2 \\ m+24 & \text{if } m = 3,4 \\ m+15 & \text{if } m \geq 5 \end{cases}$$

To prove the above we first observe, from (5), that

$$r(m,2,2) \leq m+48.$$

A simple computation yields that:

$$r(3,2,2) = 27 = (3^3) = (1^2,5)$$

$$r(4,2,2) = 28 = (1,3^3) = (1^3,5)$$

$$r(5,2,2) = 20 = (2^5) = (1^4,4).$$

Now that we have a string of two's that are optimal we cannot do better than to continually add one's to each side of the equation. Thus for $m \geq 5$

$$r(m,2,2) = m+15 = (1^{m-5},2^5) = (1^{m-1},4).$$

In general, by first finding $r(m,n,s)$ for some $m = m_0$ we then have an upper bound on $r(m,n,s)$ for all $m \geq m_0$. Next successively compute $r(m,n,s)$ for $m = m_0+1, m_0+2, \ldots$. We continue these computations until we reach a situation that cannot be improved upon (Eg., a string of two's that is optimal). Then we have a complete solution for all $m \geq m_0$, and the particular values of n and s. To illustrate this concept we have calculated the exact values of $r(m,n,s)$ for all $m \geq m_0$, and particular values of $n(2 \leq n \leq 6)$ and $s(2 \leq s \leq 4)$. All of these results are contained in the Tables which appear in the next section.

III. Tables.

Table 1 summarizes our results on $r(m,n,s)$ and gives the appropriate table in which the results can be found. Table 2 generalizes Ramanujan's remark, (6), and Table 3 generalizes Result (7).

TABLE 1: Complete Results on $r(m,n,s)$ for all $m \geq m_0$

m_0	n	s	Location of Result
2	2	2	Equation (15)
2	3	2	Table 2
2	4	2	Table 3
3	5	2	Table 4
3	6	2	Table 5
2	2	3	Table 6
3	3	3	Table 7
3	4	3	Table 8
5	5	3	Table 9
2	2	4	Table 10
3	3	4	Table 11
3	4	4	Table 12

TABLE 2: r(m,3,2)

m	r(m,3,2)	$(x_1,\ldots,x_m)$,	$(y_1,\ldots,y_m)$
2	1729	(1,12)	(9,10)
3	251	$(1,5^2)$	(2,3,6)
4	219	$(1^3,6)$	$(3,4^3)$
$5 \leq t \leq 6$	t+152	$(1^{t-3},3,4^2)$	$(1^{t-5},2^4,5)$
$7 \leq t \leq 8$	t+124	$(1^{t-1},5)$	$(1^{t-7},2^5,3,4)$
$t \geq 9$	t+63	$(1^{t-1},4)$	$(1^{t-9},2^9)$

TABLE 3: r(m,4,2)

m	r(m,4,2)	$(x_1,\ldots,x_m)$,	$(y_1,\ldots,y_m)$
2	635,318,657	(59,158)	(133,134)
3	2,673	(2,4,7)	$(3,6^2)$
$4 \leq t \leq 15$	t+255	$(1^{t-1},4)$	$(1^{t-4},2,3^3)$
$t \geq 16$	t+240	$(1^{t-3},3^3)$	$(1^{t-16},2^{16})$

TABLE 4: r(m,5,2)

m	r(m,5,2)	$(x_1,\ldots,x_m)$,	$(y_1,\ldots,y_m)$
3	1,375,298,099	(24,28,67)	(3,54,62)
4	51,445	$(4,7^3)$	$(5,6^2,8)$
$5 \leq t \leq 32$	t+4092	$(1^{t-4},4^4)$	$(1^{t-5},3^4,5)$
$t \geq 33$	t+1023	$(1^{t-1},4)$	$(1^{t-33},2^{33})$

TABLE 5: r(m,6,2)

m	r(m,6,2)	$(x_1,\ldots,x_m)$	$(y_1,\ldots,y_m)$
3	106,426,514	(3,19,22)	(10,15,23)
$4 \le t \le 5$	t+1,063,006	$(1^{t-4},2^2,9^2)$	$(1^{t-4},3,5,6,10)$
$6 \le t \le 7$	t+570,941	$(1^{t-3},6,8^2)$	$(1^{t-6},2,4^2,5^2,9)$
$8 \le t \le 9$	t+63,224	$(1^{t-5},3,5^4)$	$(1^{t-8},2^3,4^4,6)$
$10 \le t \le 12$	t+52,479	$(1^{t-9},3^8,6)$	$(1^{t-10},4^9,5)$
$13 \le t \le 17$	t+16,380	$(1^{t-4},4^4)$	$(1^{t-13},2^{12},5)$
$18 \le t \le 34$	t+13,104	$(1^{t-16},2^{13},4^3)$	$(1^{t-18},3^{18})$
$35 \le t \le 39$	t+8,190	$(1^{t-2},4^2)$	$(1^{t-35},2^{26},3^9)$
$40 \le t \le 64$	t+6,552	$(1^{t-9},3^9)$	$(1^{t-40},2^{39},4)$
$t \ge 65$	t+4,095	$(1^{t-1},4)$	$(1^{t-65},2^{65})$

TABLE 6: r(m,2,3)

m	r(m,2,3)	$(x_1,\ldots,x_m)$	$(y_1,\ldots,y_m)$	$(z_1,\ldots,z_m)$
2	325	(1,18)	(6,17)	(10,15)
3	54	(1,2,7)	$(2,5^2)$	$(3^2,6)$
$t \ge 4$	t+24	$(1^{t-1},5)$	$(1^{t-3},3^3)$	$(1^{t-4},2^3,4)$

It is interesting to observe that in Table 6, for $t \ge 8$, we pick up a fourth solution. Since:

(16) $r(8,2,3) = (1^7,5) = (1^5,3^3) = (1^4,2^3,4) = (2^8) = 32.$

As can be seen in Table 10, for s = 4, Result (16) is optimal.

TABLE 7: r(m,3,3)

m	r(m,3,3)	$(x_1,\ldots,x_m)$	$(y_1,\ldots,y_m)$	$(z_1,\ldots,z_m)$
3	5104	(1,12,15)	(2,10,16)	(9,10,15)
4	1225	(1,2,6,10)	$(3,7^2,8)$	$(4,6^2,9)$
5	766	$(1^2,2,3,9)$	$(1,4^2,5,8)$	$(2^2,4,7^2)$
$6 \le t \le 9$	t+215	$(1^{t-1},6)$	$(1^{t-4},3,4^3)$	$(1^{t-6},2^4,4,5)$
10	197	$(1^8,4,5)$	$(1^2,2^5,3,4^2)$	$(2^9,5)$
$11 \le t \le 14$	t+152	$(1^{t-3},3,4^2)$	$(1^{t-5},2^4,5)$	$(2^9,3,4)$
$t \ge 15$	t+124	$(1^{t-1},5)$	$(1^{t-7},2^5,3,4)$	$(1^{t-15},2^{14},3)$

Once again, this time at t = 9, we pick up a fourth solution

(17) $r(9,3,3) = (1^5,3,4^3) = (1^8,6) = (1^3,2^4,4,5) = (2,3^8) = 224.$

TABLE 8: r(m,4,3)

m	r(m,4,3)	$(x_1,\ldots,x_m)$	$(y_1,\ldots,y_m)$	$(z_1,\ldots,z_m)$
3	811,538	(4,23,27)	(7,21,28)	(12,17,29)
4	16,578	(1,2,9,10)	(2,5,6,11)	(3,7,8,10)
5	4,225	$(2^3,3,8)$	$(2,4^2,6,7)$	$(3,4,6^3)$
$6 \le t \le 7$	t+2670	$(1^{t-3},2,4,7)$	$(1^{t-3},3,6^2)$	$(1^{t-6},2^2,3^3,7)$
$8 \le t \le 16$	t+510	$(1^{t-2},4^2)$	$(1^{t-5},2,3^3,4)$	$(1^{t-8},2^2,3^6)$
$t \ge 17$	t+255	$(1^{t-1},4)$	$(1^{t-4},2,3^3)$	$(1^{t-17},2^{17})$

TABLE 9: r(m,5,3)

m	r(m,5,3)	$(x_1,\ldots,x_m)$	$(y_1,\ldots,y_m)$	$(z_1,\ldots,z_m)$
5	13,124,675	(1,9,10,20,25)	$(2,5,12,23^2)$	$(16,19,20^3)$
6	696,467	$(1,6,8,9^2,14)$	$(3^2,7,9,12,13)$	$(4^3,11^2,13)$
7	84,457	$(1,3,4,7^3,8)$	$(1,3,5,6^2,8^2)$	$(2,4^2,6^3,9)$
$8 \le t \le 9$	t+52,409	$(1^{t-7},4^4,6^2,8)$	$(1^{t-8},3^4,4,7^3)$	$(1^{t-8},3^4,5,6^2,8)$
$10 \le t \le 20$	t+8184	$(1^{t-8},4^8)$	$(1^{t-9},3^4,4^4,5)$	$(1^{t-10},3^8,5^2)$
$21 \le t \le 27$	t+7775	$(1^{t-1},6)$	$(1^{t-20},2^{12},3,4^7)$	$(1^{t-21},2^{12},3^5,4^3,5)$
$28 \le t \le 30$	t+7557	$(1^{t-16},2^{11},4^4,5)$	$(1^{t-17},2^{11},3^4,5^2)$	$(1^{t-28},3^{27},4)$
$31 \le t \le 35$	t+7502	$(1^{t-18},2^{11},4^7)$	$(1^{t-19},2^{11},3^4,4^3,5)$	$(1^{t-31},3^{31})$
$36 \le t \le 65$	t+4092	$(1^{t-4},4^4)$	$(1^{t-5},3^4,5)$	$(1^{t-36},2^{33},4^3)$
$t \ge 66$	t+2046	$(1^{t-2},4^2)$	$(1^{t-34},2^{33},4)$	$(1^{t-66},2^{66})$

TABLE 10: r(m,2,4)

m	r(m,2,4)	$(x_1,\ldots,x_m)$	$(y_1,\ldots,y_m)$	$(z_1,\ldots,z_m)$	$(w_1,\ldots,w_m)$
2	1105	(4,33)	(9,32)	(12,31)	(23,24)
3	129	$(1,8^2)$	$(2^2,11)$	(2,5,10)	(4,7,8)
$4 \le t \le 5$	t+48	$(1^{t-1},7)$	$(1^{t-2},5^2)$	$(1^{t-4},2,4^3)$	$(1^{t-4},3^3,5)$
$6 \le t \le 7$	t+30	$(1^{t-2},4^2)$	$(1^{t-3},2^2,5)$	$(1^{t-5},2^2,3^3)$	$(1^{t-6},2^5,4)$
$t \ge 8$	t+24	$(1^{t-1},5)$	$(1^{t-3},3^3)$	$(1^{t-4},2^3,4)$	$(1^{t-8},2^8)$

TABLE 11: r(m,3,4)

m	r(m,3,4)	$(x_1,\ldots,x_m)$	$(y_1,\ldots,y_m)$	$(z_1,\ldots,z_m)$	$(w_1,\ldots,w_m)$
3	13,896	(1,12,23)	(2,4,24)	(4,18,20)	(9,10,23)
4	1,979	$(1,5^2,12)$	(2,3,6,12)	$(5^2,9,10)$	$(6^3,11)$
5	1,252	$(1^2,5^2,10)$	(1,2,3,6,10)	$(3^2,7^2,8)$	$(3,4,6^2,9)$
6	626	$(1^3,4,6,7)$	$(1,5^5)$	$(2,3,5^3,6)$	$(3,4^4,7)$
7	470	$(1^4,5^2,6)$	$(1^3,2,3,6^2)$	$(1,3,4^3,5^2)$	$(2,3^2,4^3,6)$
8	256	$(1^6,5^2)$	$(1^5,2,3,6)$	$(1^2,2,3^2,4^3)$	$(2^5,3,4,5)$
$9 \le t \le 12$	t+215	$(1^{t-1},6)$	$(1^{t-4},3,4^3)$	$(1^{t-6},2^4,4,5)$	$(1^{t-9},2,3^8)$
$13 \le t \le 15$	t+208	$(1^{t-5},2^3,4,5)$	$(1^{t-8},3^8)$	$(1^{t-11},2^8,3,4^2)$	$(1^{t-13},2^{12},5)$
$16 \le t \le 18$	t+187	$(1^{t-2},4,5)$	$(1^{t-8},2^5,3,4^2)$	$(1^{t-10},2^9,5)$	$(1^{t-16},2^{14},3,4)$
$t \ge 19$	t+152	$(1^{t-3},3,4^2)$	$(1^{t-5},2^4,5)$	$(1^{t-11},2^9,3,4)$	$(1^{t-19},2^{18},3)$

TABLE 12: r(m,4,4)

m	r(m,4,4)	$(x_1,\ldots,x_m)$	$(y_1,\ldots,y_m)$	$(z_1,\ldots,z_m)$	$(w_1,\ldots,w_m)$
3	5,978,882	(3,40,43)	(8,37,45)	(15,32,47)	(23,25,48)
4	236,674	(1,2,7,22)	(3,6,18,19)	(7,14,16,19)	$(8,16,17^2)$
5	20,995	$(1^3,4,12)$	$(2,3^3,12)$	$(2,6,9^3)$	$(4,6,7^2,11)$
6	6,626	$(1,2^4,9)$	$(2^3,3,7,8)$	$(2,4^2,6,7^2)$	$(3,4,6^3,7)$
$7 \le t \le 8$	t+2925	$(1^{t-4},2,4^2,7)$	$(1^{t-4},3,4,6^2)$	$(1^{t-7},2^2,3^3,4,7)$	$(1^{t-7},2,3^4,6^2)$
9	2854	$(1^6,4,6^2)$	$(1^3,2^2,3^2,4,7)$	$(1^3,2,3^3,6^2)$	$(2^3,3^5,7)$
$10 \le t \le 11$	t+1610	$(1^{t-6},2^4,4,6)$	$(1^{t-7},3,4^6)$	$(1^{t-9},2^5,3^3,6)$	$(1^{t-10},2,3^4,5^5)$
$12 \le t \le 17$	t+765	$(1^{t-3},4^3)$	$(1^{t-6},2,3^3,4^2)$	$(1^{t-9},2^2,3^6,4)$	$(1^{t-12},2^3,3^9)$
$18 \le t \le 19$	t+510	$(1^{t-2},4^2)$	$(1^{t-5},2,3^3,4)$	$(1^{t-8},2^2,3^6)$	$(1^{t-18},2^{17},4)$
$20 \le t \le 31$	t+495	$(1^{t-4},3^3,4)$	$(1^{t-7},2,3^6)$	$(1^{t-17},2^{16},4)$	$(1^{t-20},2^{17},3^3)$
$t \ge 32$	t+480	$(1^{t-6},3^6)$	$(1^{t-16},2^{15},4)$	$(1^{t-19},2^{16},3^3)$	$(1^{t-32},2^{32})$

IV. Remarks and Open Problems

Most of the results in the Tables are new, especially Tables 6-12. One notes that $r(2,5,2)$ and $r(2,6,2)$ are missing from Tables 4 and 5 respectively. In [7] it is suggested that there is no solution to these equations. This still remains to be resolved. It was established in [7] that if a solution does exist it must be quite large. The next cases, $r(3,5,2)$ and $r(3,6,2)$, were previously known and appear in [7]. In fact, $r(3,6,2)$ was first discovered in 1934 by K. Subba Rao [9]. In [7], the authors observe that there are no known solutions for $r(m,n,2)$ if $2m < n$. Also, if $2m = n$ the only known solutions are $r(2,4,2)$ and $r(3,6,2)$.

In each case, the most difficult computation is $m = 2$. One observes that this value is missing from each of Tables 7,8,9,11 and 12. Once this is found we have an immediate upper bound on $r(m,n,s)$ for all $m \geq 2$:

$$(18) \qquad r(m,n,s) \leq r(2,n,s) + (m-2).$$

For the case when $s = 2$ we have an upper bound on $r(m,n,2)$ which depends only on m and n.

$$(19) \qquad r(m,n,2) \leq m+4^n-1 \text{ for } m \geq 2^n+1.$$

The above follows since

$$(20) \qquad m+4^n-1 = (1^{m-1},4) = (1^{m-2^n-1},2^{2^n+1}).$$

By observing the last line in Eq. (15), Tables 2,3,4 and 5 we see that the bound in Eq. (19) is best possible for $n = 2,3,5$ and 6. In order to further test (19) we have also computed

$$(21) \qquad r(129,7,2) = (1^{128},4) = (2^{129}) = 16,512$$

and

$$(22) \qquad r(257,8,2) = (1^{256},4) = (2^{257}) = 65,792.$$

Thus Eq. (19) is an inequality if $n = 4$ and an equality if $n \neq 4$ and $2 \leq n \leq 8$. This leads to

CONJECTURE. If $n \neq 4$ and $m \geq 2^n+1$, then

$$r(m,n,2) = m+4^n-1.$$

ACKNOWLEDGEMENT

I would like to thank James Abbott for the programming assistance which he lent to this project.

REFERENCES

1. E.T. Bell, Development of Mathematics, McGraw-Hill, New York, 1945.
2. S. Brudno, "Triples of Sixth Powers With Equal Sums," Math. Comp. 30 (1976), 646-648.
3. L.E. Dickson, History of the Theory of Numbers, Carnegie Institute of Washington, Washington, D.C., 1920
4. T. Estermann, Introduction to Modern Prime Number Theory, Cambridge Tract No. 41, Cambridge University Press, London, 1961.
5. G.H. Hardy and E.M. Wright, An Introduction to the Theory of Numbers, Oxford University Press, London, 1960.
6. L.J. Lander and T.R. Parkin, "A Counterexample to Euler's Sum of Powers Conjecture," Math. Comp. 21 (1967), 101-103.
7. L.J. Lander, T.R. Parkin and J.L. Selfridge, "A Survey of Equal Sums of Like Powers," Math. Comp. 21 (1967), 446-459.
8. J.R. Newman, World of Mathematics, Simon and Schuster, New York, 1956.
9. K.S. Rao, "On Sums of Sixth Powers," J. London Math. Soc. 9 (1934), 172-173.

THE ARITHMETIC MEAN OF THE DIVISORS OF AN INTEGER

by

Paul T. Bateman, Paul Erdös, Carl Pomerance, and E.G. Straus

Dedicated to Emil Grosswald on the occasion of his sixty-eighth birthday.

1. Introduction.

In [16], O. Ore introduced the arithmetic functions $A(n)$, $G(n)$, $H(n)$ which are respectively the arithmetic mean, the geometric mean, and the harmonic mean of the natural divisors of n. Thus one easily has

$$A(n) = \sigma(n)/d(n), \quad G(n) = \sqrt{n}\,, \quad H(n) = nd(n)/\sigma(n),$$

so that all three functions are seen to be multiplicative. We note that $G(n)$ is an integer if and only if n is a square, so that the set of n with $G(n)$ integral has density 0. In [10], Kanold showed that the set of n with $H(n)$ integral also has density 0. The corresponding statement for $A(n)$ is certainly not true, for, as Ore pointed out (in [16]), if n is odd and square-free, then $A(n)$ is integral. In fact, the set of n for which $A(n)$ is integral has density 1 (cf. [18]). In our Theorem 2.1 we study the distribution of those exceptional n for which $A(n)$ is not integral.

We show in Theorem 3.1 that the set of n for which $d(n)^2|\sigma(n)$ has asymptotic density exactly 1/2.

In our Theorem 4.1 we show that the mean value of $A(n)$ for $n \le x$ is asymptotic to $c\, x/\sqrt{\log x}$ where c is an explicit constant. Moreover, in Theorem 5.1, we show that the number of n with $A(n) \le x$ is asymptotic to a constant times $x \log x$. This last result is somewhat more difficult to establish than, say, an asymptotic formula for the number of n with $\sigma(n) \le x$ (cf. Bateman [1]).

In [18], Pomerance asked what can be said about the distribution of the distinct integral values of $A(n)$. In Theorem 6.1, we show that they have density 0 and that, in fact, the non-integral values may be thrown in as well.

At the end of the paper we briefly consider some further problems.

This paper is the long-term result of conversations which the first and fourth authors had with Herbert Wilf in the spring of 1962. We also take this opportunity to acknowledge several interesting conversations about the contents of this paper with Harold Diamond and Gábor Halász. Finally, we mention that the research of the last two authors was supported by grants from the National Science Foundation.

2. The distribution of the n for which A(n) is not integral.

THEOREM 2.1. *Let* $N(x)$ *denote the number of* $n \leq x$ *for which* $A(n) = \sigma(n)/d(n)$ *is not an integer. Then*

$$N(x) = x\cdot\exp\{-(1+o(1))\cdot 2\sqrt{\log 2}\ \sqrt{\log\log x}\ \}.$$

PROOF. We first show $N(x) \geq x\cdot\exp\{-(1+o(1))\cdot 2\sqrt{\log 2}\ \sqrt{\log\log x}\ \}$.

Let $p_0 = p_0(x)$ be the closest prime to $\sqrt{(\log\log x)/\log 2}$. (If there is a tie, choose either prime). Then $p_0 = (1+o(1))\sqrt{(\log\log x)/\log 2}$. We shall consider integers $n \leq x$ such that $2^{p_0-1}\,||\,n$, $p_0 \nmid \sigma(n)$. For such n, $d(n) \nmid \sigma(n)$. Let

$$M(y) = \#\{m \leq y : 2 \nmid m,\ p_0 \nmid \sigma(m)\}.$$

The following lemma will enable us to estimate N(x) from below.

LEMMA 2.1. *There is an absolute constant* $c > 0$ *such that for* $x^{1/2} < y < x$,

$$M(y) \geq c\ y/(\log y)^{1/(p_0-1)}.$$

Assuming Lemma 2.1, we have at once

$$N(x) \geq M(x/2^{p_0-1}) \geq c\ x/\{2^{p_0-1}\cdot(\log x)^{1/(p_0-1)}\}$$

$$= x\cdot\exp\{-(1+o(1))\cdot 2\sqrt{\log 2}\ \sqrt{\log\log x}\ \}.$$

To prove Lemma 2.1 (and several other results in this paper) we shall use the following corollary of the Siegel-Walfisz theorem.

THEOREM A. (Norton [15], Pomerance [17]). *Let* $0 < \ell < k$ *be integers with* $(\ell,k) = 1$. *Then for all* $x \geq 3$

$$\sum_{\substack{p \leq x \\ p \equiv \ell \pmod k}} 1/p = \frac{1}{\phi(k)} \log\log x + \ell^* + O\left(\frac{1}{\phi(k)} \log k\right)$$

where $\ell^* = 1/\ell$ if ℓ is prime and 0 otherwise. The constant implied by the O -notation is absolute.

PROOF OF LEMMA 2.1. Let

$$W(z) = (1/2) \prod_{\substack{q \equiv -1 \pmod{p_0} \\ q \leq z}} (1 - 1/q).$$

Then using Theorem A, we have, uniformly for $z \geq 3$,

$$W(z) = (1/2) \cdot \exp\{-\textstyle\sum' \sum_{k \geq 1} 1/(kq^k)\}$$

$$= (1/2) \cdot \exp\{-\textstyle\sum' \sum_{k \geq 2} 1/(kq^k)\} \cdot \exp\{-\textstyle\sum' 1/q\}$$

$$= \exp(O(1)) \cdot \exp(-(\log\log z)/(p_0 - 1)),$$

where $\sum'$ denotes the sum over primes $q \leq z$, $q \equiv -1 \pmod{p_0}$. Thus if

$$z_1 = y^{\exp(-(\log\log y)^{1/3})}, \quad z_2 = y^{\exp(-(\log\log y)^{1/2})},$$

then

$$c_1 y/(\log y)^{1/(p_0-1)} \leq yW(z_1) \leq yW(z_2) \leq c_2 y/(\log y)^{1/(p_0-1)} \tag{2.1}$$

where $0 < c_1 \le c_2$ are absolute constants. For $i = 1,2$ let $P_i = 2 \prod_{\substack{q \le z_i \\ q \equiv -1 \pmod{p_0}}} q$.

Since

$$M(y) \ge \#\{m \le y : (m,P_1) = 1,\ p_0 \nmid \sigma(m)\},$$

it follows that

$$M(y) \ge S_1 - S_2 - S_3 - S_4 - S_5 - S_6,$$

where

$$S_1 = \#\{n \le y : (n,P_1) = 1\},$$
$$S_2 = \#\{n \le y : (n,P_1) = 1,\ \exists q \equiv -1(p_0),\ z_1 < q < y/z_1^{10},\ q|n\}$$
$$S_3 = \#\{n \le y : (n,P_2) = 1,\ \exists q \equiv -1(p_0),\ y/z_1^{10} \le q < y/z_2^{10},\ q|n\}$$
$$S_4 = \#\{n \le y : \exists q \ge y/z_2^{10},\ q|n\}$$
$$S_5 = \#\{n \le y : (n,P_1) = 1,\ \exists q^a < y/z_1^{10},\ a > 1,\ q^{a+1} \equiv 1(p_0),\ q^a|n\}$$
$$S_6 = \#\{n \le y : \exists q^a \ge y/z_1^{10},\ a > 1,\ q^a|n\}.$$

By Theorem 2.5 in [7], we have

$$c_3 yW(z_1) \le S_1 \le c_4 yW(z_1),$$

where $0 < c_3 \le c_4$ are absolute constants. Thus by (2.1) the proof of Lemma 2.1 will be complete if we show that $S_i = o(S_1)$ for $2 \le i \le 6$. We have the following estimates (where we use Theorem A in considering S_2 and upper bound sieve results in considering S_2, S_3, and S_5)

$$S_2 \le \sum_{\substack{q \equiv -1(p_0) \\ z_1 < q < y/z_1^{10}}} \sum_{\substack{m \le y/q \\ (m,P_1) = 1}} 1$$

$$\ll yW(z_1) \sum_{\substack{q \equiv -1(p_0) \\ z_1 < q < y}} 1/q .$$

$$= yW(z_1)\cdot\{\frac{1}{p_o-1}\cdot\log\frac{\log y}{\log z_1} + O(\frac{\log p_o}{p_o})\}$$

$$<< yW(z_1)/(\log\log y)^{1/6} = o(S_1).$$

$$S_3 \le \sum_{\substack{q \equiv -1(p_o) \\ y/z_1^{10} \le q < y/z_2^{10}}} \sum_{\substack{m \le y/q \\ (m,P_2) = 1}} 1$$

$$<< yW(z_2) \sum_{y/z_1^{10} \le q < y} 1/q$$

$$<< yW(z_2)\frac{\log z_1}{\log y} = o(S_1).$$

$$S_4 \le y \sum_{y/z_2^{10} \le q \le y} 1/q << y\frac{\log z_2}{\log y} = \frac{y}{\exp(\sqrt{\log\log y})} = o(S_1).$$

$$S_5 \le \sum_{\substack{q^a < y/z_1^{10} \\ a > 1,\ q^{a+1} \equiv 1(p_o)}} \sum_{\substack{m \le y/q^a \\ (m,P_1) = 1}} 1$$

$$<< yW(z_1) \sum_{\substack{a > 1 \\ q^{a+1} \equiv 1(p_o)}} 1/q^a \le yW(z_1) \sum_{\substack{a > 1 \\ q^a > p_o^{1/2}}} 1/q^a << yW(z_1)p_o^{-1/4} = o(S_1).$$

$$S_6 \le y \sum_{\substack{q^a \ge y/z_1^{10} \\ a > 1}} 1/q^a << y/\sqrt{y/z_1^{10}} << y^{2/3} = o(S_1).$$

This completes the proof of Lemma 2.1 and thus of our estimate of N(x) from below.

We now show that $N(x) \le x\cdot\exp\{-(1+o(1))2\sqrt{\log 2}\sqrt{\log\log x}\}$. For any integer n, we may write n = sm where 4s is square-full, m is odd and square-free, and (s,m) = 1. Suppose $d(n)\nmid\sigma(n)$. Now $d(n) = d(s)d(m)$, $\sigma(n) = \sigma(s)\sigma(m)$, and

$d(m)|\sigma(m)$. Thus $d(s)\nmid\sigma(n)/d(m)$. We now divide the $n \le x$ for which $d(n)\nmid\sigma(n)$ into 3 classes:

(1) $s > \exp(4\sqrt{\log 2}\ \sqrt{\log\log x}\ )$,

(2) $s < \log\log x$,

(3) $\log\log x \le s \le \exp(4\sqrt{\log 2}\ \sqrt{\log\log x})$.

Let $N_1(x)$, $N_2(x)$, $N_3(x)$ denote, respectively, the number of n in classes 1,2,3.

Let $S(y)$ denote the number of $s \le y$ for which $4s$ is square-full. Then $S(y) \ll y^{1/2}$. Thus

$$N_1(x) \ll x\cdot\exp(-2\sqrt{\log 2}\ \sqrt{\log\log x}).$$

To consider the n in the other two classes, we shall use the following lemma.

<u>LEMMA 2.2.</u> <u>Let</u> u,v <u>be integers with</u> $(u,v) = 1$, $v \ge 3$. <u>Let</u> $N(v,u,k,y)$ <u>denote the number of integers</u> $m \le y$ <u>which are not divisible by</u> k <u>or more primes</u> $q \equiv u \pmod v$. <u>There are absolute constants</u> c and c' <u>such that</u>

$$N(v,u,k,y) \le c \sum_{i=0}^{k-1} y\{\frac{\log\log y}{\phi(v)}\}^i/\{i!(\log y)^{1/\phi(v)}\},$$

<u>provided</u> $1 \le k < \frac{1}{\phi(v)}\cdot\log\log y - c'$ <u>and</u> $y \ge 3$.

<u>PROOF</u>. This lemma follows by applying a result of Halász [6] in conjunction with Theorem A.

Now say $n = sm$ is in class 2. Since $d(s)\nmid\sigma(n)/d(m)$, it follows that m is not divisible by any prime $q \equiv -1 \pmod{2d(s)}$. From Theorem 317 in [9] we have

$$d(t) \le 2^{(1+o(1))(\log t)/\log\log t}. \tag{2.2}$$

So, since $s < \log\log x$, we have $d(s) < \frac{1}{10}\sqrt{\log\log x}$ for large x.

Applying Lemma 2.2, we have

$$N_2(x) \le \sum_{s<\log\log x} N(2d(s),-1,1,x) \le \sum_{s<\log\log x} cx/(\log x)^{1/\phi(2d(s))}$$

$$\le c\ x(\log\log x)/\exp(5\sqrt{\log\log x})\ ,$$

so that class 2 is under control.

Thus to complete the proof of Theorem 2.1, it remains only to consider the n in class 3. Suppose n = sm is in class 3. Since $d(s)\nmid\sigma(n)/d(m)$, it follows that there is a prime power p^k with $p^k|d(s)$ but $p^k\nmid\sigma(n)/d(m)$. Thus it is not the case that there are k primes in the class -1 (mod 2p) which divide m. Thus

$$N_3(x) \leq \sum_s \sum_{p^k|d(s)} N(2p, -1, k, x/s) \tag{2.3}$$

where s runs through the integers in $[\log\log x, \exp(4\sqrt{\log 2}\sqrt{\log\log x})]$ for which 4s is square-full. We shall require the following lemma.

<u>LEMMA 2.3</u>. <u>Suppose</u> t <u>is an integer and</u> $p^k|d(t)$, <u>where</u> p <u>is a prime</u>. <u>Then for all sufficiently large</u> t <u>we have</u>

$$p < 2\log t,\ k < (3/2)(\log t)/\log\log t.$$

<u>PROOF</u>. Since $p|d(t)$, there is a prime power $q^b||t$ with $p|b+1$. Then

$$p \leq b+1 \leq \frac{\log t}{\log 2} + 1 < 2\log t$$

for large t. From (2.2) we have

$$p^k \leq d(t) \leq 2^{(1+o(1))(\log t)/\log\log t}.$$

so that

$$k \leq (1+o(1))(\log t)/\log\log t < (3/2)(\log t)/\log\log t$$

for large t. Thus Lemma 2.3 is proved.

In view of (2.3) and Lemma 2.3, we have for all large x

$$N_3(x) \leq \sum_{\substack{p<\sqrt{8\log\log x}\\ k<\sqrt{6\log\log x}}} \sum_{p^k|d(s)} N(2p,-1,k,x/s),$$

where the inner sum is taken over those s such that 4s is square-full and $d(s) \equiv 0 \pmod{p^k}$. Let $f(p^k)$ denote the inner sum. Thus

$$N_3(x) \leq 48 \log\log x \cdot \max\{f(p^k) : p < 8\sqrt{\log\log x},\ k < 6\sqrt{\log\log x}\}\ .$$

Thus to prove Theorem 1, it is enough to show that each

$$f(p^k) \leq x \cdot \exp\{-(1+o(1))\ 2\sqrt{\log 2}\ \sqrt{\log\log x}\}\ . \tag{2.4}$$

By Lemmas 2.2 and 2.3, we have

$$f(p^k) \leq \frac{c\ x}{(\log x)^{1/\phi(2p)}} \cdot \sum_{p^k | d(s)} \frac{1}{s} \sum_{i=0}^{k-1} \{\frac{1}{\phi(2p)} \log\log x\}^i / i!\ . \tag{2.5}$$

The following two lemmas will enable us to prove (2.4) and thus complete the proof of the upper estimate of $N(x)$.

LEMMA 2.4. *Suppose the prime power* $p^k | d(t)$ *where* $p-1 > \frac{1}{12}\sqrt{\log\log x}$ *and* $t \leq \exp(4\sqrt{\log 2}\ \sqrt{\log\log x})$. *Then for all sufficiently large* x, *we have* $k < 60$.

PROOF. Say $q^b || t$ where $p | b+1$. If $p^2 | b+1$, then

$$t \geq q^b \geq 2^{p^2-1} > 2^{\frac{1}{144} \log\log x} > t$$

for all large x. Thus we may assume $p || b+1$. Hence there are k distinct primes $q_1, \ldots, q_k$ with $(q_1 \ldots q_k)^{p-1} | t$. Thus $2^{k(p-1)} \leq t$, so that $k < 48/\sqrt{\log 2} < 60$. Thus Lemma 2.4 is proved.

LEMMA 2.5. *Let* p *be a prime and let* S *be a set of integers whose elements satisfy* $p | d(s)$ *and* $4s$ *is square-full. Then there is an absolute constant* c *such that*

$$\sum_{s \in S} 1/s \leq c \cdot 2^{-p}.$$

PROOF. We may assume p is odd. If $p | d(s)$, there is a prime power $q^b || s$ with $p | b+1$. Let t be the product of all such q^b in s and write $s = ut$. Then

$$\sum_{s \in S} 1/s \leq (\sum 1/u)(\sum 1/t)$$

where u runs through all integers for which 4u is square-full and t runs through all integers > 1 which are (p-1)-full. Thus

$$\sum_{s \in S} 1/s \ll \sum 1/t \ll \zeta(p-1)-1 \ll 2^{-p},$$

where ζ is Riemann's function. Thus Lemma 2.5 is proved.

We now show that (2.4) holds if $p-1 > \frac{1}{12}\sqrt{\log\log x}$. In this case, Lemma 2.4 implies we may assume $k < 60$. Thus by (2.5) and Lemma 2.5,

$$f(p^k) \le c\cdot 60 \cdot \frac{x(\log\log x)^{59}}{(\log x)^{1/(p-1)}} \cdot \sum_{p^k | d(s)} 1/s$$

$$\ll \frac{x(\log\log x)^{59}}{(\log x)^{1/(p-1)}\cdot 2^{p-1}} .$$

Applying the inequality of the arithmetic and geometric means to the logarithm of the last denominator, we have (2.4).

Now suppose $p-1 \le \frac{1}{12}\sqrt{\log\log x}$. If $p^k | d(s)$, since $s \le \exp(4\sqrt{\log 2}\sqrt{\log\log x})$, Lemma 2.3 implies for all large x

$$k < 12\sqrt{\log 2}\sqrt{\log\log x}/\log\log\log x < 10\sqrt{\log\log x}/\log\log\log x.$$

Thus by (2.5) we have

$$f(p^k) \le ecx\cdot(\log\log x)^{10\sqrt{\log\log x}/\log\log\log x}(\log x)^{-1/\phi(2p)} \sum_{p^k | d(s)} 1/s$$

$$\ll x\exp(-2\sqrt{\log\log x}),$$

so that a stronger result than (2.4) holds in this case. Thus Theorem 2.1 is completely proved.

REMARK. Although giving the approximate rate of growth of N(x), the estimate in Theorem 2.1 is not an asymptotic formula. We believe an asymptotic formula for N(x) could be established along the general lines of our proof, but it appears that certain strong conjectures about the distribution of prime numbers in short intervals (specifically, in a short interval centered at $\sqrt{\log\log x}/\sqrt{\log 2}$) would have to be assumed.

3. The n for which $d(n)^2|\sigma(n)$.

For every positive real number β, let $< n^\beta > = \prod_{p^a||n} p^{[a\beta]}$. Thus if β is a positive integer, $< n^\beta > = n^\beta$.

THEOREM 3.1. *For any* ε *in* $(0,2)$, *the set of* n *for which* $< d(n)^{2-\varepsilon} > |\sigma(n)$ *has asymptotic density* 1, *the set of* n *for which* $< d(n)^{2+\varepsilon} > |\sigma(n)$ *has asymptotic density* 0, *and the set of* n *for which* $d(n)^2|\sigma(n)$ *has asymptotic density* 1/2.

PROOF. Write $d(n) = a(n)b(n)$, where $a(n)$ is odd and $b(n)$ is a power of 2. We first note that the set of n for which $< a(n)^\beta >$ divides $\sigma(n)$ has asymptotic density 1, no matter what positive value we choose for β. This can be seen as follows. Let ε be an arbitrary positive number. If we write $n = sm$ where $(s,m) = 1$, s is square-full, and m is square-free, then it is easy to see that there is a positive integer K such that the set of n whose square-full part s exceeds K has asymptotic density $< \varepsilon$. (Indeed, this follows at once from the fact that the sum of the reciprocals of the square-full numbers is convergent.) Hence we may consider only those n whose square-full part s does not exceed K. Since $a(n)|d(s)$, we accordingly know that $< a(n)^\beta > \le K^\beta$. If $p_1, p_2, \ldots$ are the primes congruent to -1 modulo $[K^\beta]!$, let P be the set of positive integers n for which there is no i such that $p_i||n$; since $\sum 1/p_i$ diverges (according to Dirichlet), the asymptotic density of the set P is

$$\prod_{i=1}^{\infty} \left(1 - \frac{1}{p_i} + \frac{1}{p_i^2}\right) = 0.$$

But if n has square-full part $\le K$ and if $p_i||n$ for some i, then $< a(n)^\beta > |(p_i+1)|\sigma(n)$. Thus the set of n for which $< a(n)^\beta > \nmid \sigma(n)$ is contained in the union of P and the set of integers n with $s(n) > K$; accordingly the set of n for which $< a(n)^\beta > \nmid \sigma(n)$ has density less than ε. Since ε is arbitrary, the set of n for which $< a(n)^\beta > |\sigma(n)$ has asymptotic density 1.

Since $< a(n)^{\beta} >$ and $< b(n)^{\beta} >$ are relatively prime and their product is $< d(n)^{\beta} >$, we need only be concerned with the divisibility of $\sigma(n)$ by $< b(n)^{\beta} >$ for $\beta = 2-\varepsilon$, 2, and $2+\varepsilon$.

Let $v_2(n)$ denote the exponent on 2 (possibly 0) in the prime factorization of n. Let $g_\beta(n) = v_2(\sigma(n))-\beta v_2(d(n))$. Note that g_β is an additive function. Moreover $< b(n)^{\beta} > |\sigma(n)$ if and only if $g_\beta(n) > -1$. We shall require the following lemma.

LEMMA 3.1. *If $\beta \neq 2$, the normal value of $g_\beta(n)$ is $(2-\beta)\log\log n$; that is, if $\varepsilon > 0$, the set of n with $(2-\beta-\varepsilon)\log\log n < g_\beta(n) < (2-\beta+\varepsilon)\log\log n$ has asymptotic density 1. For every real number u, the set of n with $g_2(n) \leq u\sqrt{2\log\log n}$ has asymptotic density $(2\pi)^{-1/2}\int_{-\infty}^{u} e^{-v^2/2}\,dv \overset{\text{def}}{=} G(u)$.*

We can see how the Theorem is a corollary of the Lemma. Indeed $g_{2-\varepsilon}(n) > -1$ for all n but for a set of asymptotic density 0, $g_{2+\varepsilon}(n) \leq -1$ for all n but for a set of asymptotic density 0, and $g_2(n) > -1$ for a set of n of asymptotic density $G(0) = 1/2$.

PROOF OF THE LEMMA. If g(n) is any real valued additive function, let $A(x) = \sum_{p \leq x} g(p)/p$, and let $B^2(x) = \sum_{p \leq x} g^2(p)/p$. We shall use the following generalization of the Erdös-Kac Theorem (see Kubilius [12] or Shapiro [21]): *If $B(x) \to \infty$ as $x \to \infty$ and if for every $\eta > 0$,*

$$\sum_{\substack{p \leq x \\ |g(p)| > \eta B(x)}} g^2(p)/p = o(B^2(x)),$$

then for every real number u,

$$\lim_{x \to \infty} \frac{1}{x} \sum_{\substack{n \leq x \\ (g(n)-A(x))/B(x) \leq u}} 1 = G(u).$$

We apply this theorem to the functions $g_\beta(n)$.

Using Theorem A, we have for any β,

$$A_\beta(x) \overset{\text{def}}{=} \sum_{p \le x} g_\beta(p)/p = \sum_{i=1}^{\infty} \sum_{\substack{p \le x \\ p \equiv 2^i-1 \pmod{2^{i+1}}}} i/p \; -\beta \sum_{p \le x} 1/p$$

$$= \sum_{i=1}^{\infty} i\cdot 2^{-i} \log\log x + O(\sum_{i=1}^{\infty} i^2\cdot 2^{-i}) - \beta \log\log x + O(1)$$

$$= (2-\beta) \log\log x + O(1).$$

Moreover,

$$B_\beta^2(x) \overset{\text{def}}{=} \sum_{p \le x} g_\beta^2(p)/p = \sum_{i=1}^{\infty} \sum_{\substack{p \le x \\ p \equiv 2^i-1 \pmod{2^{i+1}}}} (i-\beta)^2/p$$

$$= \sum_{i=1}^{\infty} (i-\beta)^2.2^{-i}\log\log x + O(\sum_{i=1}^{\infty} i(i-\beta)^2.2^{-i})$$

$$= (6-4\beta+\beta^2)\log\log x + O(1).$$

Now for every $\eta > 0$, let $i_\beta = i_\beta(\eta,x) = \eta B_\beta(x) + \beta$. We have (if we assume x large enough so that $\beta < \eta B_\beta(x)$)

$$\sum_{\substack{p \le x \\ |g_\beta(p)| > \eta B_\beta(x)}} g_\beta^2(p)/p = \sum_{i > i_\beta} \sum_{\substack{p \le x \\ p \equiv 2^i-1 \pmod{2^{i+1}}}} (i-\beta)^2/p$$

$$= \sum_{i > i_\beta} (i-\beta)^2.2^{-i}\log\log x + O(\sum_{i > i_\beta} i(i-\beta)^2.2^{-i})$$

$$= O\left(\frac{\eta^2 B_\beta^2(x)\log\log x}{2^{\eta B_\beta(x)}}\right) = o(1) = o(B_\beta^2(x))$$

by our estimate for $B_\beta^2(x)$.

Hence the generalization of the Erdös-Kac Theorem quoted above is applicable. Thus the normal value of $g_\beta(n)$ for $\beta \neq 2$ is $(2-\beta)\log\log n$ and, if $\rho(n)$ tends to infinity with n, we have for all n

$$-\rho(n)(\log\log n)^{1/2} < g_\beta(n)-(2-\beta)\log\log n < \rho(n)(\log\log n)^{1/2}$$

except for a set of asymptotic density zero. Moreover, since $A_2(x) = O(1)$ and since $B_2(x)$ is indistinguishable from $B_2(n)$ for n near x, we have our assertion about $g_2(n)$. This completes the proof of Lemma 3.1 and accordingly Theorem 3.1 is established.

4. The mean value of A(n).

Let g(s) be the sum of the Dirichlet Series $\sum c(n)n^{-s}$ whose Euler product is

$$\prod_p (\{1-p^{-s}\}^{1/2}\{1 + \tfrac{1}{2}(1 + \tfrac{1}{p})p^{-s} + \tfrac{1}{3}(1 + \tfrac{1}{p} + \tfrac{1}{p^2})p^{-2s} + \ldots\}),$$

the square-root being the principal branch. Clearly both series and product converge absolutely for Re s > $\frac{1}{2}$, since the general term of the product has the form

$$1 + \frac{1}{2p}\, p^{-s} + \sum_{k=2}^{\infty} \varepsilon_k(p)p^{-ks},$$

where $|\varepsilon_k(p)| < 1$. The following result indicates that A(n) behaves like $g(1)n(\pi \log n)^{-1/2}$ on average.

THEOREM 4.1. *As* $x \to \infty$ *we have*

$$\sum_{n \le x} A(n) \sim \frac{g(1)}{2\pi^{1/2}} \frac{x^2}{(\log x)^{1/2}}. \tag{4.1}$$

PROOF. For Re s > 1 we have

$$\sum_{n=1}^{\infty} n^{-1}A(n)n^{-s} = \prod_p \{1 + \tfrac{1}{2}(1 + \tfrac{1}{p})p^{-s} + \tfrac{1}{3}(1 + \tfrac{1}{p} + \tfrac{1}{p^2})p^{-2s} + \ldots\} \tag{4.2}$$

$$= \zeta(s)^{1/2}g(s),$$

where $\zeta(s)^{1/2}$ is real for positive real s greater than 1. From (4.2) it is possible to deduce in various ways that

$$B(x) = \sum_{n \le x} n^{-1}A(n) \sim \frac{g(1)}{\pi^{1/2}} \frac{x}{(\log x)^{1/2}}; \tag{4.3}$$

in fact we shall sketch below several different methods of going from (4.2) to (4.3). The partial summation formula

$$\sum_{n \le x} A(n) = \int_{1-}^{x} t\, dB(t) = x\, B(x) - \int_1^x B(t)dt$$

then readily enables one to deduce (4.1) from (4.3).

Here are five methods I-V for deducing (4.3) from (4.2).

I One can use the classical method of contour integration (cf. Landau [13], Landau [14], Wilson [24], Stanley [22], and Hardy [8].)

II One can appeal to general theorems established by Kienast [11] and Dixon [4] by the method of contour integration.

III One can start with the result

$$D_{1/2}(x) = \sum_{n \le x} d_{1/2}(n) = \frac{x}{(\pi \log x)^{1/2}} + O\left(\frac{x}{(\log x)^{3/2}}\right)$$

of Selberg [22] or Diamond [3], where

$$\sum_{n=1}^{\infty} d_{1/2}(n)n^{-s} = \zeta(s)^{1/2} \qquad (\mathrm{Re}\ s > 1),$$

and then use the identity

$$\sum_{n \le x} n^{-1}A(n) = \sum_{n \le x} c(n)\, D_{1/2}(x/n),$$

where as above $\sum c(n)n^{-s} = g(s)$.

IV One can use the following Tauberian theorem of Delange:

LEMMA 4.1. *Suppose* $1 \le h_1 < h_2 < \cdots$ *and* $h_n \to +\infty$. *Suppose the Dirichlet series* $\sum a_n h_n^{-s}$ *has non-negative coefficients and converges for* Re $s > 1$ *to a sum* $f(s)$. *Suppose there is a real number* r *which is not a negative integer such that*

$$f(s) = (s-1)^{-r-1}h(s) + k(s),$$

where h *and* k *are holomorphic functions on some domain containing the closed half-plane* Re $s \ge 1$ *and* $h(1) \ne 0$. *Then as* $x \to +\infty$ *we have*

$$\sum_{h_n \le x} a_n \sim \frac{h(1)}{\Gamma(r+1)}\, x(\log x)^r.$$

Lemma 4.1 is a special case of Theorem 3 of Delange [2]. To obtain (4.3) from (4.2) by using Lemma 4.1 we need only take $r = -1/2$, $h_n = n$, $a_n = A(n)/n$, $h(s) = \{(s-1)\zeta(s)\}^{1/2}g(s)$, and $k(s) = 0$.

V One can apply the Tauberian theorem of Delange quoted as Lemma 5.1 in the next section to

$$\sum(\log n)\frac{A(n)}{n}\frac{1}{n^s} = -\frac{d}{ds}\{\zeta(s)^{1/2}g(s)\}.$$

This approach gives

$$\sum_{n\leq x}(\log n)\frac{A(n)}{n} \sim \frac{g(1)}{\pi^{1/2}}x(\log x)^{1/2}$$

from which (4.3) follows by partial summation.

Any of the first three of the above five methods will in fact give the more precise result

$$\sum_{n\leq x} A(n) = \frac{x^2}{(\log x)^{1/2}}\{\frac{g(1)}{2\pi^{1/2}} + \frac{b_1}{\log x} + \frac{b_2}{\log^2 x} +\ldots+ \frac{b_m}{\log^m x} + O(\frac{1}{\log^{m+1} x})\}$$

for any positive integer m and suitable constants $b_1, b_2, \ldots, b_m$.

Instead of starting with (4.2) it would also be possible to begin with

$$\sum_{n=1}^{\infty} n^{-1}A(n)n^{-s} = \sum_{n=1}^{\infty}\frac{1}{d(n)}n^{-s} \cdot \sum_{n=1}^{\infty}\ell(n)n^{-s},$$

where $\sum \ell(n)n^{-s}$ has abscissa of convergence less than 1, and then use the result of Wilson [24] that

$$\sum_{n\leq x}\frac{1}{d(n)} \sim \frac{cx}{(\log x)^{1/2}}$$

for a certain positive constant c.

5. The number of n with $A(n) \leq x$.

We begin by quoting (as Lemma 5.1) another Tauberian theorem of Delange which is somewhat more powerful than Lemma 4.1. While it requires the parameter r to be non-negative, it requires a much less stringent condition than the analyticity of the functions h and k which was imposed in Lemma 4.1. If $r \geq 0$, Lemma 4.1 clearly follows from Lemma 5.1; if r is negative but not an integer, Lemma 4.1 can be readily deduced from Lemma 5.1 by repeated differentiation and partial summation.

LEMMA 5.1. *Suppose* $1 \le h_1 < h_2 < \dots$ *and* $h_n \to +\infty$. *Suppose the Dirichlet series* $\sum a_n h_n^{-s}$ *has non-negative coefficients and converges for* Re $s > 1$ *to a sum* $f(s)$. *Suppose*

$$\lim_{\substack{s \to 1+iy \\ \mathrm{Re}\, s > 1}} f(s)$$

exists for each non-zero y. *And suppose there exist real numbers* A, r, θ *with*

$$A > 0,\ r \ge 0,\ 0 \le \theta < 1$$

such that

$$f(s) - A(s-1)^{-r-1} = O(|s-1|^{-r-\theta})$$

for Re $s > 1$, $|s-1| < 1$. *Then as* $x \to +\infty$ *we have*

$$\sum_{h_n \le x} a_n \sim \frac{A}{\Gamma(r+1)}\, x(\log x)^r.$$

Lemma 5.1 is a special case of Theorem 1 of Delange [2].

The following theorem not only provides valuable information about the distribution of the values $A(n)$, but also is interesting in that it is a clearcut instance where Lemma 5.1 appears to be needed rather than the easier Lemma 4.1.

THEOREM 5.1. *As* $x \to +\infty$ *we have*

$$\#\{n : A(n) \le x\} = \sum_{A(n) \le x} 1 \sim \lambda\, x \log x,$$

where

$$\lambda = \prod_p \left\{\left(1 - \frac{1}{p}\right)^2 \left(1 + \frac{2}{p+1} + \frac{3}{p^2+p+1} + \frac{4}{p^3+p^2+p+1} + \dots\right)\right\}.$$

PROOF. For $\sigma = \mathrm{Re}\, s > 1$ we have

$$(5.1) \qquad \sum A(n)^{-s} = \prod_p \{1 + A(p)^{-s} + A(p^2)^{-s} + A(p^3)^{-s} + \dots\}$$

$$= \zeta(s)^{2^s} G(s),$$

where

$$\zeta(s)^{2^s} = \exp\{2^s \sum_p \sum_{m=1}^{\infty} m^{-1}p^{-ms}\}$$

and

$$(5.2)\quad G(s) = \prod_p \left(\{1-\frac{1}{p^s}\}^{2^s}\{1+\frac{2^s}{(p+1)^s}+\frac{3^s}{(p^2+p+1)^s}+\frac{4^s}{(p^3+p^2+p+1)^s}+\ldots\}\right).$$

For $\sigma = \text{Re } s > 1/2$ we have the estimates

$$1+\frac{2^s}{(p+1)^s}+\frac{3^s}{(p^2+p+1)^s}+\frac{4^s}{(p^3+p^2+p+1)^s}+\ldots = 1+\frac{2^s}{(p+1)^s}+O(\frac{4^\sigma}{p^{2\sigma}})$$

and

$$(1-\frac{1}{p^s})^{2^s} = \exp\{2^s \log(1-\frac{1}{p^s})\} = \exp\{-\frac{2^s}{p^s}+O(\frac{2^\sigma}{p^{2\sigma}})\},$$

so that

$$(5.3)\quad \{1-\frac{1}{p^s}\}^{2^s}\{1+\frac{2^s}{(p+1)^s}+\frac{3^s}{(p^2+p+1)^s}+\frac{4^s}{(p^3+p^2+p+1)^s}+\ldots\} =$$

$$1-\frac{2^s}{p^s}+\frac{2^s}{(p+1)^s}+O(\frac{4^\sigma}{p^{2\sigma}}).$$

Since

$$|-\frac{2^s}{p^s}+\frac{2^s}{(p+1)^s}| = |-2^s \int_p^{p+1} \frac{s}{u^{s+1}}\,du| \leq \frac{2^\sigma|s|}{p^{\sigma+1}},$$

(5.2) and (5.3) show that G is holomorphic for $\sigma = \text{Re } s > 1/2$.

Put

$$H(s) = G(s)\{(s-1)\zeta(s)\}^{2^s}$$

in some domain containing $\text{Re } s \geq 1$ in which ζ has no zeros. Then for Re s >

we obtain from (5.1)

$$(5.4)\quad \sum_{n=1}^{\infty} A(n)^{-s} = (s-1)^{-2^s}H(s)$$

$$= (s-1)^{-2}H(1) + (s-1)^{-2}\{H(s)-H(1)\}$$

$$+ H(s)(s-1)^{-2}(\exp\{(2-2^s)\log(s-1)\}-1),$$

where the logarithm is the principal branch. Hence if Re s > 1 and $|s-1| < 1$, we have

$$|\exp\{(2-2^s)\log(s-1)\} -1| \leq \exp|(2-2^s)\log(s-1)| -1$$
$$\leq C_1|(2-2^s)\log(s-1)|$$
$$\leq C_2|(s-1)\log(s-1)|$$
$$\leq C_2|s-1|(\log|s-1|^{-1}+\pi)$$
$$\leq C_3|s-1|^{1-\varepsilon},$$

for any fixed positive $\varepsilon < 1$ and suitable constants C_1,C_2,C_3. Thus (5.4) gives

$$\left| \sum_{n=1}^{\infty} \frac{1}{A(n)^s} - \frac{H(1)}{(s-1)^2} \right| \leq C_4(s-1)^{-1-\varepsilon} \quad (\text{Re } s > 1,\ |s-1| < 1)$$

for a suitable constant C_4. We may therefore apply Lemma 5.1 with h_n = the n-th distinct value in the range of A, a_n = the number of times the value h_n is taken on by A, $r = 1$, $\theta = \varepsilon$, and

$$A = H(1) = G(1) = \prod_p\{(1-\frac{1}{p})^2(1+\frac{2}{p+1}+\frac{3}{p^2+p+1}+\frac{4}{p^3+p^2+p+1}+\ldots)\}.$$

Thus the result of Theorem 5.1 follows.

6. <u>The distribution of the numbers</u> A(n).

<u>THEOREM 6.1</u>. <u>There is a positive constant</u> ν <u>such that the number of distinct rationals of the form</u> $\sigma(n)/d(n)$ <u>not exceeding</u> x <u>is</u> $O(x/(\log x)^\nu)$.

We shall use the following result of Erdös and Wagstaff [5]: <u>There is a positive constant</u> μ <u>such that the number of</u> $n \leq x$ <u>such that</u> n <u>has a divisor</u> p+1 <u>with</u> p <u>a prime</u>, $p > T$, <u>is</u> $O(x/(\log T)^\mu)$ <u>uniformly for all</u> $x \geq 1$, $T \geq 2$. Actually Erdös-Wagstaff prove this for p-1 in place of p+1, but the proof is identical.

PROOF OF THE THEOREM. Let $\varepsilon > 0$ be arbitrarily small. Let x be large and let $S = (\log x)^{2\mu}$, where μ is the constant in the Erdös-Wagstaff theorem. Any integer $n > 0$ can be written uniquely in the form $n = s(n)\cdot m(n) = sm$ where $(s,m) = 1$, m is odd and square-free, and 4s is square-full. Let

$$N_1 = \#\{n > x(\log x)^4 : \quad \sigma(n)/d(n) \le x\} ,$$

$$N_2 = \#\{n \le x(\log x)^4 : \ P(n) \le x^{1/\log\log x}\},$$

$$N_3 = \#\{n \le x(\log x)^4 : P(n) > x^{1/\log\log x},\ P(n)^2 | n\} ,$$

$$N_4 = \#\{r \le x : r = \sigma(n)/d(n) \text{ for some } n \text{ with } P(n) > x^{1/\log\log x}, P(n) \| n, s(n) \le S\},$$

$$N_5 = \#\{r \le x : r = \sigma(n)/d(n) \text{ for some } n \text{ with } s(n) > S\},$$

where P(n) denotes the largest prime factor of n. Thus, if f(x) denotes the number of distinct rationals not exceeding x and having the form $\sigma(n)/d(n)$, we clearly have

$$f(x) \le N_1 + N_2 + N_3 + N_4 + N_5, \tag{6.1}$$

so that it remains to estimate these 5 quantities. Note that in the definitions of N_1, N_2, and N_3 we are counting the number of positive integers n satisfying the conditions in question, but that in the definitions of N_4 and N_5 we are counting only distinct values of the ratio $\sigma(n)/d(n)$ arising from at least one n satisfying the conditions mentioned.

We have (see p. 240 of [24])

$$N_1 \le \sum_{\substack{n > x \log^4 x \\ d(n) \ge n/x}} 1 \le \sum_{i=0}^{\infty} \sum_{\substack{2^i x \log^4 x < n \le 2^{i+1} x \log^4 x \\ d(n) > 2^i \log^4 x}} 1$$

$$\le \sum_{i=0}^{\infty} \frac{1}{2^{2i} \log^8 x} \sum_{n \le 2^{i+1} x \log^4 x} d^2(n)$$

$$\ll \sum_{i=0}^{\infty} \frac{1}{2^{2i}\log^8 x} \cdot 2^{i+1} x \log^4 x \cdot (i+2\log x)^3$$

(6.2) $$\ll \frac{x}{\log^4 x} \sum_{i=0}^{\infty} \frac{(i+2\log x)^3}{2^i} \ll \frac{x}{\log x}\,.$$

From Rankin [12]

(6.3) $$N_2 \ll x/\log x.$$

Clearly

(6.4) $$N_3 \le \sum_{d > x^{1/\log\log x}} x(\log x)^4/d^2 \ll x/\log x.$$

We use the Erdös-Wagstaff theorem to estimate N_4. If $\sigma(n)/d(n)$ is counted by N_4, then $\sigma(n)/d(n) = (\sigma(n)/d(m))/d(s)$ where $\sigma(n)/d(m)$ is an integer and for each $\varepsilon > 0$ (cf. (2.2))

$$d(s) \le Z \overset{\text{def}}{=} \max\{d(s) : s \le S\} \le (\log x)^{\varepsilon}$$

for all large x depending on the choice of ε. Thus the integer $2\sigma(n)/d(m)$ is at most $2Zx$ and is divisible by a $p+1$ where p is prime, $p > x^{1/\log\log x}$. By the Erdös-Wagstaff theorem, we thus have

(6.5) $$N_4 \ll xZ^2/(\log(x^{1/\log\log x}))^{\mu} \ll x/(\log x)^{\mu-3\varepsilon}.$$

Note that if $\sigma(n)/d(n)$ is counted by N_5 and $n = sm$, then $\sigma(n)/d(n) = (\sigma(m)/d(m))(\sigma(s)/d(s))$, so that $(\sigma(m)/d(m))\sigma(s) \le xd(s) \le xs^{\varepsilon}$ for large x. But $\sigma(m)/d(m)$ is an integer. Thus for each fixed $s > S$ for which $4s$ is square full, the number of $\sigma(n)/d(n) \le x$ with $s(n) = s$ is at most the number of multiples of $\sigma(s)$ below xs^{ε}. Thus

(6.6) $$N_5 \le \sum_{s > S} xs^{\varepsilon}/\sigma(s) \le \sum_{s > S} x/s^{1-\varepsilon} \ll x/(\log x)^{\mu-2\varepsilon\mu}.$$

Our theorem now follows from (6.1), (6.2), (6.3), (6.4), (6.5), (6.6).

Note that if q runs through the primes not exceeding 2x-1, the $\pi(2x-1)$ numbers $\sigma(q)/d(q) = (q+1)/2$ lie in $[1,x]$ and are all distinct. Thus the constant ν in Theorem 6.1 cannot be larger than 1. In fact, by more complicated arguments we can prove that, if f(x) is as above, then for every positive ε and every positive integer k we have

$$\frac{x}{\log x}(\log\log x)^k \ll f(x) \ll \frac{x}{(\log x)^{1-\varepsilon}}.$$

7. Other problems.

In this section we shall state some further results, giving either sketchy proofs or no proofs at all.

THEOREM 7.1. *There is a constant* c *so that*

$$\{\prod_{i=1}^{n} A(i)\}^{1/n} \sim c\, n/(\log n)^{\log 2}.$$

Since there is a known asymptotic formula for $\{\prod_{i=1}^{n} d(i)\}^{1/n}$ due to Ramanujan (cf. Wilson [24]), Theorem 7.1 can be proved by establishing an asymptotic formula for $\{\prod_{i=1}^{n} \sigma(i)\}^{1/n}$. It is possible to give the constant c explicitly and also to give arbitrarily many secondary terms.

THEOREM 7.2. *The set of integers* n *with an integral arithmetic mean for the divisors* d *of* n *with* $1 \le d < n$ *has density* 0.

That is, Theorem 7.2 asserts that the set of n for which $d(n)-1 \mid \sigma(n)-n$ has density 0. We now sketch a proof for square-free n. The non-square-free case is much harder.

Assume n is square-free and K is large. All but a density 0 of n have $2^p-1 \mid \sigma(n)$ for every prime $p \le K$. For each prime p, $2 \le p \le K$, the n with $p \mid \omega(n)$ have relative density 1/p. (Here $\omega(n)$ is the number of distinct prime factors of n.) In fact, the relative density of the square-free n for which

(7.1) $$p \mid \omega(n),\ 2^p-1 \nmid n,\ 2^p-1 \mid \sigma(n)$$

is $p^{-1}(1-(2^p-1)^{-1})$. But such an n has $d(n)-1 \nmid \sigma(n)-n$ since $2^p-1 \mid d(n)-1$, $2^p-1 \nmid \sigma(n)-n$. The events (7.1) for different primes p are independent. Thus the relative density of the square-free n for which $d(n)-1 \mid \sigma(n)-n$ is at most

$$\prod_{2 \le p \le K} \{1-\frac{1}{p}(1-\frac{1}{2^p-1})\}.$$

Letting $K \to \infty$, this product goes to 0 and we have our result for square-free n.

The above heuristic argument can be made the backbone of a rigorous proof. Roughly the same idea can be used in the more general case when d(n) is a power of 2.

Recall that $\sigma_i(n) = \sum_{d|n} d^i$. Thus $\sigma(n) = \sigma_1(n)$, $d(n) = \sigma_0(n)$.

THEOREM 7.3. *Let $\delta_{i,j}$ denote the asymptotic density of the set of n for which $\sigma_i(n)|\sigma_j(n)$, where i,j are integers and $0 \le i < j$. Then*

$$\delta_{i,j} = \begin{cases} 1, & \text{if } j/i \text{ is an odd integer;} \\ 1, & \text{if } i = 0,\ j \text{ is odd;} \\ 0, & \text{if } i \ge 1,\ j/i \text{ is not an odd integer.} \end{cases}$$

Moreover if $i = 0$, j is even, then $\delta_{i,j}$ exists and $0 < \delta_{i,j} < 1$.

REFERENCES

1. P.T. Bateman, The distribution of values of the Euler function, Acta Arith. 21 (1972), 329-345.
2. Hubert Delange, Généralisation du theorème de Ikehara, Ann. Sci. École Norm. Sup. (3) 71 (1954), 213-242.
3. Harold G. Diamond, Interpolation of the Dirichlet divisor problem, Acta Arithmetica 13 (1967), 151-168.
4. Robert D. Dixon, On a generalized divisor problem, J. Indian Math. Soc. (N.S.) 28 (1964), 187-196.
5. P. Erdös and S.S. Wagstaff, Jr., The fractional parts of the Bernoulli numbers, Illinois J. Math. 24 (1980), 104-112.

6. G. Halász, Remarks to my paper "On the distribution of additive and the mean value of multiplicative arithmetic functions", Acta Math. Acad. Sci. Hungar. 23 (1972), 425-432.

7. H. Halberstam and H.-E. Richert, Sieve Methods, London,Academic Press, 1974.

8. G.H. Hardy, Ramanujan, Cambridge, The University Press, 1940, particularly §4.4-4.8.

9. G.H. Hardy and E.M.Wright, An Introduction to the Theory of Numbers, London, Oxford University Press, 1960.

10. H.-J. Kanold, Über das harmonische Mittel der Teiler einer natürlichen Zahl, Math. Ann. 133 (1957), 371-374.

11. Alfred Kienast, Über die asymptotische Darstellung der summatorischen Funktion von Dirichletreihen mit positiven Koeffizienten, Math. Zeit. 45 (1939), 554-558.

12. J.P. Kubilius, Probabilistic Methods in the Theory of Numbers, Translations of Mathematical Monographs, Vol. 11, American Math. Soc., Providence 1964.

13. Edmund Landau, Über die Einteilung der positiven ganzen Zahlen in vier Klassen nach der Mindestzahl der zu ihrer additiven Zussamensetzung erforderlichen Quadrate, Archiv der Mathematik und Physik (Ser. 3) 13 (1908), 305-312.

14. Edmund Landau, Handbuch der Lehre von der Verteilung der Primzahlen, Leipzig, Teubner, 1909, particularly §§176-183.

15. K.K. Norton, On the number of restricted prime factors of an integer. I, Illinois J. Math. 20 (1976), 681-705.

16. O. Ore, On the averages of the divisors of a number, Amer. Math. Monthly 55 (1948), 615-619.

17. C. Pomerance, On the distribution of amicable numbers, J. reine angew. Math. 293/4 (1977), 217-222.

18. C. Pomerance, Problem 6144, Amer. Math. Monthly 84 (1977), 299-300.

19. R.A. Rankin, The difference between consecutive prime numbers, J. London Math. Soc. 13 (1938), 242-247.

20. Atle Selberg, Note on a paper by L.G. Sathe, J. Indian Math. Soc. (N.S.) 18 (1954), 83-87.

21. H.N. Shapiro, Distribution functions of additive arithmetic functions, Proc. Nat. Acad. Sci. U.S.A. 42 (1956), 426-430.

22. G.K. Stanley, Two assertions made by Ramanujan, J. London Math. Soc. 3 (1928), 232-237 and 4 (1929), 32.

23. D.V. Widder, The Laplace Transform, Princeton University Press, 1941.

24. B.M. Wilson, Proofs of some formulae enunciated by Ramanujan, Proc. London Math. Soc. (2) 21 (1923), 235-255.

University of Illinois, Urbana, Illinois

Hungarian Academy of Science, Budapest, Hungary

University of Georgia, Athens, Georgia

University of California at Los Angeles

THE NEXT PELLIAN EQUATION

Harvey Cohn*

Mathematics Department, City College of New York

New York, 10031, N.Y.

Affectionately dedicated to Emil Grosswald

in appreciation of his enthusiasm

for concrete results

Abstract. The pellian equations $x^2-dy^2 = -4$ in Z or $\xi^2-\partial\eta^2 = 4i$ in Z[i] both have similar criteria of solvability according to factors of 2 or 4 in the class number for $\mathbb{Q}(\sqrt{-p})$, when a prime p = d or $N\partial$. The next pellian equation leads to a tower of pellian equations whose height limits the power of 2 dividing that class number.

INTRODUCTION

The pellian equation refers variously to equations over Z of the form

(1a) $$x^2-dy^2 = \pm 1, \pm 4$$

where d > 0 and is not a perfect square. It has a gaussian analogue over Z[i]

(1b) $$\xi^2-\partial\eta^2 = 1, 4, i, 4i$$

where now ∂ is not a perfect square in Z[i]. (The option of a $\pm$ sign is not important since $i^2 = -1$). The use of the factor of 4 is not a major difficulty, as we note later, but the nonsquare unit -1 in (1a) or i in (1b) is a matter of deeply theoretical concern. For instance, referring just to the rational case (1a), for d = p(prime), the best-known result is that (1a) is solvable if and only if p ≡ 1 mod 4, when -1 or -4 is used on the right. Yet if d is composite

* Research supported by NSF Grant MCS 7903060.

there is no "satisfactory" answer, although conditions can be found [7] on the level of sophistication of reciprocity in algebraic number fields. The concept of "satisfactory" is somewhat elusive, since a more modern approach would favor computational complexity [5] over class field theory. For instance, one might ask if (1a) is numerically solvable in essentially the amount of time required to read d into a computer.

If we follow the classical route, we restrict ourselves to $p \equiv 1 \bmod 4$ and decompose

(2a) $$p = a^2+b^2,\ a(\text{odd}),\ b(\text{even}) \in Z.$$

Then we factor

(2b) $$p = \pi\pi',\ (\pi = a+bi,\ \pi = a-bi)$$

and, according to a result of Dirichlet [4], the equation

(3) $$\alpha^2-\pi\beta^2 = 4i$$

is solvable for α, β in $Z[i]$ if and only if

(4a) $$p \equiv 1 \bmod 8$$

(as we henceforth assume). The above relation can also be written

(4b) $$b \equiv 0 \bmod 4.$$

Under the assumption (4a), it will be convenient to choose the sign of b in the manner of [2], by first using the fundamental unit of $\mathbb{Q}(\sqrt{p})$

(4c) $$\varepsilon = s+t\sqrt{p} \qquad s > 0 \qquad t > 0 \qquad s,t \in Z$$

which itself satisfies a pellian equation (without "denominator")

(4d) $$s^2-t^2p = -1.$$

What we do is choose the sign of b so that (1+si)/(a+bi) is a perfect square of a gaussian integer γ

(4e) $$\gamma^2\pi = 1+si, \ \gamma \in Z[i]$$

(4f) $$\gamma = z_1+z_2 i, \ z_1 > 0, \ z_2 > 0$$

(4f) $$t = z_1^2+z_2^2 .$$

(Thus by taking the norm of (4e), we obtain (4d)). The relation (4e) is convenient from the Galois group point of view because it is the square of the relation

(4g) $$(\sqrt{\varepsilon}-1/\sqrt{\varepsilon})/(1-i) = \gamma\sqrt{\pi}$$

since from (4d)

(4h) $$\varepsilon' = -1/\varepsilon = s-t\sqrt{p}.$$

Now the choices of sign of $\sqrt{-1}$, $\sqrt{p}$ and $\sqrt{\varepsilon}$, ($\sqrt{-p} = \sqrt{-1}\ \sqrt{-p}$), determine that of $\sqrt{\pi}$.

We note a simplification of the pellian equation (3) when

(5a) $$\pi \equiv \pm 1 \bmod 4(1+i).$$

This assumption was made by Lakein in his useful table [6], for algorithmic convenience, but we need it later on (see (8d) below). The congruence condition (5a) characterized all odd gaussian squares, thus we can simplify (3) as follows:

(5b) $$\alpha = (1-i)\alpha_0, \ \beta = (1-i)\beta_0,$$

(5c) $$\alpha_0^2-\beta_0^2\pi = -2,$$

We easily recognize a common feature of (1a) and (1b) namely the form

(6a) $$N_{K/k}\Theta_K = \Theta_k$$

where $K = k(\sqrt{\Delta})$ is a quadratic extension and θ_K as well as θ_k are units of the corresponding fields. Thus in (1ab)

(6b) $$\Theta_K = (x+y\sqrt{d})/2 \text{ or } (\xi+\eta\sqrt{\partial})/2$$

and the norm $N_{K/k}\Theta_K$ is merely $\Theta_K\Theta_K'$, where

(6c) $$\Theta_K' = (x-y\sqrt{d})/2 \text{ or } (\xi-\eta\sqrt{\partial})/2.$$

Of course, we specialize in (1ab) to $d = p$ or $\partial = \pi$, and $\Theta_k = -1$ or i. Also the denominator of (6bc) (or the 4 in (1ab)) creates no problem. Units are closed under multiplication so no higher denominator occurs, while often the denominator is cancelled out, (as in (4d) or (5c)).

We might be tempted to look for pellian equations over higher fields to give us solvability criteria depending on p. This can not be done "cheaply", however. For instance if Θ_k is a unit in the field k, and $i \in k$, then the pellian equation

(7a) $$N_{K/k}\Theta_K = \Theta_k$$

will have the almost trivial solution

(7b) $$\Theta_K = \sqrt{(i\Theta_k)}$$

if we choose K as $k(\Theta_K)$. The way to avoid this embarassment is to insist that K/k is only oddly ramified (as we shall require).

THE PELLIAN TOWER

Rational number theory can easily give us the wrong clue if we expect the next pellian equation to depend on a further congruence condition on p (say "$p \equiv 1 \bmod 16$"). For the correct clue we must define

(8a) $$h = h(-4p) = \text{class number of } \mathbb{Q}(\sqrt{-p})$$

or the number of forms of discriminant -4p, (associated with x^2+py^2). It is then classically known that

(8b) $$p \equiv 1 \bmod 4 \iff 2|h$$

(8c) $$p \equiv 1 \bmod 8 \iff 4|h.$$

We therefore look for a pellian equation solvable only when $8|h$. The condition for this is $p = A^2+32B^2$ or (see [1])

(8d) $$\pi \equiv \pm 1 \bmod 4(1+i) \iff 8|h.$$

More generally we search for a pellian equation solvable when

(9) $$M = 2^T|h, \; T \geq 2.$$

<u>MAIN THEOREM.</u> <u>Let</u> $M = 2^T$ (≥ 4) <u>divide</u> h. <u>Then a tower of</u> T <u>fields</u>

(10a) $$\mathbb{Q}(i) = K_2 \subset K_4 \subset \ldots \subset K_m$$

<u>(of degrees shown) can be constructed with odd ramifications such that</u>

(10b) $$K_{2m} = K_m(\sqrt{\Delta_m}), \; 2 \leq m \leq M/2,$$

<u>for</u> Δ_m <u>and integer of</u> K_m <u>of absolute norm</u> $p^{(odd)}$. <u>Moreover, (showing norms by the degrees of the fields),</u>

(10c) $$N_{2m/m}\Delta_{2m} = \Delta_m'^{(odd)}$$

<u>for</u> Δ_m', <u>the conjugate of</u> Δ_m <u>in</u> K_m <u>over</u> $K_{m/2}$. <u>Furthermore, each</u> K_m <u>has a unit</u> Θ_m <u>satisfying the chain of pellian equations</u>

(10d) $$N_{2m/m}\Theta_{2m} = \Theta_m, \; 2 \leq m \leq M/2$$

<u>starting with</u> $\Delta_2 = \pi$, $\Theta_2 = i$ <u>and</u> $\Theta = (\alpha+\beta\sqrt{\pi})/2 = \Theta_4$.

The proofs will not be repeated here, (see [3, §5] for details). The origin of the fields K_m is that the tower of fields $k_{2m} = K_m(\sqrt{-p})$ over $k_2 = \mathbb{Q}(\sqrt{-p})$ is a "2-cyclic class field tower" of unramified extensions which can exist if and only if $M|h$. Nevertheless, the solutions of the pellian equations required for constructing this tower k_{2m} are rather special and it is not conversely obvious that if the K_m tower exists as in (10a-d), then that tower (or some "better" K_m tower) could be pushed up to a class field tower k_{2m}. So generally, the main theorem would seem only to give a necessary condition for $M|h$ (although it would be no surprise if the condition of the main theorem were also sufficient).

The source [3] is couched in terms of relative discriminants $D_m = \text{disc } K_{2m}/K_m$, which are ideals (not numbers) in K_m of norm p. To translate to the present context, note that all the class numbers of the K_m are odd [3, §3] so that $D_m^{(odd)} = (\Delta_m)$, and taking odd powers does not destroy the chain (10c).

THE NEXT PELLIAN EQUATION

If we restrict our attention to just the case (8d), or M = 8, then we can explicitly calculate the pellian equation

(11a) $$N_{8/4}\Theta_8 = \Theta_4$$

for Θ_4 taken from (3) or (5c), namely,

(11b) $$\Theta_4 = (\alpha+\beta\sqrt{\pi})/2 = (\alpha_0+\beta_0\sqrt{\pi})/(1+i),$$

under circumstances seemingly satisfied within the range of computation, and we even can give a perhaps weaker condition for solvability of the pellian equation to be sufficient for $8|h$.

First we consider the class field tower k_{2m} over $k_2 = \mathbb{Q}(\sqrt{-p})$, given by $k_{2m} = K_m(\sqrt{-p})$. Thus, $K_4 = K_2(\sqrt{\pi})$,

(12a) $$k_8 = k_2(\Theta_4) = K_4(\sqrt{-p}) = \mathbb{Q}(i, p, \sqrt{\varepsilon}) = \mathbb{Q}(\sqrt{\pi}, i),$$

(12b) $$\text{Gal } k_8/k_2 = \langle c|c^4 = 1\rangle.$$

Here the action of c preserves $\sqrt{-p}(=(\sqrt{-p})^c)$, but also

(13a) $$i^c = -i,\ (\sqrt{p})^c = -\sqrt{p},\ (\varepsilon^c = \varepsilon',\ \pi^c = \pi').$$

Thus, since $\sqrt{\varepsilon}\sqrt{\varepsilon'} = i$,

(13b) $$(\sqrt{\varepsilon})^c = i/\sqrt{\varepsilon},\ (\sqrt{\varepsilon})^{cc} = -\sqrt{\varepsilon},$$

(13c) $$(\sqrt{\pi})^{cc} = -\sqrt{\pi}$$

but $(\sqrt{\pi})^c$ must be unraveled from (4g), better written as

(13d) $$\gamma\sqrt{\pi} = (\sqrt{\varepsilon}+(\sqrt{\varepsilon})^{1/c})/(1-i).$$

We now excerpt from the basic construction [3, Cor. 5.15] the analogue of (13d). There it is seen that

(14a) $$K_8 = K_4(\Gamma_8)$$

(14b) $$\Gamma_8 = (H_8+H_8^{1/c})/(1-i)$$

where H_8 is either root in

(14c) $$H_8^2 = \sqrt{\varepsilon}\Theta_4\Theta_4^c.$$

We can now define Δ_4 to within a superfluous square by

(15a) $$\Delta_4^* = \Gamma_8^2 = \Delta_4\Lambda^2,$$

(15b) $$\Delta_4^* = \Theta_4(1-(\Theta_4/\sqrt{\varepsilon}+\sqrt{\varepsilon}/\Theta_4)/2)^c$$

(15c) $$\Delta_4^* = \Theta_4(1 - \beta_0\pi'\gamma'/2 + i\gamma\alpha_0'\sqrt{\pi}/2).$$

Here Δ_4^* was calculated by the rules (13a-d). It has the factor Λ^2 which can be removed (in some cases) so that

(16a) $$N_{4/2}\Delta_4^* = \pi'\lambda^2,$$

(16b) $$N_{4/2}\Lambda = \lambda = (\gamma'-\beta_0')/(1-i),$$

(16c) $$\Delta_4 = \Delta_4^*/\Lambda^2, \; N_{4/2}\Delta_4 = \pi'.$$

Unfortunately the factor Λ is always present since λ is always divisible by 1+i. Nevertheless, some choice of sign $\pm\beta_0$ will guarantee that 1+i is the only even factor of λ.

Assume π' has a principal factor as it splits in K_4. Then whenever $8|h$ the equation

(17) $$X^2 - Y^2\Delta_4 = 4\Theta_4$$

is solvable (in integers X, Y of K_4) for θ_4 some unit of K_2 of type (11b).

Conversely, if (17) is solvable for $N_{4/2}\Delta_4 = \pi'$, and Y is odd, then $8|h$.

For proof, note that if π' has a principal factor in K_4 this factor (or its conjugate over K_2) divides Δ_4^* of (15c). The remaining factor of Δ_4^* is an ideal square, which is also principal hence the square of a principal ideal (since the class number of K_4 is odd). In symbols, $\Delta_4^* = \Delta_4\Lambda^2$. Since Δ_4 generates disc K_8/K_4, all the integers of K_4 easily lie in the module $[1, \sqrt{\Delta_4}]/2$ over the integers of K_2. Thus the unit

(18a) $$\Theta_8 = (X+Y\sqrt{\Delta_4})/2$$

satisfies (17), or

(18b) $$N_{8/4}\Theta_8 = \Theta_4.$$

To see the converse part, note that with Y odd and Θ_8 an integer, disc K_8/K_4 is odd. But it we examine the tower of $k_{2m} = K_m(\sqrt{-p})$ for m = 2, 4, 8 we see each stage is also generated by the adjunction of a unit, and in particular (see [3, §5]) $k_{16} = k_8(H_8)$ from (14c). Thus k_{16}/k_8 is also at most even ramified, and therefore unramified. Thus the unramified tower extends all the way from k_2 to k_{16}, and consequently $8|h$.

ILLUSTRATION

We take the first case where $8|h(-4p)$. The class number of K_4 is unity here:

<u>$p = 41$</u>

$$\varepsilon = 32+5\sqrt{41},\ \pi = 5-4i,\ \gamma = 1+2i,\ K_4 = \mathbb{Q}(\sqrt{\pi})$$

$$\Theta_4(=\Theta) = (2-i+\sqrt{\pi})/(1+i),\ (\alpha_0 = 2-i,\ \beta_0 = 1)$$

$$\Delta_4^* = \Theta_4((-11+6i)/2 - 5\sqrt{\pi}/2),\ \lambda = 1-i$$

$$\Lambda = i\Theta_4(1+\sqrt{\pi})/2,\ N_{4/2}\Lambda = \lambda$$

$$\Delta_4(=\Delta) = \Delta_4^*/\Lambda^2 = -2+2i+\sqrt{\pi},\ N_{4/2}\Delta_4 = \pi'(= 5+4i)$$

$$K_8 = K_4(\sqrt{\Delta_4})$$

$$\Theta_8 = i(1+\Theta_4\sqrt{\Delta_4})/2,\ N_{8/4}\Theta_8 = \Theta_4$$

pellian equations:

$$s^2-t^2 41 = -1 \quad \ldots \quad (11b)$$

$$\alpha_0^2-\beta_0^2(5-4i) = -2 \quad \ldots \quad (14c)$$

$$X^2-Y^2(-2+2i+\sqrt{\pi}) = 4\Theta_4 \quad \ldots \quad (9b)$$

$$(X = i,\ Y = i\Theta_4).$$

Further efforts at computation suggest that the problem is numerically very challenging. For $p = 257$, e.g., $16|h$, so here Θ_{16} is definable (with norm Θ_8) but is very hard to compute.

A REAL PELLIAN TOWER

Incidentally, a real analogue of the pellian tower does exist; it consists of the (unique) tower of real subfields of k_{2M}:

(18a) $$\mathbb{Q} = L_1 \subset \mathbb{Q}(\sqrt{p}) = L_2 \subset L_4 \subset \ldots \subset L_M$$

(18b) $$L_{2m} = L_m(B_{2m})$$

(18c) $$N_{2m/m}B_{2m} = -B_m, \quad 1 \le m \le M/2.$$

For a general formula for the B_m we refer to [3], but we list the first few cases,

(18d) $$B_2 = \varepsilon,\ B_4 = \sqrt{\varepsilon},\ B_8 = H_8/\Theta_8\Theta_8^c\ .$$

We are dealing this time with wild ramifications, which present a more difficult theory. Indeed, it is "not natural" to explain why $\sqrt{\varepsilon}$ is acceptable for B_4 while $\sqrt{\sqrt{\varepsilon}}$ is not acceptable for B_8, (which exists only when $8|h$)!

REFERENCES

1. Barrucand, P. and Cohn, H., Note on primes of type x^2+32y^2, class number and residuacity, J. reine angew. Math. 238(1969) 67-70.
2. Cohn, H., Cyclic-sixteen class fields for $\mathbb{Q}(-p)^{1/2}$ by modular arithmetic, Math. of Comp. 33(1979) 1307-1316.
3. Cohn, H., The explicit Hilbert 2-cyclic class field for $Q(-p)^{1/2}$, J. reine angew. Math. 321(1981) 64-77.
4. Dirichlet, P.G.L., Recherches sur les formes quadratiques à coéfficients et à indéterminées complexes, J. reine angew. Math. 24(1842) 291-371.
5. Lagarias, J.C., On the computational complexity of determining the solvability of the Diophantine equation $x^2-Dy^2 = -1$, Trans. Amer. Math. Soc. (to appear).
6. Lakein, R., Class number and units of complex quartz fields, in Computers in Number Theory, pp. 167-172, Academic Press, 1971.
7. Scholz, A., Über die Lösbarkeit der Gleichung $t^2-Du^2 = -4$, Math Zeit. 39 (1934) 93-111.

BEST DIOPHANTINE APPROXIMATIONS
FOR TERNARY LINEAR FORMS, II

T.W. Cusick

Dedicated to Emil Grosswald.

1. Introduction

We consider approximations to zero by the linear form $x + \alpha y + \beta z$, where x, y, z are integers and α, β are real irrational numbers. In this situation we say that $x_0 + y_0 + \beta z_0$ is a <u>best approximation</u> if

$$|x_0 + \alpha y_0 + \beta z_0| \le |x + \alpha y + \beta z|$$

for all integer triples x, y, z with $0 < \max(|y|, |z|) \le \max(|y_0|, |z_0|)$.

Very little is known about these best approximations. In a recent paper [2], I gave a method which enables one to calculate all of the best approximations to zero by a form $x + \alpha y + \beta z$, where $[1, \alpha, \beta]$ is a basis for a nontotally real cubic field F. In the present paper I use this method to construct two examples showing that the best approximations need not have very nice properties.

As in [2], we assume for simplicity that $[1, \alpha, \beta]$ is an <u>integral</u> basis for F. We let θ, $0 < \theta < 1$, denote the fundamental unit of F. If δ is any element of a real cubic field, then $\delta, \delta', \delta''$ denote the conjugates of δ; $N(\delta)$ denotes the norm $\delta\delta'\delta''$ of δ.

We say that a number $a_0 + \alpha a_1 + \beta a_2$ is a <u>better approximation</u> than $b_0 + \alpha b_1 + \beta b_2$ if both $|a_0 + \alpha a_1 + \beta a_2| < |b_0 + \alpha b_1 + \beta b_2|$ and $\max(|a_1|, |a_2|) \le \max(|b_1|, |b_2|)$ hold. Clearly $b_0 + \alpha b_1 + \beta b_2$ is a best approximation if and only if there do not exist any numbers $x + \alpha y + \beta z$ which are better approximations than $b_0 + \alpha b_1 + \beta b_2$.

The method of [2] is based on the fact that for any best approximation $\delta = x + \alpha y + \beta z$ (where $[1, \alpha, \beta]$ is a basis for a nontotally real cubic field), $N(\delta)$ is bounded by a constant depending only on α and β [2, formula (6)]. Given an integral basis $[1, \alpha, \beta]$ for a nontotally real cubic field F with fundamental unit θ, let B denote the set of all of the best approximations $x + \alpha y + \beta z$. For each positive integer k, let B_k denote the set of all best approximations δ with $|N(\delta)| \leq k$, and let S_k denote the set of all numbers $\delta = x + \alpha y + \beta z$ such that $|N(\delta)| \leq k$ and no number $x + \alpha y + \beta z$ which also satisfies this norm inequality is a better approximation than δ. Thus we have $S_k \supseteq B_k \supseteq B_{k-1}$ for each k. Since the norms of the best approximations are bounded, there exists an integer n such that $S_n = B_n = B$.

In [2] it is shown that the set B_1 is infinite. A simple conjecture one might make is that the set S_1 contains all of the units. Theorem 1 below shows this need not be true; it follows that all units need not be best approximations.

Theorem 1 involves the linear form $x + \gamma^2 y + (\gamma^2 - \gamma)z$, where $\gamma^3 - \gamma - 1 = 0$. The real cubic field F with discriminant -23 is generated by the real root $\gamma = 1.3247...$ of $x^3 - x - 1 = 0$. The fundamental unit θ, $0 < \theta < 1$, of F is given by $\theta = \gamma^{-1} = \gamma^2 - 1 = .7549...$, and $[1, \gamma, \gamma^2]$ is an integral basis for F. Thus $[1, \gamma^2, \gamma^2 - \gamma]$ is also an integral basis. We define a_m, b_m, c_m by

(1) $$\theta^m = a_m + \alpha b_m + \beta c_m$$

where $\alpha = \gamma^2$ and $\beta = \gamma^2 - \gamma$. Our result is:

<u>THEOREM 1.</u> The set S_1 for the linear form $x + \gamma^2 y + (\gamma^2 - \gamma)z$ is made up of those units (1) which satisfy $b_m c_m \neq 0$ and either

(2) $$b_m c_m > 0 \text{ and } |a_m| > \max(|b_m|, |c_m|)$$

or

(3) $$b_m c_m < 0,\ |a_m| > \max(|b_m|, |c_m|) \text{ and } |a_m + b_m + c_m| > |b_m| .$$

The conditions (2) and (3) are equivalent to

(4) $$b_m c_m > 0 \text{ and } \gamma - 1 < b_m/c_m$$

and

(5) $$b_m c_m < 0 \text{ and } \gamma^2(\gamma^2 - \gamma + 1)^{-1} < |b_m/c_m| < \gamma + 1 ,$$

respectively.

A second conjecture one can make, about the distribution of norms of best approximations, is that there exists a number M such that for every possible value $t \leq M$ of $|N(x + \alpha y + \beta z)|$, there is a best approximation with $|N(\delta)| = t$. Our next theorem disproves this conjecture by considering the linear form $x + \eta y + (\eta^2 - \eta)z$, where $\eta^3 = 2$. The real cubic field F with discriminant -108 is generated by the real root $\eta = 2^{1/3} = 1.2599...$ of $x^3 - 2 = 0$. The fundamental unit θ, $0 < \theta < 1$, of F is given by $\theta = \eta - 1 = .2599...$, and $[1, \eta, \eta^2]$ is an integral basis for F; thus $[1, \eta, \eta^2 - \eta]$ is also an integral basis. The values of the norm of the linear form include all positive integers ≤ 5, but there are no best approximations of norm ± 4, as is shown in the following theorem.

THEOREM 2. The best approximations to zero by $x + \eta y + (\eta^2 - \eta)z$ all have norm ≤ 5 in absolute value. There are no best approximations or norm ± 4. Let $\theta^m = a_m + \eta b_m + (\eta^2 - \eta)c_m$. The best approximations of norms 1, 2, 3, 5 are given by the sets:

$M_1 = \{\theta^m: m \geq 2\}$,

$M_2 = \{\theta^m \eta: m \geq 1, b_m c_m > 0 \text{ and } (\eta^2 - \eta + 1)(2 - \eta)^{-1} < b_m/c_m\}$,

$M_3 = \{\theta^m (1 + \eta): m \geq 1, b_m c_m > 0 \text{ and } (\eta^2 - \eta)(2 - \eta)^{-1} < b_m/c_m < 1\}$,

$M_5 = \{\theta^m(1 + \eta^2): m \geq 1, b_m c_m > 0 \text{ and } (-\eta^2 + \eta + 2)\eta^{-1} < b_m/c_m < 2\}$,

respectively.

REMARK. It is not hard to see that there are infinitely many units not satisfying the conditions of Theorem 1, and that there infinitely many elements in each of the sets M_2, M_3 and M_5 in Theorem 2. This follows from the fact that the numbers b_m/c_m in both theorems are dense in the real numbers.

2. Determining the set S_1

We define the integers a_m, b_m, c_m by equation (1) and let $V_m = \max(b_m^2, c_m^2)$. The subsequence $V_{d(1)}, V_{d(2)},...$ of successive minima of the sequence $V_1, V_2,...$

is defined as follows: We define d(1) to be the largest integer such that $V_{d(1)} \leq V_m$ for all $m \leq d(1)$. By induction, d(j) for $j \geq 2$ is defined to be the largest integer such that $V_{d(j)} \leq V_m$ for all m with $d(j - 1) < m < d(j)$. Thus $\theta^{d(1)} > \theta^{d(2)} > \ldots$ are the members of S_1.

If the sequence $V_1, V_2, \ldots$ is monotone increasing, then S_1 is simply the set of all θ^m, $m \geq 1$. This is the situation in the example of [2, Theorem 3], for instance. If the V_i are not monotone increasing, then we must compute the successive minima. The proof of Theorem 1 illustrates how this is done.

The following lemma is the key step in the proof of Theorem 1. The point is that the differences d(j) - d(j - 1) are bounded.

LEMMA 1. For the linear form of Theorem 1, the integers V_i satisfy $V_k > V_m$ for all $m \geq 1$ and all $k \geq m + 4$.

PROOF. The lemma will follow if we prove

(6) $$V_{m+j} > V_m \text{ for } j = 4, 5, 6, 7 \text{ and all } m > 1.$$

To prove this, it is convenient to have expressions for the coefficients of θ^{m+j} (j = 4, 5, 6, 7) in terms of the coefficients of θ^m; these are given in Table 1 (with the subscripts m omitted, so $\theta^m = a + b\gamma^2 + c(\gamma^2 - \gamma)$).

j	1	γ^2	$\gamma^2 - \gamma$
0	a	b	c
1	-a - c	a + b + c	-b - c
2	a + b + 2c	-c	-a
3	-b - 2c	b + c	a + c
4	-a + b + c	a	-a - b - 2c
5	2a + c	-a - c	b + 2c
6	-2a - b - 3c	a + b + 2c	a - b - c
7	a + 2b + 4c	-b - 2c	-2a - c

TABLE 1. Coefficients of θ^{m+j}

According to Table 1, the case $j = 4$ of (6) follows at once if $|a| > \max(|b|, |c|)$. Hence we may assume that

$$|a| \leq \max(|b|, |c|) \quad . \tag{7}$$

We have

$$a \approx \gamma^2 b - (\gamma^2 - \gamma)c \tag{8}$$

and $\gamma^2 = 1.75\ldots$, $\gamma^2 - \gamma = .43\ldots$. Hence if $bc \geq 0$, then (7) and (8) imply that $\max(|b|, |c|) = |c|$, and in fact that

$$\left|\frac{c}{b}\right| \geq \frac{\gamma^2}{1 - (\gamma^2 - \gamma)} > 3 \quad .$$

Now it follows from

$$|-a - b - 2c| \approx |(\gamma^2 - 1)b + (\gamma^2 - \gamma - 2)c|$$

that we have

$$|-a - b - 2c| > \max(|b|, |c|)$$

when $bc \geq 0$. Similar calculations using (7) and (8) show that the above inequality also holds if $bc < 0$. By Table 1, this proves the case $j = 4$ of (6).

The cases $j = 5, 6, 7$ of (6) are proved by similar calculations, making use of (8) and Table 1. Similar calculations are given in more detail in [2, Sections 3 to 5].

PROOF OF THEOREM 1. By Lemma 1, the unit θ^m belongs to S_1 if and only if none of θ^{m+j} with $j = 1, 2$ or 3 is a better approximation than θ^m. By using (8) and Table 1, it is easy to describe the conditions under which one of these units θ^{m+j} is a better approximation than θ^m.

For example, θ^{m+1} is better than θ^m if and only if

$$\max(|a + b + c|, |b + c|) \leq \max(|b|, |c|).$$

This inequality is never true if $bc > 0$, but is always true if $b = 0$ or $c = 0$. If

$bc < 0$, then calculation shows that the inequality is true if and only if

$$\gamma + 1 < |b/c| \quad \text{or} \quad |b/c| < -\gamma^2 + \gamma + 1,$$

that is

$$(9) \qquad |a + b + c| \le \max(|b|,|c|) \quad .$$

Similarly, we find that θ^{m+2} is better than θ^m if and only if

$$(10) \qquad |a| \le \max(|b|,|c|)$$

and θ^{m+3} is better than θ^m if and only if

$$(11) \qquad bc < 0 \text{ and } |a + c| \le \max(|b|,|c|) \quad .$$

By (10), we see that if $bc > 0$, then (2) is necessary and sufficient for θ^m to be in S_1. If $bc < 0$, then combining (9), (10) and (11) gives the condition (3) for θ^m to be in S_1. Calculation shows that (2) and (3) are equivalent to (4) and (5), respectively. This completes the proof of Theorem 1.

3. PROOF OF THEOREM 2.

The proof of Theorem 2 follows exactly the same lines as that of [2, Theorem 3]. The only point where particular care in the calculations is required is in the proof that there are no best approximations of norm $\pm$ 4. All values of the linear form with norm 4 have the shape

$$\theta^m\eta^2 = 2(b_m - c_m) + \eta(a_m + 2c_m) + (\eta^2 - \eta)a_m$$

and we have $\theta^{m-1} > \theta^m\eta^2 > \theta^m$. Thus it suffices to show that θ^m is a better approximation than $\theta^m\eta^2$, that is

$$(12) \qquad \max(|b_m|,|c_m|) \le \max(|a_m + 2c_m|,|a_m|) \quad .$$

If $b_mc_m < 0$, then $|a_m + 2c_m|$ exceeds the left hand side of (12). However, if $b_mc_m > 0$, then (12) is a sharp inequality in the sense $|a_m| \ge \max(|b_m,c_m|)$ if and only if

$$(13) \qquad |b_m/c_m| \ge (-\eta^2 + \eta + 1)\eta^{-1} \pm \theta^m(\eta|c_m|)^{-1}$$

and $|a_m + 2c_m| \geq \max(|b_m|,|c_m|)$ if and only if the inequality sign in (13) is reversed. Here the upper sign in the $\pm$ holds if $c_m > 0$, and the lower sign holds if $c_m < 0$. In this calculation we have used the equality

$$a_m = \theta^m - \eta b_m - (\eta^2 - \eta)c_m$$

instead of neglecting the small term θ^m, which can always be safely done except in cases like (12) where a delicate inequality is involved.

4. Applications and open questions.

It is well known that the simple continued fraction algorithm provides all the best approximations to zero by a linear form $x + \alpha y$. Much effort has been expended on the search for an algorithm which will furnish best approximations in higher dimensions in some reasonably simple way (see Cusick [1,2] and Szekeres [3] for some references and discussion). The method of [2] can be used to provide lists of known best approximations against which the efficacy of any proposed "higher dimensional continued fraction" can be tested (in dimension 2).

In particular, the algorithm proposed by Szekeres [3] can be tested against various examples in dimension 2. Szekeres [3, p.117] conjectured that his algorithm would always give all of the best approximations. This is certainly false, because in Theorem 2 above the best approximation $\theta^2 = \eta^2 - \eta - (\eta - 1)$ is omitted by the Szekeres algorithm applied to the linear form $x + \eta y + (\eta^2 - \eta)z$.

One can weaken the conjecture to assert that all but finitely many best approximations are produced by the Szekeres algorithm. However, I believe that this weaker conjecture is also false, and that a counterexample could be found using the method of [2].

I conclude with two other open questions: Can one find a linear form such that infinitely many members of S_1 are not in B? Do the sets $\{m: \theta^m \text{ is in } S_1\}$ and $\{m: \theta^m \text{ is in B}\}$ have an asymptotic density?

REFERENCES

1. T.W. Cusick, The Szekeres multidimensional continued fraction, Math. Comput.

31 (1977), 280-317.

2. T.W. Cusick, Best Diophantine approximations for ternary linear forms, J. reine angew. Math. 315 (1980), 40-52.
3. G. Szekeres, Multidimensional continued fractions, Ann. Univ. Sci. Budapest Eötvös Sect. Math. 13 (1970), 113-140.

STATE UNIVERSITY OF NEW YORK AT BUFFALO

CONSTRUCTIVE ELEMENTARY ESTIMATES FOR M(x)

Harold G. Diamond and Kevin S. McCurley

Dedicated to Professor Emil Grosswald on the occasion of his retirement.

1. Introduction.

Let $M(x)$ denote the summatory function of the Möbius μ-function. The estimate $M(x) = o(x)$ is a familiar result "equivalent" to the prime number theorem (cf. [3], [4], [6]). Here we shall describe a systematic constructive method of estimating $|M|$ and give two examples and a theoretical result.

The first example involves a rather simple calculation and gives the estimate

$$\limsup_{x \to \infty} |M(x)|/x < 1/32; \tag{1}$$

the other one requires far more calculation and yields

$$\limsup_{x \to \infty} |M(x)|/x < 1/105. \tag{2}$$

We shall also show that for any positive ε our method applied with a finite amount of computation (which increases with $1/\varepsilon$) can be used to establish the bound

$$\limsup_{x \to \infty} |M(x)|/x < \varepsilon.$$

Our method is inspired by Chebyshev's use of approximate Möbius inversion for elementary estimates of the prime counting function $\pi(x)$ [1]. The result (2) improves somewhat the elementary bound 1/80 of MacLeod [5]. Our method could be adapted to improve the prime counting estimates of Chebyshev and Sylvester [7], but it is most naturally suited to estimating $|M(x)|$.

In [2] the analogous problem of estimating $\pi(x)$ was treated. The methods used in the two articles differ principally in the way the approximate Möbius

* Research supported in part by a grant from the National Science Foundation.

inversion is carried out. We shall note these differences later. While such elementary estimates are of course superseded by the prime number theorem and the bound $M(x) = o(x)$, we wanted to see how far one can get with arguments akin to Chebyshev's.

Our starting point is the formula

$$(3) \qquad E(x) \stackrel{\Delta}{=} \sum_{n \leq x} \mu(n)\left[\frac{x}{n}\right] = \begin{cases} 1, & \text{if } x \geq 1 \\ 0, & \text{if } 0 \leq x < 1. \end{cases}$$

This follows from the Möbius inversion formula

$$\sum_{d|n} \mu(d) = \begin{cases} 1 & \text{if } n = 1 \\ 0 & \text{if } n = 2,3,\ldots \end{cases}$$

upon summing over all positive integers $n \leq x$ and reordering the sum.

We shall construct a periodic function

$$(4) \qquad F(x) \stackrel{\Delta}{=} c_1[x/a_1] + \ldots + c_r[x/a_r]$$

which is "usually" equal to $E(x)$ for $x \geq 1$. We combine this function with the relation

$$(5) \qquad |M(x)| \leq \sum_{n \leq x} \left|1-F\left(\frac{x}{n}\right)\right| |\mu(n)| + O(1),$$

which will be shown below, to give our estimate of $|M(x)|$.

We shall choose certain <u>rational</u> numbers for the sequence $a_1, a_2, \ldots, a_r$ in such a way that they approximate the first several square free numbers quite accurately. This will allow us to approximate $E(x)$ quite accurately and systematically by $F(x)$.

2. Construction of a "nearly constant" periodic function.

Let K be a positive integer and let $1 = a_1 < a_2 < \ldots < a_r$ be rational numbers for which

$$(6) \qquad K/a_i \in \mathbb{Z}, \quad 1 \leq i \leq r,$$

and let $c_1, c_2, \ldots, c_r$ be real numbers such that

(7) $$c_1/a_1 + c_2/a_2 + \ldots + c_r/a_r = 0.$$

LEMMA 1. *Let* $F(x)$ *be defined on* R *by* (4) *with* $K, a_1, \ldots, a_r$, *and* $c_1, \ldots, c_r$ *satisfying* (6) *and* (7). *Then* F *has period* K.

PROOF. We have

$$F(x+K) = \sum_{i=1}^{r} c_i\left[\frac{x+K}{a_i}\right] = \sum_{i=1}^{r} c_i\left[\frac{x}{a_i}\right] + K\sum_{i=1}^{r} \frac{c_i}{a_i} = F(x).$$

In our examples the period K will be chosen as a multiple of the first several primes, so that we can use those primes and their products as a_i's. Beyond that we will choose K so that several of the subsequent a_i's will be very near primes. We shall discuss the construction of F further after considering the following example.

EXAMPLE. Suppose we choose $K = 30$. Then 2,3, and 5 divide K, and moreover we have $30/4 = 7\frac{1}{2} \approx 7$. We take

(8) $$F_1(x) = [x] - [x/2] - [x/3] - [x/5] + [x/6] - [4x/30].$$

This function satisfies (6) and (7), so it has period 30. Also $F_1(x) = 1$ for $0 \leq x < 30$ except for the intervals $[7, 7\frac{1}{2})$, $[13,15)$, $[19,20)$, and $[29,30)$, where it has the value 2 and the intervals $[0,1)$, $[10,11)$, $[15,17)$, and $[22\frac{1}{2}, 23)$, where its value is zero.

One might ask why the terms $[x/10]$, $[x/15]$, and $[x/30]$ did not appear in $F_1(x)$. We did not include these terms because the sum of the Möbius function near 10,15, and 30 is zero, or "nearly so." That is,

$$\mu(10) + \mu(11) = 0,$$

$$\mu(13) + \mu(14) + \mu(15) + \mu(17) = 0,$$

and

$$\mu(22) + \mu(23) + \mu(26) + \mu(29) + \mu(30) = -1.$$

(We might have included a term $-[x/30]$, except that (7) was satisfied the way F_1 was chosen.)

3. Application to estimate $M(x)$.

Let $F(x)$ satisfy (4), where as usual $1 = a_1 < a_2 < \ldots < a_r \leq K$. Möbius inversion gives

$$\sum_{n \leq x} F(\tfrac{x}{n})\mu(n) = \sum_{i=1}^{r} c_i \sum_{n \leq x/a_i} [\tfrac{x}{na_i}]\, \mu(n) = \sum_{a_i \leq x} c_i .$$

Thus, if $x \geq K$ we have

$$M(x) = \sum_{i=1}^{r} c_i + \sum_{n \leq x} (1-F(\tfrac{x}{n}))\mu(n).$$

This establishes relation (5). We use this relation to prove

LEMMA 2.

(9) $$\limsup_{x \to \infty} \frac{|M(x)|}{x} \leq \frac{6}{\pi^2} \int_1^{\infty} \frac{|1-F(u)|}{u^2}\, du.$$

PROOF. Let $B_1 = |c_1+\ldots+c_r|$,

$$B_2 = \sup_{u \in \mathbb{R}} |1-F(u)|,$$

and take A to be a large number. We have

$$|M(x)| \leq B_1+B_2\, x/A + \sum_{x/A < n \leq x} |1-F(\tfrac{x}{n})||\mu(n)|.$$

Now let $1 = b_1 < b_2 < \ldots$ be the points at which $F(x)$ jumps, arranged in ascending order. We may assume that A is one of the b_i. If we organize together those terms n in the last sum for which x/n lies in an interval $[b_i, b_{i+1})$, we obtain the estimate

$$|M(x)| \leq B_1+B_2\, x/A+ \sum_{b_i < A} |1-F(b_i)| \sum_{\frac{x}{b_{i+1}} < n \leq \frac{x}{b_i}} |\mu(n)|.$$

The elementary estimate of the counting function of square free numbers implies that the last sum is asymptotic to $(6/\pi^2)(x/b_i - x/b_{i+1})$ as $x \to \infty$ (for fixed A). Thus we have

$$|M(x)| \le \frac{6x}{\pi^2} \sum_{b_i < A} |1-F(b_i)| \left(\frac{1}{b_i} - \frac{1}{b_{i+1}}\right) + B_1 + B_2\, x/A + \sum_{b_i < A} B_2 o(x)$$

or

$$\limsup_{x \to \infty} \frac{|M(x)|}{x} \le \frac{6}{\pi^2} \sum_{b_i < A} |1-F(b_i)| \int_{b_i}^{b_{i+1}} \frac{du}{u^2} + \frac{B_2}{A}$$

$$\le \frac{6}{\pi^2} \int_1^A |1-F(u)| \frac{du}{u^2} + \frac{B_2}{A} .$$

Since A is arbitrary, the lemma is proved.

For the function F_1 defined by (8) we estimate the right side of (9) by making a 5¢ computer calculation over 100 periods and a separate estimate for the tail. For the mth period of the tail we use the facts that $|1-F_1(u)| \le 1$ and length of the intervals in $[0,30)$ for which $1-F(u) \neq 0$ is 9. Thus

$$\int_{30m}^{30(m+1)} |1-F(u)| \frac{du}{u^2} \le \frac{9}{(30m)^2} = \frac{1}{100\, m^2} .$$

It follows that

$$\int_{3000}^{\infty} |1-F(u)| \frac{du}{u^2} \le \sum_{m=100}^{\infty} \frac{1}{100\, m^2} < \frac{1}{100} \int_{99}^{\infty} \frac{du}{u^2} = \frac{1}{9900} .$$

Combining the two parts we obtain

$$\limsup_{x \to \infty} \frac{|M(x)|}{x} < \frac{6}{\pi^2} (.051237 + .000102) < .03122 < \frac{1}{32} .$$

4. More accurate approximations.

The following identity (cf. [7], p. 14, footnote) enables us to use the values of F on the interval (0,K/2) to evaluate F on (K/2,K). This will be convenient for treating functions F having a large period.

LEMMA 3. *Let* $K, a_1,\dots,a_r, c_1,\dots,c_r$, *and* $F(x)$ *be as in Lemma 1. Then*

$$F(x) + F(K-x) = -(c_1+\dots+c_r)$$

for all x *such that* $x/a_i \notin \mathbb{Z}$ *for* $1 \le i \le r$.

PROOF. By periodicity $F(K-x) = F(-x)$. Also $[y] + [-y] = -1$ for $y \notin \mathbb{Z}$. Thus we have

$$F(x) + F(K-x) = F(x)+F(-x) = \sum_{i=1}^{r} c_i([\tfrac{x}{a_i}] + [\tfrac{-x}{a_i}]) = -\sum_{i=1}^{r} c_i.$$

In the preceding example $-\sum c_i = 2$. Thus F_1 has an average value of 1 and the intervals where F_1 equals 0 pair up with the intervals where it is equal to 2. (In the argument of Chebyshev the function

$$F_0(x) = [x] - [x/2] - [x/3] - [x/5] + [x/30]$$

was used. In this case $-\sum c_i = 1$ and F_0 assumes the values 1 or 0. Consideration of Chebyshev's argument shows that this function is very good for the purpose, because it is equal to 1 for $1 \le x < 6$ and is frequently zero thereafter.)

Now we shall construct a function $F_2(x)$ having the form (4) which more closely approximates $E(x)$ than F_1 did, and we shall apply (9) to estimate $|M(x)|/x$. The calculations here are rather extensive and were done mainly by computer. The most costly operation was sorting the points at which F_2 has discontinuities.

We take for the period $K = 4200$. Since

$$4200 \equiv 0 \pmod{m}, \quad m = 2,3,5,7,$$

we can take 2,3,5, and 7 as values of a_i. Moreover,

$$4200 \equiv -2 \pmod{11}$$
$$4200 \equiv 1 \pmod{m}, \quad m = 13,17,19,$$

and we may take as "approximate primes" the numbers

$$\frac{4200}{382} \doteq 10.995,\ \frac{4200}{323} \doteq 13.003,\ \frac{4200}{247} = 17.004,\ \frac{4200}{221} \doteq 19.005.$$

Subsequent "approximate primes", such as $4200/183 \doteq 22.951$ are generally not as close to their exact counterparts, but the influence of these numbers is less than that of the initial ones.

Introducing a slight change in notation, we take

$$F_2(x) = \sum_{i=1}^{4200} c_i \left[\frac{ix}{4200}\right],$$

where all c_i are zero except for the values given in the following table:

i	4200	2100	1400	840	700	600	420	382	323	300
c_i	1	-1	-1	-1	1	-1	1	-1	-1	1

i	280	247	221	200	191	183	162	145	140	135	127	124	120
c_i	1	-1	-1	1	1	-1	1	-1	-1	-1	1	1	1

i	114	111	108	102	100	98	91	89	82	79	76	74	72	71	69
c_i	-1	1	1	-1	-1	-1	1	-1	1	-1	1	1	1	-1	-1

i	68	65	64	63	61	60	59	58	57	55	54	53	49	48	47	46
c_i	1	1	-1	-1	1	-1	-1	-1	1	1	-1	-1	2	1	-1	1

i	45	44	43	42	41	40	39	37	36	35	34	33	32	31	29	28	26	25
c_i	2	1	-1	-1	-2	-1	-1	-2	2	1	2	-1	1	-1	3	-2	1	-2

i	24	22	21	20	18	16	15	14	13	12	10	9	8	7	6	5	4	3
c_i	-2	-2	2	6	-3	-2	-4	5	3	-2	-4	1	7	-8	3	1	-2	1

The nonzero values c_i for $53 \leq i \leq 4200$ were chosen to make $c_i = \mu(k)$, where k is the nearest integer to $4200/i$. For $i < 53$ there is generally more than one square free integer near $4200/i$. For a preliminary trial we grouped the integers near each number $4200/i$ into blocks $B(i)$ and chose $c_i = \sum_{\ell \in B(i)} \mu(\ell)$. We computed the resulting function $F_2^*(x)$ and then adjusted some of the c_i with small indices to satisfy the conditions $\sum ic_i = 0$ and $\sum c_i = -2$ and to improve the proximity of $F_2(x)$ to 1 in certain large ranges of x. The largest i for which c_i needed adjustment from our initial guess was $i = 40$.

A computer calculation of F_2 over ten period, with a crude error analysis, shows that

$$\int_1^{42000} |1-F(u)| \frac{du}{u^2} < .015584.$$

For the tail we use the estimate

$$\int_{4200k}^{4200(k+1)} |1-F(u)| \frac{du}{u^2} \leq \frac{1}{(4200k)^2} \int_0^{4200} |1-F(u)|du.$$

The last integral is less than 12186, as shown by a computer calculation. Thus

$$\int_{42000}^{\infty} \frac{|1-F(u)|}{u^2} du \leq \frac{12186}{(4200)^2} \sum_{k=10}^{\infty} \frac{1}{k^2} < 7.7 \times 10^{-5},$$

and so

$$\limsup_{x \to \infty} \frac{|M(x)|}{x} \leq \frac{6}{\pi^2} (.015661) < .009521 < 1/105.$$

5. <u>Theoretical estimates</u>.

In this section we describe the construction of a periodic function F of the type (4) which approximates (1) with desired accuracy.

<u>THEOREM</u>. <u>Given</u> $\varepsilon > 0$ <u>there are positive integers</u> K <u>and</u> r <u>and real</u> r <u>tuples</u> $a_1, \ldots, a_r$, $c_1, \ldots, c_r$ <u>satisfying</u> (6) <u>and</u> (7) <u>and a function</u> F <u>of period</u> K <u>satisfying</u> (4) <u>such that</u>

$$\int_1^{\infty} |1-F(u)|u^{-2}du < \varepsilon. \tag{10}$$

<u>This estimate can be verified with a finite number of computations</u>.

Taken together, (9) and (10) assure that Chebyshev type methods are capable in principle of producing an estimate

$$\limsup_{x \to \infty} |M(x)|/x < \varepsilon$$

for any $\varepsilon > 0$. However, the guarantee that (10) is attainable appears to depend in turn on the estimate $M(x) = o(x)$, so our argument does not offer a new proof of this result.

LEMMA 4. Let $m(x) = \sum_{n \leq x} \frac{\mu(n)}{n}$. Then $m(x)$ changes sign infinitely often.

PROOF. The Mellin transform

$$\int_1^\infty x^{-s} m(x)dx = 1/\{(s-1)\zeta(s)\}$$

converges for $\sigma = \mathrm{Re}\ s > 1$ since m is bounded.

If m were always of one sign, then the transform would converge for $\sigma > -2$ by Landau's theorem, since $1/\{(s-1)\zeta(s)\}$ has no real singularities to the right of $s = -2$. It would then follow that $1/\{(s-1)\zeta(s)\}$ would be analytic in the half plane $\sigma > -2$. This is impossible since $\zeta(s)$ has an infinite number of zeros in the strip $0 < \sigma < 1$.

COROLLARY. There exists a sequence of integers $Q \to \infty$ such that

$$|m(Q-1)| < \frac{1}{2Q} + \frac{1}{Q^2} .$$

PROOF. By the preceding lemma there exist a sequence of integers k satisfying $m(k)m(k+1) \leq 0$. Then $|m(k)| \leq 1/(2k)$ or $|m(k-1)| \leq 1/(2k)$, for otherwise

$$\frac{|\mu(k)|}{k} = |m(k)-m(k-1)| > \frac{1}{k} .$$

If $|m(k)| \leq |m(k-1)|$, then take $Q = k+1$ and we have

$$|m(Q-1)| \leq \frac{1}{2Q-2} < \frac{1}{2Q} + \frac{1}{Q^2} .$$

If $|m(k)| > |m(k-1)|$, then take $Q = k$ and obtain a slightly better estimate.

Let R be a large integer and Q a larger integer satisfying the preceding corollary and such that the bound $|M(y)| < y/R$ holds for $Q \leq y \leq RQ$. The existence of numbers Q satisfying the last condition is guaranteed by the prime number theorem; for given R the relation may be verified for a suitable Q in a finite number of calculations. Let $K = R^2Q^2$, and then define

$$F(x) = \sum_{n \le Q-1} \mu(n)[[\tfrac{K}{n}] \tfrac{x}{K}] + C(Q)[\tfrac{x}{Q}] \tag{11}$$

where $C(Q)$ has been chosen so that $F(K) = 0$, i.e.

$$\sum_{n \le Q-1} \mu(n)[\tfrac{K}{n}] + C(Q)R^2Q = 0. \tag{12}$$

<u>LEMMA 5</u>. <u>If</u> $R > 2$ <u>then</u> $|C(Q)| < 1$.

<u>PROOF</u>. We have

$$K \sum_{n \le Q-1} \frac{\mu(n)}{n} = \theta(\frac{K}{2Q} + \frac{K}{Q^2}), \text{ where } |\theta| \le 1.$$

Subtracting this relation from (12) we obtain

$$\sum_{n \le Q-1} \mu(n)(\tfrac{K}{n} - [\tfrac{K}{n}]) - C(Q)R^2Q = \theta(\frac{R^2Q}{2} + R^2).$$

Thus

$$C(Q)R^2Q = \sum_{n \le Q-1} \mu(n) \{\tfrac{K}{n}\} - (\frac{R^2Q}{2} + R^2)\theta,$$

where $\{y\} = y-[y]$, and so

$$|C(Q)| \le \frac{1}{2} + \frac{1}{Q} + \frac{1}{R^2} < 1.$$

<u>LEMMA 6</u>. <u>If</u> $F(x)$ <u>is defined as in</u> (11), <u>then</u> $|F(x)| \le Q-1$ <u>for all</u> x.

<u>PROOF</u>. Multiplying (12) by x/K and subtracting from (11) we obtain the representation

$$F(x) = -\sum_{n \le Q-1} \mu(n) \{[\tfrac{K}{n}] \tfrac{x}{K}\} - C(Q) \{\tfrac{x}{Q}\}.$$

Thus

$$|F(x)| \le \sum_{n \le Q-1} |\mu(n)| + |C(Q)| \le Q-1 \text{ if } Q \ge 5.$$

We now show that $\int_1^\infty |1-F(u)| \frac{du}{u^2}$ can be made arbitrarily small by choosing R and Q large. Write

$$\int_1^\infty = \int_1^{\sqrt{K}} + \int_{\sqrt{K}}^\infty = I_1 + I_2.$$

We treat I_2 first. Since $|1-F(u)| \leq Q$,

$$I_2 \leq \int_{\sqrt{K}}^\infty \frac{Q}{u^2}\, du \leq \frac{Q}{\sqrt{K}} = \frac{1}{R},$$

which is small for R sufficiently large.

For $x \geq 1$ we have from (3) the representation $1 = \sum_{n \leq x} \mu(n)[\frac{x}{n}]$. Thus for $1 \leq x \leq \sqrt{K}$ we have

$$|F(x)-1| \leq |\sum_{n \leq Q-1} \mu(n)([[\tfrac{K}{n}]\tfrac{x}{K}] - [\tfrac{x}{n}])| + |C(Q)|\,[\tfrac{x}{Q}]$$

$$+ |\sum_{Q \leq n \leq x} \mu(n)\,[\tfrac{x}{n}]|.$$

Note that only the first sum appears if $x < Q$. Thus

$$I_1 \leq \int_1^{\sqrt{K}} \sum_{n \leq Q-1} ([\tfrac{x}{n}] - [[\tfrac{K}{n}]\tfrac{x}{K}])\frac{dx}{x^2}$$

$$+ \int_Q^{\sqrt{K}} \frac{|C(Q)|}{Q}\frac{dx}{x} + \int_Q^{\sqrt{K}} |\sum_{Q \leq n \leq x} \mu(n)\,[\tfrac{x}{n}]|\,\frac{dx}{x^2}$$

$$= I_3 + I_4 + I_5, \text{ say.}$$

For $1 \leq x \leq \sqrt{K}$ we have

$$\frac{x}{n} - [\tfrac{K}{n}]\frac{x}{K} = \frac{x}{K}\{\tfrac{K}{n}\} \leq \frac{1}{\sqrt{K}} < 1,$$

so $[\frac{x}{n}] - [[\frac{K}{n}]\frac{x}{K}]$ is either zero or one. Furthermore the last expression equals one precisely when there is a positive integer ℓ such that $\frac{x}{n} \geq \ell > [\frac{K}{n}]\frac{x}{K}$, or $\ell n \leq x < \frac{\ell K}{[K/n]}$. Thus we have

$$I_3 \le \sum_{n \le Q} \sum_{\ell \le \sqrt{K}/n} \int_{\ell n}^{\ell K/[K/n]} \frac{dx}{x^2}$$

$$= \frac{1}{K} \sum_{n \le Q} \sum_{\ell \le \sqrt{K}/n} \frac{1}{\ell} \{\frac{K}{n}\} \le \frac{1}{K} \sum_{n \le Q} \sum_{\ell \le \sqrt{K}/n} \frac{1}{\ell}$$

$$\le \frac{Q}{K} \sum_{\ell \le QR} \frac{1}{\ell} \le \frac{1+\log QR}{R^2 Q},$$

which is small for R large.

We estimate I_4 with the aid of Lemma 5. We obtain

$$I_4 \le \frac{1}{Q} \int_Q^{\sqrt{K}} \frac{dx}{x} = \frac{\log R}{Q} < \frac{\log R}{R}; .$$

which is small if R is large.

It remains then to estimate I_5, and this is the only point at which we must rely on the prime number theorem. We have

$$\sum_{Q \le n \le x} \mu(n)[\frac{x}{n}] = \sum_{Q \le n \le x} \mu(n) \sum_{m \le x/n} 1 = \sum_{m \le x/Q} \sum_{Q < n \le x/m} \mu(n)$$

$$= \sum_{m \le x/Q} (M(\frac{x}{m})-M(Q)) = M(x)+M(\frac{x}{2})+\ldots+M(\frac{x}{[x/Q]}) - [\frac{x}{q}]M(Q).$$

Therefore our assumption that $|M(y)| \le y/R$ on the finite range $Q \le y \le \sqrt{K}$ yields

$$|\sum_{Q < n \le x} \mu(n)[\frac{x}{n}]| \le \frac{x}{R} (1+ \frac{1}{2} + \frac{1}{3} +\ldots+ \frac{1}{[x/Q]}) + \frac{x}{R}$$

$$\le \frac{x}{R} (2+ \frac{1}{2} + \frac{1}{3} +\ldots+ \frac{1}{R}) \le \frac{x}{R} (2+\log R),$$

and so

$$I_5 \le \int_Q^{\sqrt{K}} (\frac{2+\log R}{R}) \frac{dx}{x} = (\frac{2+\log R}{R}) \log R$$

which can be made small for R large.

Thus we have shown that, for any given positive ε, knowledge of the Möbius function on a finite interval can be used to prove that

$$\limsup_{x \to \infty} |M(x)|/x < \varepsilon.$$

6. Comparison with other methods.

We compare the present method with that of Diamond-Erdös [2] and with elementary proofs of the prime number theorem (P.N.T.). Also we shall discuss some connections with the Riemann zeta function.

Recall that we have constructed a function F of period K such that $F(x)$ is usually 1 for $1 \leq x \leq \sqrt{K}$. For the region $x > \sqrt{K}$ we have an easily established absolute estimate $|F(x)| < \sqrt{K}/R$, R large. An appeal was made to the P.N.T. to show that F(x) is close to 1 on average for $\sqrt{K}/R \leq x \leq \sqrt{K}$.

In the Diamond-Erdös article a function G is constructed which plays an analogous role to our F. For T a suitable large parameter, $G(x) = 1$ for $1 \leq x < T$. G has period K equal to the least common multiple of $1,2,\dots,T-1$, i.e. $K = \exp(T+o(T))$, which is huge compared with the parameters of F. The P.N.T. is used to obtain an absolute bound on G.

The present method and that of [2] involve a one step operation. The behavior of M on a fixed bounded interval is used to give estimates of $M(x)/x$ or $\psi(x)/x-1$ for all large values of x. The method in each case leads to a somewhat worse estimate for $M(x)/x$ or $\psi(x)/x-1$ for large values of x than what one had in the initial range, so further iteration is not useful.

In contrast, elementary proofs of the P.N.T. use the Selberg formula (or some analogue), which involves expressions of greater order of magnitude than ψ or M. This is helpful for making an infinite recursion which leads to successively smaller estimates of $M(x)/x$ or $\psi(x)/x-1$.

Finally, we note some connections between the functions occurring in our method and the Riemann zeta function. If we form the Mellin transforms of the functions E and F defined by (3) and (4), we obtain for $\sigma = \text{Re } s > 1$

$$1 = \int_0^\infty x^{-s}\, dE(x) = (1/\zeta(s))\zeta(s),$$

$$\int_0^\infty x^{-s}\, dF(x) = s \int_0^\infty x^{-s-1} \sum_{n \le r} c_n[\frac{x}{a_n}]dx =$$

$$(\sum_{n \le r} c_n a_n^{-s})\ (s \int_0^\infty u^{-s-1}\,[u]du) = P(s)\zeta(s).$$

Since F is constructed to emulate E, the polynomial P should be an approximation to $1/\zeta$. We show a few aspects of this behavior. The condition (7) can be restated as $P(1) = 1/\zeta(1)$. We had found it convenient in §4 to require that $\sum c_i = -2$; this is the condition $P(0) = 1/\zeta(0)$. The following table compares a few values of P_1, P_2, and $1/\zeta$, where P_1 and P_2 are the polynomials associated with the functions F_1 and F_2.

s	1/2	2	1/2 + 14.1347...i
$1/\zeta(s)$	-.6848	.607927	∞
$P_1(s)$	-.6886	.608889	3.310 + .615i
$P_2(s)$	-.6865	.607932	8.724 - .806i

REFERENCES

1. P.L. Chebyshev, Mémoire sur les nombres premiers, J. Math. Pure Appl. 17 (1852), 366-390. Also appears in Mémoires présentés à l'Académie Imperial des Sciences de St.-Pétersbourg par divers Savants et lus dans ses Assemblées 7 (1854), 15-30. Also in Oeuvres, V. 1 (1899), 49-70.

2. H.G. Diamond and P. Erdös, On sharp elementary prime number estimates, to appear in l'Enseignement Math.

3. W.J. Ellison in collaboration with M. Mendès France, Les nombres premiers, Herrmann, Paris, 1975.

4. E. Landau, Handbuch der Lehre von der Verteilung der Primzahlen, Band 2, Teubner, Leipzig, 1909. Reprinted with an appendix by P.T. Bateman, Chelsea, New York, 1953.

5. R.A. MacLeod, A new estimate for the sum $M(x) = \sum_{n \leq x} \mu(n)$, Acta Arith. 13 (1967), 49-59, Errata, Ibid. 16 (1969), 99-100.

6. L. Schoenfeld, An improved estimate for the summatory function of the Möbius function, Acta Arith. 15 (1968/69), 221-233.

7. J.J. Sylvester, On arithmetical series, II, Messenger of Math (2) 21 (1892), 87-120.

ON THE SECOND LARGEST PRIME DIVISOR OF AN ODD PERFECT NUMBER

Peter Hagis, Jr.
Temple University, Philadelphia, Pa. 19122

With gratitude, respect and affection this paper is dedicated to Emil Grosswald, my teacher, my colleague and my friend.

1. Introduction.

A positive integer n is said to be perfect if n is equal to the sum of its proper divisors, so that

(1) $$\sigma(n) = 2n$$

where σ is the familiar divisor sum function. Over a period of time spanning more than two thousand years only twenty-seven perfect numbers have been found, all of them even. The largest of these, $(2^p-1)2^{p-1}$ where $p = 44497$, was discovered in 1979 [8]. While no one knows whether or not any odd perfect numbers exist their properties have been the subject of many papers. Suppose that n is an odd perfect number and that P is the largest prime factor of n. In 1944 Kanold [4] showed that $P \geqq 61$, and in 1975 Hagis and McDaniel [2] proved that

(2) $$P \geqq 100129 .$$

More recently Condict [1] has shown that $P > 300000$. If s is the second largest prime factor of n Pomerance [7] has proved that $s \geqq 139$. The purpose of the present paper is to show that $s > 1000$. While the method of proof differs from that of Pomerance there are many similarities.

It should also be pointed out that in all likelihood our lower bound on s could be significantly improved (say to 50000) by anyone willing to expend the necessary effort and computer time. All of the computations and searches described in the sequel were carried out on the CDC CYBER 174 at the Temple University Computer Center.

2. Some Preliminaries.

In what follows a,b,c,... will represent non-negative integers with p and q denoting odd primes. n will be an odd perfect number with largest prime factor P and second largest prime factor s. M will denote an odd integer with the property that if $p|M$ then $p \geqq 100129$. The dth cyclotomic polynomial will be symbolized by F_d, so that $F_p(x) = 1 + x + x^2 + \dots + x^{p-1}$. Finally, if $(m,p) = 1$ then $h = h(m;p)$ will denote the exponent to which m belongs modulo p; and if $q^c||m$ we shall write $v_q(m) = c$.

We now recall some facts concerning odd perfect numbers and cyclotomic polynomials. According to Theorem 3.4 in [6]:

(3) $$\sigma(p^a) = \prod F_d(p) \text{ where } d|(a + 1) \text{ and } d > 1.$$

From Theorems 94 and 95 in [5]:

(4) $q|F_k(p)$ if and only if $k = q^{\beta} \cdot h(p;q)$. If $\beta > 0$, then $q||F_k(p)$; if $\beta = 0$, then $q \equiv 1 \pmod{k}$.

From (4) we see easily that:

(5) If $q_1|F_{k_1}(p)$ and $q_2|F_{k_2}(p)$ $(k_1 > k_2 \geq 2)$ where $q_i \equiv 1 \pmod{k_i}$ then $q_1 \neq q_2$.

From (21) in [3] we have:

(6) If $k \geq 3$ then $F_k(p)$ has at least one prime factor q such that $q \equiv 1 \pmod{k}$.

If n is odd and perfect, Euler showed that

(7) $$n = p_0^{a_0} p_1^{a_1} \cdots p_t^{a_t} \text{ where } p_0 \equiv a_0 \equiv 1 \pmod 4 \text{ and } 2|a_i \text{ if } i > 0.$$

Since $\sigma(n) = 2n$ it follows from (3) that

(8) $$2n = \prod_{i=0}^{t} \prod F_d(p_i) \text{ where } d|(a_i + 1) \text{ and } d > 1.$$

Of course, the set of p_i in (7) is identical with the set of odd prime factors of the $F_d(p_i)$ in (8).

We conclude this section by stating two lemmas. The first follows easily from (3) and (4).

LEMMA 1. If $p^b||\sigma(q^a)$ and $h = h(q;p)$ then: (i) $b = 0$ if $h \nmid (a+1)$; (ii) $b = v_p(a+1)$ if $h = 1$; (iii) $b = v_p(a+1) + v_p[F_h(q)]$ if $h > 1$ and $h|(a+1)$.

LEMMA 2. Suppose that $\sigma(p^c) = Kq^e$ where $e \geq 1$ and $(K,q) = 1$. Let h and H be the exponents to which q and K, respectively, belong modulo p. Also, suppose that $K^H - 1 = wp + vp^2$ where $0 \leq w < p$. Then $p||F_h(q)$ if $p \nmid (H-w)$.

PROOF. If we raise both sides of the equation $Kq^e = \sigma(p^c)$ to the power H and use the fact that $K^H = 1 + wp + vp^2$ we see that $q^{eH} \equiv 1 + (H - wq^{eH})p \pmod{p^2}$. Therefore, $q^{eH} \equiv 1 \pmod p$ so that $h|eH$ and $q^{eH} \equiv 1 + (H-w)p \pmod{p^2}$. It follows that if $p \nmid (H-w)$ then $p||(q^{eH} - 1)$ so that $p||(q^h - 1)$ and (from (3) and (4)) $p||F_h(q)$.

COROLLARY (Pomerance [7]). If $\sigma(p^c) = q^e$ where $e \geq 1$ and h is the exponent to which q belongs modulo p then $p||F_h(q)$.

3. Acceptable Positive Integers.

We now state formally the result we wish to prove.

THEOREM. If s is the second largest prime factor of an odd perfect number then $s > 1000$.

Our proof of this theorem will be given in Section 5 and will be by reductio ad absurdum. Thus, we shall assume the existence of an odd perfect number with exactly one prime divisor $P > 1000$ (by (2), $P \geq 100129$) and show that this assumption is untenable.

Now, consider the set $S = \{4,127,151,331,31,19,7,97,61,11,13,3,29,5,307,17\}$ which is ordered from left to right so that 4 precedes 127, 127 precedes 151, 31 precedes 7, etc. Suppose that $p \in S$ and let q be an odd prime such that $q < 1000$. (There are 167 odd primes less than 1000.) If $h = h(q;p)$ is the exponent to which q belongs modulo p we shall say that the positive integer k is pq-acceptable if each of the following is true:

(I) $k + 1 = hp^{\alpha}$ where $\alpha \geqq 0$ and $4 \nmid h$.

(II) If u precedes p in S then $u \nmid \sigma(q^k)$.

(III) $\sigma(q^k)$ has no prime factor between 1000 and 100128.

(IV) $\sigma(q^k)$ has at most one prime factor greater than 100128.

(V) If $\sigma(q^k)$ has exactly one prime factor $Q \geq 100129$ then for some prime t (with t=2 being considered only if $Q \equiv 1 \pmod 4$)) all of the prime factors of $F_t(Q)$ are less than 1000 and none precedes p in S. (Note that, according to (6), $F_t(Q)$ has a prime factor greater than 1000 if $t > 500$.)

If β is the smallest integer such that $hp^{\beta} \geqq 1000$ then from (6),(5),(3),(III),(IV), it follows that k is not pq-acceptable if $k + 1 = hp^{\alpha}$ where $\alpha > \beta$. Thus, the set of pq-acceptable integers is finite. For each p in S and $3 \leqq q \leqq 997$, $\sigma(q^k)$ is given in Table 1 for each pq-acceptable value of k which is even. For odd k at most one value of $\sigma(q^k)$ is listed for each p, that for which $v_p(\sigma(q^k))$ is maximal (and k is odd and pq-acceptable). k may or may not be pq-acceptable for those entries in Table 1 such that $M_i | \sigma(q^k)$, as the search for prime factors of M_i was not pursued beyond 10^8.

TABLE 1

p	$\sigma(q^k)$
127	$\sigma(19^2) = 3 \cdot 127$, $\sigma(107^2) = 7 \cdot 13 \cdot 127$,
	$\sigma(389^6) = 127 \cdot 337 \cdot 659 \cdot 827 \cdot 148933$,
	$\sigma(761) = 2 \cdot 3 \cdot 127$.

TABLE 1 (cont.)

p	$\sigma(q^k)$
151	$\sigma(269^2) = 13 \cdot 37 \cdot 151$, $\sigma(571^2) = 3 \cdot 7 \cdot 103 \cdot 151$, $\sigma(787^2) = 3 \cdot 37^2 \cdot 151$, $\sigma(19^4) = 151 \cdot 911$, $\sigma(59^4) = 11 \cdot 41 \cdot 151 \cdot 181$.
331	$\sigma(31^2) = 3 \cdot 331$, $\sigma(293^{10}) = 67 \cdot 331M_1$, $\sigma(773^{10}) = 331 \cdot 727 \cdot 991M_2$, $\sigma(661) = 2 \cdot 331$.
31	$\sigma(5^2) = 31$, $\sigma(67^2) = 3 \cdot 7^2 \cdot 31$, $\sigma(149^2) = 7 \cdot 31 \cdot 103$, $\sigma(191^2) = 7 \cdot 13^2 \cdot 31$, $\sigma(211^2) = 3 \cdot 13 \cdot 31 \cdot 37$. $\sigma(439^2) = 3 \cdot 31^2 \cdot 67$, $\sigma(521^2) = 31^2 \cdot 283$, $\sigma(811^2) = 3 \cdot 31 \cdot 73 \cdot 97$, $\sigma(997^2) = 3 \cdot 13 \cdot 31 \cdot 823$, $\sigma(157^4) = 11 \cdot 31 \cdot 1793161$, $\sigma(281^4) = 5 \cdot 31 \cdot 271 \cdot 148961$, $\sigma(373^{30}) = 31M_3$, $\sigma(61) = 2 \cdot 31$.
19	$\sigma(7^2) = 3 . 19$, $\sigma(11^2) = 7 . 19$, $\sigma(83^2) = 19 . 367$, $\sigma(163^2) = 3 . 7 . 19 . 67$, $\sigma(277^2) = 3 . 7 . 19 . 193$, $\sigma(653^2) = 7 . 13^2 . 19^2$, $\sigma(809^2) = 7 . 13 . 19 . 379$, $\sigma(919^2) = 3 . 7 . 13 . 19 . 163$, $\sigma(191^{18}) = 19M_4$, $\sigma(571^{18}) = 19.457M_5$, $\sigma(647^{18}) = 19M_6$, $\sigma(761^{18}) = 19M_7$, $\sigma(37) = 2 \cdot 19$.
7	$\sigma(23^2) = 7 \cdot 79$, $\sigma(37^2) = 3 \cdot 7 \cdot 67$, $\sigma(53^2) = 7 \cdot 409$, $\sigma(79^2) = 3 \cdot 7^2 \cdot 43$, $\sigma(109^2) = 3 \cdot 7 \cdot 571$, $\sigma(137^2) = 7 \cdot 37 \cdot 73$, $\sigma(263^2) = 7^2 \cdot 13 \cdot 109$, $\sigma(359^2) = 7 \cdot 37 \cdot 499$, $\sigma(373^2) = 3 \cdot 7^2 \cdot 13 \cdot 73$, $\sigma(431^2) = 7 \cdot 67 \cdot 397$, $\sigma(499^2) = 3 \cdot 7 \cdot 109^2$, $\sigma(821^2) = 7 \cdot 229 \cdot 421$, $\sigma(823^2) = 3 \cdot 7 \cdot 43 \cdot 751$, $\sigma(877^2) = 3 \cdot 7 \cdot 37 \cdot 991$, $\sigma(947^2) = 7 \cdot 277 \cdot 463$,

TABLE 1 (cont.)

p	$\sigma(q^k)$
	$\sigma(977^2) = 7 \cdot 136501$, $\sigma(991^2) = 3 \cdot 7 \cdot 13^2 \cdot 277$, $\sigma(97) = 2 \cdot 7^2$.
97	$\sigma(61^2) = 3 \cdot 13 \cdot 97$, $\sigma(229^2) = 3 \cdot 97 \cdot 181$, $\sigma(193) = 2.97$.
61	$\sigma(13^2) = 3 \cdot 61$, $\sigma(47^2) = 37 \cdot 61$, $\sigma(379^2) = 3 \cdot 61 \cdot 787$, $\sigma(131^4) = 5 \cdot 61 \cdot 973001$, $\sigma(733^{60}) = 61M_8$.
11	$\sigma(3^4) = 11^2$, $\sigma(5^4) = 11 \cdot 71$, $\sigma(199^{10}) = 11M_9$, $\sigma(331^{10}) = 11 \cdot 23 \cdot 89M_{10}$, $\sigma(617^{10}) = 11 \cdot 67M_{11}$, $\sigma(991^{10}) = 11 \cdot 23M_{12}$, $\sigma(241) = 2 \cdot 11^2$.
13	$\sigma(3^2) = 13$, $\sigma(29^2) = 13 \cdot 67$, $\sigma(113^2) = 13 \cdot 991$, $\sigma(139^2) = 3 \cdot 13 \cdot 499$, $\sigma(971^2) = 13 \cdot 79 \cdot 919$. $\sigma(131^{12}) = 13 \cdot 79M_{13}$, $\sigma(313^{12}) = 13M_{14}$, $\sigma(443^{12}) = 13M_{15}$. $\sigma(677^{12}) = 13M_{16}$, $\sigma(911^{12}) = 13 \cdot 53 \cdot 937M_{17}$, $\sigma(937^{12}) = 13 \cdot 53 \cdot 599M_{18}$, $\sigma(337) = 2 \cdot 13^2$.
3	$\sigma(43^2) = 3 \cdot 631$, $\sigma(181^2) = 3 \cdot 79 \cdot 139$, $\sigma(283^2) = 3 \cdot 73 \cdot 367$, $\sigma(307^2) = 3 \cdot 43 \cdot 733$, $\sigma(313^2) = 3 \cdot 181^2$, $\sigma(337^2) = 3 \cdot 43 \cdot 883$, $\sigma(547^2) = 3 \cdot 163 \cdot 613$, $\sigma(631^2) = 3 \cdot 307 \cdot 433$, $\sigma(809) = 2 \cdot 3^4 \cdot 5$.
29	$\sigma(523^{28}) = 29 \cdot 59M_{19}$.

TABLE 1 (cont.)

p	$\sigma(q^k)$
5	$\sigma(229) = 2 \cdot 5 \cdot 23.$
307	$\sigma(17^2) = 307,\ \sigma(331^{16}) = 307M_{20},$ $\sigma(421^{16}) = 103 \cdot 307M_{21},\ \sigma(587^{16}) = 103 \cdot 307M_{22},$ $\sigma(883^{16}) = 239 \cdot 307M_{23},\ \sigma(887^{16}) = 307 \cdot 409M_{24},$ $\sigma(613) = 2 \cdot 307.$
17	$\sigma(103^{16}) = 17M_{25},\ \sigma(137^{16}) = 17 \cdot 103M_{26},$ $\sigma(239^{16}) = 17M_{27},\ \sigma(577) = 2 \cdot 17^2.$

4. Feasible Primes.

If $p \in S$ we shall say that the prime r is p-feasible if each of the following is true:

(A) If u precedes p in S then $u \nmid F_r(p)$.

(B) $F_r(p)$ has no prime factor between 1000 and 100128.

(C) $F_r(p)$ has at most one prime factor greater than 100128.

(D) If $F_r(p)$ has exactly one prime factor $Q \geq 100129$ then for some prime t (with $t = 2$ being considered only if $Q \equiv 1 \pmod 4$)) all of the prime factors of $F_t(Q)$ are less than 1000 and none precedes p in S.

Now suppose that r is a prime such that (A) and (B) are satisfied and suppose also that $q | F_r(p)$ where $q \geq 100129$. Then $F_r(p) = KM$ where every prime factor of K is less than 1000 and every prime factor of M exceeds 100128. If K belongs to the exponent H modulo p and $K^H - 1 = wp + vp^2$ (where $0 \leq w < p$) then we shall say that r is a p-nasty prime if $p | (H-w)$. Either M has two distinct prime factors which exceed 100128 (so that, by (C), r is not p-feasible) or $F_r(p) = Kq^e$. In the latter case, if $h = h(q;p)$, then by Lemma 2 $p || F_h(q)$ if r is not p-nasty (so that $p \nmid (H-w)$).

If $r > 500$ and $p \in S$ then it follows easily from (2) and (4) that either (B) is not satisfied (so that r is not p-feasible) or $F_r(p) = M$. In the latter case, since $H-w = 1 - 0 = 1$, r is not p-nasty. Thus, for each p in S only a search of finite length is required to find both: (i) a prime $L(p)$ such that r is not p-feasible if $r < L(p)$; (ii) the set $N(p)$ of all p-nasty primes r such that $r \geq L(p)$ and (A) and (B) are satisfied. In Table 2, $L(p)$ and $N(p)$ are given for each p in S.

TABLE 2

p	L(p)	N(p)
127	11	ϕ
151	11	ϕ
331	11	ϕ
31	17	ϕ
19	17	ϕ
7	29	ϕ
97	11	ϕ
61	13	ϕ
11	19	{29,179}
13	19	ϕ
3	43	ϕ
29	11	ϕ
5	37	{71,359}
307	11	ϕ
17	17	{97}

5. The Proof of Our Theorem.

We now assume the existence of an odd perfect number n whose second largest prime factor is less than 1000. Thus,

(9) $$n = P^g \prod_{i=1}^{u} q_i^{a_i} \quad \text{where } P \geqq 100129 \text{ and } 3 \leqq q_i \leqq 997.$$

Our first objective is to show that if $p \in S$ then $p \nmid n$. Suppose that $p|n$ where $p \in S$, but n is not divisible by any prime in S which precedes p.

REMARK 1. If $p|\sigma(q_i^{a_i})$ then Lemma 1 and a few moments of reflection convince us that $(k+1)|(a_i+1)$ where k is pq_i-acceptable. Therefore, $\sigma(q_i^k)|\sigma(n)$.

REMARK 2. If $p^a||n$ and t is the smallest prime factor of $a + 1$ then $t \geqq L(p)$ so that $a \geqq L(p) - 1$ (see (3) and Section 4).

REMARK 3. A computer search revealed that if t is a prime and $t \geqq L(p)$ then $F_t(p)$ has a prime factor which exceeds 100128. It follows that if $p^a||n$ then $P|\sigma(p^a)$ (see Remark 2).

REMARK 4. If $p^b||\sigma(P^g)$ and $p^f < 500$ but $p^{f+1} > 500$ then $b \leqq f + v_p[F_h(P)]$ where $h = h(P;p)$ (see Lemma 1, (3), (6)). (If $p > 19$ then $f = 1$.)

REMARK 5. If b and f are as given in Remark 4 and N(p) is empty then $b \leqq f + 1$

(see Section 4 and Lemma 2 , and recall Remark 3).

In what follows $\underline{a}$ denotes a positive integer. Suppose that $127^a || n$. From Table 1 and Remark 1 at most $127^4 | \sigma(n/P^g)$, and from Table 2 and Remark 5 at most $127^2 | \sigma(P^g)$. Therefore, $127^7 \nmid \sigma(n)$. But, from Table 2 and Remark 2, $a \geqq 10$. This contradicts (1), and we conclude that $127 \nmid n$.

Similarly, if $151^a || n$ (but $127 \nmid n$) then $a \geqq 10$ but $151^8 \nmid \sigma(n)$. Again (1) is contradicted so that $151 \nmid n$. If $331^a || n$ then $a \geqq 10$ but $331^7 \nmid \sigma(n)$. Therefore, $331 \nmid n$. If $31^a || n$ then $a \geqq 16$ but $31^{16} \nmid \sigma(n)$ (note that 1793161,148961 and M_3 are relatively prime in pairs and that n has exactly one prime factor which exceeds 100128 so that at most $31^{13} | \sigma(n/P^g)$). Therefore, $31 \nmid n$. If $19^a || n$ then $a \geqq 16$ but $19^{16} \nmid \sigma(n)$ (note that M_5, M_6, M_7 are relatively prime in pairs since $M_5 \doteq 5 \cdot 10^{45}$, $M_6 \doteq 2 \cdot 10^{49}$, $M_7 \doteq 4 \cdot 10^{50}$ and every prime factor of M_i exceeds 10^8) so that $19 \nmid n$. If $7^a || n$ then $a \geqq 28$ but $7^{27} \nmid \sigma(n)$. If $97^a || n$ then $a \geqq 10$ but $97^6 \nmid \sigma(n)$. If $61^a || n$ then $a \geqq 12$ but $61^8 \nmid \sigma(n)$. Therefore $(7 \cdot 97 \cdot 61, n) = 1$.

Now suppose that $11^a || n$. Since $N(11) = \{29,179\}$ we see from Remark 4 and the discussion immediately preceding Table 2 that either $29|(a+1)$ or $179|(a+1)$ or at most $11^3 | \sigma(P^g)$. $F_{29}(11) = 523M_{28}$ and $F_{179}(11) = 359M_{29}$ where every prime factor of M_i exceeds 10^8. Therefore, either $F_{29}(11) = 523P^e$ or $F_{179}(11) = 359P^d$ or $11^4 \nmid \sigma(P^g)$. If $1 + 11 + 11^2 + \ldots + 11^{28} = 523P^e$ then $133 \equiv 523P^e \pmod{11^3}$, and it follows that $111 \equiv 133^{10} \equiv 523^{10}P^{10e} \equiv -10P^{10e} \pmod{11^3}$. If $h = h(P;11)$ then, since $h|10e$, we see that $P^h \not\equiv 1 \pmod{11^3}$ and it follows that at most $11^2 | F_h(P)$. Similarly, if $F_{179}(11) = 359P^d$ then at most $11^2 | F_h(P)$. By Remark 4 at most $11^4 | \sigma(P^g)$. But, by Remark 1 and Table 1, at most $11^9 | \sigma(n/P^g)$ so that $11^{14} \nmid \sigma(n)$. Since, by Remark 2 and Table 2, $a \geqq 18$ we have a contradiction and must conclude that $11 \nmid n$.

If $13^a || n$ then $a \geqq 18$ but $13^{17} \nmid \sigma(n)$. If $3^a || n$ then $a \geqq 42$ but $3^{19} \nmid \sigma(n)$. If $29^a || n$ then $a \geqq 10$ but $29^4 \nmid \sigma(n)$. Therefore, $(13 \cdot 3 \cdot 29, n) = 1$.

Suppose that $5^a || n$. Since $N(5) = \{71,359\}$ either $71|(a + 1)$ or $359|(a + 1)$ or at most $5^4 | \sigma(P^g)$. $F_{71}(5) = 569M_{30}$ and $F_{359}(5) = 719M_{31}$ where every prime factor of M_i exceeds 10^8. It follows that either $F_{71}(5) = 569P^e$ or $F_{359}(5) = 719P^d$ or $5^5 \nmid \sigma(P^g)$. If $1 + 5 + 5^2 + \ldots + 5^{358} = 719P^d$ then $151 \equiv 719P^d \pmod{5^4}$, and it follows that $271 \equiv 156^4 \equiv 719^4P^{4d} \equiv 521P^{4d} \pmod{5^4}$. If $h = h(P;5)$ then, since $h|4d$, we see that $P^h \not\equiv 1 \pmod{5^4}$ so that at most $5^3 | F_h(P)$. By Remark 4 at most $5^6 | \sigma(P^g)$. By a similar argument $5^7 \nmid \sigma(P^g)$ if $F_{71}(5) = 569P^e$. Since $5^2 \nmid \sigma(n/P^g)$ we see that $5^8 \nmid \sigma(n)$. Since $a \geqq 36$ we conclude that $5 \nmid n$.

$307 \nmid n$ since if $307^a || n$ then $a \geqq 10$ while $307^{10} \nmid \sigma(n)$. If $17^a || n$ then either $97 | (a + 1)$ or $17^4 \nmid \sigma(P^g)$. Therefore, since $F_{97}(17) = 389M_{32}$ either $M_{32} = P^e$ or $17^4 \nmid \sigma(P^g)$. If $1 + 17 + 17^2 + \ldots + 17^{96} = 389P^e$ then $307 \equiv 389P^e (\mathrm{mod}\ 17^3)$, and it follows that $273 \equiv 307^{16} \equiv 389^{16}P^{16e} \equiv 2585P^{16e} (\mathrm{mod}\ 17^3)$. If $h = h(P;17)$ then, since $h | 16e$, we see that $P^h \not\equiv 1 (\mathrm{mod}\ 17^3)$ so that $17^3 \nmid F_h(P)$. By Remark 4, $17^5 \nmid \sigma(P^g)$ and, since $17^6 \nmid \sigma(n/P^g)$, it follows that $17^{10} \nmid \sigma(n)$. Since $a \geqq 16$ we see that $17 \nmid n$.

It is now a simple matter to complete the proof of our theorem. For if n is an odd perfect number with second largest prime factor s where $s < 1000$ then n has the form (9) where none of the q_i is an element of S. Since $1 < \sigma(q^a)/q^a = (q^{a+1} - 1)/q^a(q - 1) < q/(q - 1)$ and since $x/(x - 1)$ is monotonic decreasing for $x \geqq 2$ we see that

$$\sigma(n)/n < P/(P - 1) \prod_{i=1}^{u} q_i/(q_i - 1) \leqq (100129/100128)(23/22) \prod_{q=37}^{997} q/(q - 1) < 2.$$

This contradiction to (1) shows that $s > 1000$.

6. Some Concluding Remarks.

While our bound on s may not be of any great importance or interest in itself, it should prove very useful in investigations concerning either the number of distinct prime factors an odd perfect number must possess or its magnitude. For example, around 1950 it was proved by Kühnel and Webber [9] (independently) that every odd perfect number has at least six distinct prime factors. This result is now almost trivial. For if n is odd and perfect then it is well known that $3\cdot5\cdot7 \nmid n$ and $3\cdot5^2\cdot11 \nmid n$. From (2) and our theorem it follows that if n has less than six prime factors then $\sigma(n)/n < (3/2)(6/5)(11/10)(1009/1008)(100129/100128) < 2$, in contradiction to (1).

Two final comments. L(p), as described in Section 4, is not uniquely defined, and the values given in Table 2 were chosen with Remarks 2 and 3 of Section 5 in mind. [1] has not been refereed and is not readily available. Therefore, the result contained therein, that $P > 300000$, was not used in the present paper although it would have been helpful to do so.

REFERENCES

1. J.T. Condict, On An Odd Perfect Number's Largest Prime Divisor, Senior Thesis, Middlebury College, May 1978.

2. P. Hagis, Jr. and W.L. McDaniel, "On the Largest Prime Divisor of an Odd Perfect Number. II," Math. Comp., v. 29, 1975, pp. 922-924.

3. H.-J. Kanold, "Untersuchungen über ungerade vollkommene Zahlen," J. Reine Angew. Math., v. 183, 1941, pp. 98-109.

4. H.-J. Kanold, "Folgerungen aus dem Vorkommen einer Gausschen Primzahl in der Primfaktorenzerlegung einer ungeraden vollkommenen Zahl," J. Reine Angew. Math., v. 186, 1944, pp. 25-29.

5. T. Nagell, Introduction to Number Theory, Wiley, New York, 1951.

6. I. Niven, Irrational Numbers, Wiley, New York, 1956.

7. C. Pomerance, "The Second Largest Prime Factor of An Odd Perfect Number," Math. Comp., v. 29, 1975, pp. 914-921.

8. D. Slowinski, "Searching For the 27th Mersenne Prime," J. Recr. Math., v. 11, 1978-79, pp. 258-261.

9. G.C. Webber, "Non-existence of Odd Perfect Numbers of the Form $3^{2\beta}p^{\alpha}s_1^{2\beta_1}s_2^{2\beta_2}s_3^{2\beta_3}$," Duke Math. J., v. 18, 1951, pp. 741-749.

A COMPLEMENT TO RIDOUT'S P-ADIC GENERALIZATION OF THE THUE-SIEGEL-ROTH THEOREM

J.C. Lagarias

Bell Laboratories
Murray Hill, New Jersey 07974

Dedicated to Emil Grosswald on the occasion of his 68th birthday.

0. Abstract

Ridout's theorem asserts that for algebraic numbers $\alpha_0,\ldots,\alpha_N$ with $\alpha_0 \varepsilon \mathbb{R}$, $\alpha_i \varepsilon Q_{p_i}$ the p_i-adic field and any $\varepsilon > 0$ there are only finitely many rationals $\frac{h}{q}$ with $q > 0$ such that

$$\min(1,|q\alpha_0 - h|) \prod_{i=1}^{N} \min(1,||q\alpha_i - h||_{p_i}) < H^{-1-\varepsilon}$$

where $H = \max(|h|, q)$. The complementary problem is that of finding algebraic numbers which can be well-approximated by rationals. For a single valuation $||\cdot||$, whether real or p-adic, it is not known whether there exists an algebraic number α_0 for which there is an increasing function $f(x)$ such that $f(x) \to \infty$ as $x \to \infty$ and such that

$$||q\alpha_0 - h|| < H^{-1}(f(H))^{-1}$$

has an infinite number of integer solutions (h,q). This paper gives an improvement in the complementary direction when more than one valuation is involved. Let $\log_i x = \log(\log_{i-1} x)$ and $\log_1 x = \log x$ so that $\log_2 x = \log\log x$, and so on. We show that for any positive integer N and any set of N primes $p_1,\ldots,P_N$ there exist algebraic $\alpha_0 \varepsilon \mathbb{R}$, $\alpha_i \varepsilon Q_{p_i}$ for $1 \le i \le N$ such that

$$\min(1, |q\alpha_0 - h|) \prod_{i=1}^{N} \min(1, ||q\alpha_i - h||_{p_i}) \ll H^{-1}(f(H))^{-1}$$

has an infinite number of solutions, where

$$f(H) = \prod_{i=1}^{N} \log_i H.$$

1. Introduction

The Thue-Siegel-Roth theorem asserts that real algebraic numbers cannot be too well approximated by rational numbers.

THEOREM (Thue-Siegel-Roth) Let $\alpha \in \mathbb{R}$ be a real algebraic number. Then for any fixed $\varepsilon > 0$ there are only finitely many rational numbers $\frac{h}{q}$ such that

$$|q\alpha - h| < q^{-1-\varepsilon} \quad . \tag{1.1}$$

In the complementary direction, it is well known that for any irrational real number θ the inequality

$$|q\theta - h| < q^{-1}$$

has an infinite number of integer solutions (h,q). Furthermore for a real irrational number θ it is known that the following properties are equivalent.

(i) There exists an increasing unbounded function $f(t)$ such that

$$|q\theta - h| < q^{-1}(f(q))^{-1} \tag{1.2}$$

has an infinite number of integer solutions (h,q).

(ii) The ordinary continued fraction expansion of θ has unbounded partial quotients.

It has never been proved that there exists any real irrational algebraic number with unbounded partial quotients in its continued fraction expansion, although as far as is known all real algebraic irrationals other than real quadratic irrationals could have this property. This is an apparently very difficult open problem.

In passing we remark that the Thue-Siegel-Roth theorem does imply restrictions on the rate of growth of the partial quotients of the continued fraction expansion of a real algebraic number, but these restrictions are not very strong. For example, the real number α given by the continued fraction expansion $\alpha = [a_0, a_1, a_2, \ldots]$ with $a_j = 2^{j}$ can be shown to have the property that for each $\varepsilon > 0$ (1.1) has only finitely many solutions, so that the Thue-Siegel-Roth theorem does not prevent α from being an algebraic number. However the Thue-Siegel-

Roth theorem does guarantee that $\alpha = [a_0, a_1, a_2, \ldots]$ with $a_j = 2^{2^j}$ is transcendental.

D. Ridout [7] gave a p-adic strengthening of the Thue-Siegel-Roth Theorem. To state his result, let Q_p denote the p-adic numbers, with the multiplicative p-adic valuation $|| \cdot ||_p$ normalized so that $||p||_p = \frac{1}{p}$. The height H of a rational number $r = \frac{h}{q}$ with $q > 0$ and $(h,q) = 1$ is given by

$$H = H(r) = \max(|h|, q).$$

An algebraic number in Q_p is one that satisfies a polynomial equation with rational integer coefficients. Ridout's theorem asserts the following (Lang [2], p.93).

<u>THEOREM</u>. <u>(Ridout) Let</u> $\alpha_0, \alpha_1, \ldots, \alpha_N$ <u>be algebraic numbers with</u> $\alpha_0 \in \mathbb{R}$ <u>and the other</u> $\alpha_i \in Q_{p_i}$ <u>distinct</u> p_i<u>-adic fields. Then for any</u> $\varepsilon > 0$ <u>there are only finitely many rational numbers</u> $\frac{h}{q}$ <u>such that</u>

$$\min(1, |q\alpha_0 - h|) \prod_{i=1}^{N} \min(1, ||q\alpha_i - h||_{p_i}) < H^{-1-\varepsilon}$$

<u>where</u> $H = \max(|h|, q)$.

The known results in the complementary direction for a single p-adic algebraic number are as limited as that for a single real algebraic number. By a pigeonhole principle argument one can show for any irrational $\alpha \in Q_p$ that

$$||q\alpha - h||_p < H^{-1}$$

has an infinite number of relatively prime integer solutions (h,q). On the other hand, it has not been shown for even a single irrational algebraic number $\alpha \in Q_p$ that there is an increasing unbounded function f(t) such that

$$||q\alpha - h||_p < H^{-1}(f(H))^{-1}$$

has an infinite number of solutions (h,q).

The object of this paper is to show that an improvement in the complementary direction to Ridout's theorem is possible when more than one valuation is involved. We use the notation $\log_1 x = \log x$ and $\log_j x = \log(\log_{j-1} x)$ so that $\log_2 x = \log\log x$ and so on.

THEOREM A. <u>Given distinct primes</u> $p_1,\ldots,p_N$ <u>there exist algebraic numbers</u> $\alpha_0 \in \mathbb{R}$ <u>and</u> $\alpha_i \in Q_{p_i}$ <u>the</u> p_i<u>-adic field and a positive constant</u> $c_0=c_0(\alpha_0,\alpha_1,\ldots,\alpha_N)$ <u>such that</u>

$$|q\alpha_0 - h| \prod_{i=1}^{N} (||q\alpha_i - h||_{p_i}) < c_0 H^{-1}(f(H))^{-1} \tag{1.3}$$

<u>has infinitely many integer solutions</u> (h,q), <u>where</u>

$$f(H) = \prod_{i=1}^{N} \log_i H. \tag{1.4}$$

<u>Similarly given distinct primes</u> $p_0,\ldots,p_N$ <u>there exist algebraic numbers</u> $\alpha_i \in Q_{p_i}$ <u>and a constant</u> $c_1 = c_1(\alpha_0,\ldots,\alpha_N)$ <u>such that</u>

$$\prod_{i=0}^{N} (||q\alpha_i - h||_{p_i}) < c_1 H^{-1}(f(H))^{-1} \tag{1.5}$$

<u>has infinitely many integer solutions</u> (h,q).

Note that (1.3) implies that

$$\min(1, |q\alpha_0 - h|) \prod_{i-1}^{N} \min(1, ||q\alpha_i - h||_{p_i}) < c_0 H^{-1}(f(H))^{-1} \quad ,$$

which may be compared to the result of Ridout's theorem.

From the viewpoint of the metric theory of Diophantine approximation, the inequality (1.3) has an infinite number of solutions for almost all $N + 1$ tuples $(\alpha_0,\alpha_1,\ldots,\alpha_N) \in \mathbb{R} \times Q_{p_1} \times \ldots \times Q_{p_N}$. In fact the stronger inequality

$$|q\alpha_0 - h| < H^{-1}(f(H))^{-1}$$

with $f(H)$ given by (1.4) has an infinite number of solutions for almost all $\alpha_0 \in \mathbb{R}$ in the sense of Lebesgue measure. This is a special case of Khinchin's divergence theorem (Lang [3], p. 24) which asserts that if $f(t)$ is a positive function for which $\sum_{j=1}^{\infty} f(j)$ diverges, then

$$|q\alpha_0 - h| < H^{-1}(f(H))^{-1}$$

has infinitely many solutions for almost all $\alpha_0 \in \mathbb{R}$ in the sense of Lebesgue measure. The inequality (1.5) also has infinitely many solutions for almost all $(\alpha_0,\ldots,\alpha_N) \in Q_{p_0} \times Q_{p_1} \times \ldots \times Q_{p_N}$, cf. Cantor [2]. Thus Theorem A does not exhibit a set of algebraic numbers that differs metrically from the bulk of all non-algebraic numbers.

We mention some related work. Bumby [1] observed that

$$\log(\max(|x|,|y|))|x|\cdot|y| \; ||x||_2 \; ||y||_3 \; ||x-y||_5 < c$$

has infinitely many solutions for $c = 2/5 \log 2$, while a generalization of Roth's theorem (see Mahler [5]) asserts that

$$(\max(|x|,|y|))^{\varepsilon} \; |x|\cdot|y| \; ||x||_2 \; ||y||_3 \; ||x-y||_5 < c$$

has only finitely many solutions for any $\varepsilon > 0$, $c > 0$. Peck [6] considered simultaneous approximation of sets of algebraic numbers. His results imply that if $(\beta_1,\ldots,\beta_N)$ are such that $[1, \beta_1,\ldots,\beta_N]$ is a Q-basis of a real algebraic number field, then

$$\prod_{i=1}^{N} |q\beta_i - p_i| < q^{-1}(\log q)^{-1}$$

has an infinite number of solutions. We remark that the construction of Theorem A can be extended to the case of simultaneous real and p-adic approximations to a set of algebraic numbers.

I am indebted to E.G. Straus, whose astute observation at the Grosswald conference led to the extension of Theorem A from the case N = 2 to the general case, and to W.W. Adams for comments on a draft of this paper.

2. Proof of Theorem A.

In what follows we use $||\cdot||_i$ as shorthand for the p-adic valuation $||\cdot||_{p_i}$.

We deal first with the case that a real valuation is present. Let $p_1,\ldots,p_N$

be given and pick a real quadratic field $Q(\sqrt{D})$ in which the prime ideals (p_1) all split completely, i.e. for which $\sqrt{D}$ is in Q_{p_i} for all i. Such a D exists using Dirichlet's theorem on primes in arithmetic progression, by taking D to be a prime $D \equiv 1 \pmod 4$ which is a quadratic residue (mod p_i) for all i. Then all the (p_i) split completely in $Q(\sqrt{D})$ by the quadratic reciprocity law.

Let ε be the smallest unit of the field $Q(\sqrt{D})$ with $\varepsilon > 1$ and $\bar{\varepsilon} > 0$, where $\bar{\varepsilon}$ is the algebraic conjugate of ε. Hence $\bar{\varepsilon} = 1/\varepsilon$. We choose $\alpha_0 = \varepsilon$ and we set

$$h(k) = \varepsilon^{k+1} + \bar{\varepsilon}^{k+1}$$

$$q(k) = \varepsilon^{k} + \bar{\varepsilon}^{k}.$$

Since $\varepsilon > 1$, the height H of $\frac{h(k)}{q(k)}$ is

$$H = H_k = \max(h(k),q(k)) = h(k).$$

In what follows we omit the subscript of H where it can be inferred from context. It is well-known that

$$|q(k)\alpha_0 - h(k)| < a_0 H^{-1} \tag{2.1}$$

where $a_0 = 2\varepsilon|\varepsilon - \bar{\varepsilon}|$ is a positive constant. Indeed

$$|q(k)\varepsilon - h(k)| = |\varepsilon^{-k}(\varepsilon - \bar{\varepsilon})| \leq \frac{a_0}{2} \varepsilon^{-k-1}$$

while $H < 2\varepsilon^{k+1}$.

Thus, (2.1) shows that the set

$$S_0 = \{(h(k),q(k)) \mid k = 0,1,2,\dots\}$$

produces a contribution of H^{-1} in Ridout's theorem from the real valuation alone. The main device of the proof is to extract subsequences $s_i(k)$ of the integers k such that the sets

$$S_i = \{(h(s_i(K)),q(s_i(k)) \mid k = 0,1,2,\dots\}$$

have

$$S_0 \supset S_1 \supset \cdots \supset S_N \tag{2.2}$$

and such that the elements of the subsequence S_i converge p_i-adically to an algebraic number α_i at a rate such that

$$||q(s_i(k))\alpha_i - h(s_i(k))||_i \leq a_i(\log_i H)^{-1} \tag{2.3}$$

for a constant a_i independent of k. Then conditions (2.1)-(2.3) immediately imply (1.3) of Theorem A, using the sequence S_N and taking $c_0 = \prod_{i=0}^{N} a_i$.

The subsequences $s_i(k)$ are defined recursively as follows. Let $s_0(k) = k$ and

$$s_i(k) = s_{i-1}(f_i p_i^{\,k}) \tag{2.4}$$

where f_i is to be specified. Let $\phi(n)$ be Euler's totient function and set

$$f_i = \phi(p_i)\phi(p_{i+1})\cdots\phi(p_N)\ . \tag{2.5}$$

We sum up the results we need to prove in the following lemma.

LEMMA 2.1. *Let* $\mu_i = s_{i-1}(0)$ *and set*

$$\alpha_i = \frac{\varepsilon^{\mu_{i+1}} + \bar{\varepsilon}^{\mu_{i+1}}}{\varepsilon^{\mu_i} + \bar{\varepsilon}^{\mu_i}} \tag{2.6}$$

for $1 \leq i \leq N$. *Then*

$$||q(s_i(k))\alpha_i - h(s_i(k))||_{p_i} \leq a_i\ (\log_i\ H)^{-1} \tag{2.7}$$

where $H = h(s_i(k))$ *and the* $a_i = a_i(\varepsilon, p_1, \ldots, p_N)$ *are positive constants.*

PROOF. We will establish the following claims.

CLAIM 1. *For all* α *that are* p_i*-adic units. i.e. for all* $\alpha \in U_i = Z_{p_i} - p_i Z_{p_i}$, *we have*

$$\alpha^{s_i(k)} \equiv \alpha^{s_{i-1}(0)} \pmod{p_i^{\,k}}\ . \tag{2.8}$$

CLAIM 2. There are positive constants $b_i = b_i(\varepsilon, p_1, \ldots, p_N)$ such that

$$\log_i H = \log_i h(s_i(k)) \leq b_i p_i^{\,k} . \tag{2.9}$$

Assuming that the claims are proved, since $\sqrt{D} \in Q_{p_i}$ the units ε, $\bar{\varepsilon} \in U_i$ so that by Claim 1,

$$q(s_i(k)) = \varepsilon^{s_i(k)} + \bar{\varepsilon}^{s_i(k)}$$

$$= \varepsilon^{s_{i-1}(0)} + \bar{\varepsilon}^{s_{i-1}(0)} \pmod{p_i^{\,k}}$$

$$= \varepsilon^{\mu_i} + \bar{\varepsilon}^{\mu_i} \pmod{p_i^{\,k}} .$$

Similarly

$$h(s_i(k)) = \varepsilon^{\mu_{i+1}} + \bar{\varepsilon}^{\mu_{i+1}} \pmod{p_i^{\,k}} .$$

This implies

$$||q(s_i(k))\alpha_i - h(s_i(k))||_i \leq (p_i)^{d-k} = b_i^* \, (p_i)^{-k} \tag{2.10}$$

where $p_i^{-d} = ||\varepsilon^{\mu_i} + \bar{\varepsilon}^{\mu_i}||_i$ is a constant. (Note that d is finite since $\varepsilon^{\mu_i} + \bar{\varepsilon}^{\mu_i} = Tr(\varepsilon^{\mu_i}) \neq 0$ in Q_p since it is a nonzero integer in $\mathbb{Z}$.) By Claim 2

$$(\log_i H)^{-1} \geq b_i^{-1}(p_i)^{-1}$$

and together with (2.10) this establishes the lemma, where we take $a_i = b_i b_i^*$.

To prove Claim 1, we first introduce some notation. Let $\underline{v}$, $\underline{w}$ be vectors with all entries nonnegative and with the same number j of components. Let $(\underline{v},\underline{w})$ denote the iterated exponential expression

$$(\underline{v},\underline{w}) = v_1 w_1^{\,v_2 w_2^{\,\cdot^{\cdot^{\cdot^{\,v_{j-1} w_{j-1}^{\,v_j w_j}}}}}}$$

For such a vector $\underline{v}$ and for $i \leq j$, we use the notation

$$\underline{v}^{(i)} = (v_i, \ldots, v_j).$$

To begin the proof, let

$$\underline{v}_k = (1,f_1,f_2,\ldots,f_i,k) ,$$

$$\underline{w} = (\alpha,p_1,p_2,\ldots,p_i,1) .$$

By definition of $s_i(k)$ we have

$$(\underline{v}_k,\underline{w}) = \alpha^{s_i(k)} .$$

Next we set

$$\underline{x} = (1,f_1,\ldots,f_{i-2},f_{i-1}) ,$$

$$\underline{y} = (\alpha,p_1,\ldots,p_{i-2},1) ,$$

and note that

$$(\underline{x},\underline{y}) = \alpha^{s_{i-1}(0)} .$$

We now establish that

$$(\underline{v}_k^{(m)}, \underline{w}^{(m)}) = (\underline{x}^{(m)},\underline{y}^{(m)}) \pmod{p_i^{\,k}} \qquad (2.11)$$

by induction on decreasing m for $i \geq m \geq 1$. For m=i we have

$$\begin{aligned}(\underline{v}_k^{(i)},\underline{w}^{(i)}) &= f_{i-1}p_{i-1}^{\;f_ip_i^{\,k}} \\ &= f_{i-1}(1 + c_ip_i^{\,k}) \\ &= f_{i-1} \pmod{p_i^{\,k}} \\ &= (\underline{x}^{(i)},\underline{y}^{(i)}) \pmod{p_i^{\,k}} .\end{aligned}$$

In the second line above c_i is an integer, which follows from the fact that

$$\phi(p_i^{\,k}) = \phi(p_i)p_i^{\,k-1} \mid f_ip_i^{\,k} ,$$

using (2.5). For the induction step, suppose it's true for m + 1. Then

$$(\underline{v}_k^{(m)},\underline{w}^{(m)}) = f_{m-1}p_{m-1}^{(\underline{v}_k^{(m+1)},\underline{w}^{(m+1)})}$$

$$= f_{m-1}p_{m-1}^{(\underline{x}^{(m+1)},\underline{y}^{(m+1)})(1+c_{m+1}p_i^k)}$$

$$= f_{m-1}p_{m-1}^{(\underline{x}^{(m+1)},\underline{y}^{(m+1)})}[1+c_m p_i^k]$$

$$\equiv (\underline{x}^{(m)},\underline{y}^{(m)}) \pmod{p_i^k} , \qquad (2.12)$$

where c_m is an integer. The step from the second to third line of (2.12) is justified by observing that

$$\phi(p_i) \mid f_m \quad \text{and} \quad f_m \mid (\underline{x}^{(m+1)},\underline{y}^{(m+1)}) ,$$

using definition (2.5) of f_m, so that

$$\phi(p_i^k) \mid (\underline{x}^{(m+1)},\underline{y}^{(m+1)})p_i^k .$$

Thus (2.12) completes the induction step. The case m = 1 is Claim 1. □

To prove Claim 2, we use the fact that

$$H_k = h(k) < (2\varepsilon)\varepsilon^k. \qquad (2.13)$$

The proof is by induction on i. For i=1 we have

$$\log H < \log (2\varepsilon) + s_1(k)\log \varepsilon$$

$$\leq b_1 p_1^k$$

where $b_1 = f_1 \log \varepsilon + \log 2\varepsilon$. For the general step, using the induction hypothesis we obtain

$$\log_i H = \log_i [h(s_i(k))]$$

$$= \log [\log_{i-1}(h(s_{i-1}(f_i p_i^k)))]$$

$$\leq \log [b_{i-1}(p_{i-1})^{f_i p_i^k}]$$

$$\leq f_i p_i^k \log p_{i-1} + \log b_{i-1}$$

$$\leq b_i p_i^{\,k}$$

where $b_i = f_i \log p_{i-1} + \log b_{i-1}$. This proves Claim 2. □

To complete the proof of Theorem A, we must consider the case that all valuations are p-adic. The proof is similar to the preceding case, with the following modifications. We take $\alpha_0 = 0$ and

$$h(k) = (p_0)^k$$

$$q(k) = 1.$$

We use the identical set of subsequences $s_i(k)$ constructed earlier, but take

$$\alpha_i = (p_0)^{s_{i-1}(0)}$$

for $1 \leq i \leq N$. Theorem A is proved. □

REFERENCES

1. R. Bumby, An elementary example in p-adic Diophantine approximation, Proc. Amer. Math. Soc. 15(1964), 22-25.
2. D. Cantor, On the elementary theory of Diophantine approximation over the ring of Adeles I, Illinois J. Math 9 (1965), 667-700.
3. S. Lang, Diophantine Geometry, John Wiley and Sons, New York 1962.
4. S. Lang, Introduction to Diophantine Approximation, Addison-Wesley, Reading, Mass. 1966
5. K. Mahler, Lectures on Diophantine Approximation. Part 1: g-adic numbers and Roth's theorem, Cushing-Malloy, Ann Arbor, Michigan 1951.
6. L.G. Peck, Simultaneous rational approximation to algebraic numbers. Bull. A.M.S. 67 (1961), 197-201.
7 D. Ridout, The p-adic generalization of the Thue-Siegel-Roth theorem, Mathematika 5 (1958), 40-48.

<u>Note added in proof</u>: (August 1981) C.L. Stewart has independently proved a similar result, stated in a somewhat different form, in: "A note on the product of consecutive integers," to appear in the Proceedings of the Number Theory Colloquium, Budapest, Hungary, July 1981.

CYCLOTOMY FOR NON-SQUAREFREE MODULI

D.H. and Emma Lehmer

Dedicated to Emil Grosswald

Cyclotomy is to be regarded not as an incidental application, but as the natural and inherent centre and core of the arithmetic of the future.

J.J. Sylvester, 1879.

ABSTRACT

Kummer considered cyclotomy for a composite modulus n and defined the periods by

$$\eta_k = \sum_{v=0}^{f-1} \zeta_n^{kq^v} \qquad k \not\equiv q^j \pmod n,\ 0 < j < \phi(n)$$

where $\zeta_n = \exp(2\pi i/n)$ and $f=t(n)$ is the order of q modulo n. He stated that, except when all the η's vanish, they satisfy an irreducible monic period equation with integer coefficients of degree $\phi(n)/t(n)$. Soon thereafter Fuchs gave a necessary and sufficient condition for the vanishing of the η's, namely: $\eta_k = 0$ if and only if $t(n)=pt(n/p)$ for some prime p dividing n. A modern proof for a generalized period was recently given by R.J. Evans.

The present paper examines the differences between Kummer's periods for composite n, when n is not squarefree, and the classical Gaussian periods for n a prime. Besides the fact that for a given non-squarefree n there exist values of q for which the η's vanish, the cyclotomy may no longer be unique. These and other distinctions are illustrated by examples.

Just as the Gaussian periods give rise to families of difference sets, the Kummer periods can be used to explain some of the known Singer difference sets arising from finite projective geometry. It is hoped that the further study of Kummer periods may bring to light a new family of difference sets.

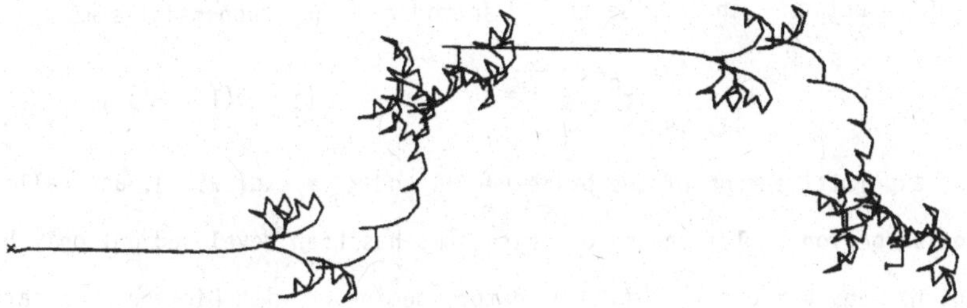

FIGURE 1

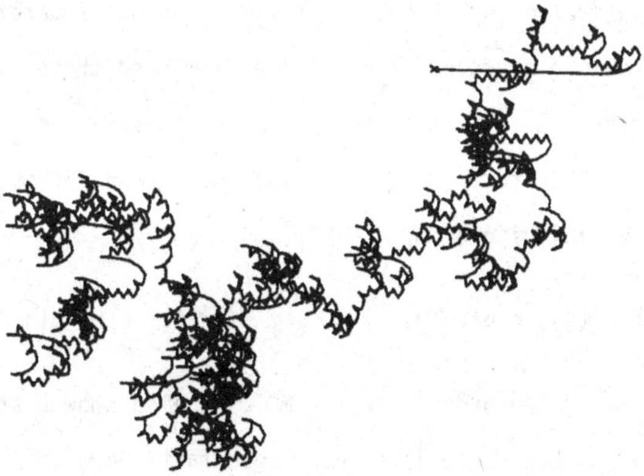

FIGURE 2

Introduction. In 1801 Gauss [4] introduced the exponential sums

(1) $$\eta_j = \sum_{\nu=0}^{f-1} \zeta_p^{g^{e\nu + j}} \qquad (j = 0(1)\ e-1) ,$$

where g is a primitive root of the prime p=ef+1 and $\zeta_p = \exp(2\pi i/p)$, and called them f-nomial periods. The theory of these sums has been developed not only by Gauss, but by many writers of the 20th century beginning with Dickson. In particular cyclotomy for most values of $e \leq 24$ has been largely explored and applied to the problem of finding difference sets.

In 1856 Kummer proposed a generalization in which p was replaced by a general number n and g^e by a generator q of order f modulo n. He used the notation

(2) $$\eta_k = \sum_{\nu=0}^{f-1} \zeta_n^{kq^\nu} \qquad k \not\equiv q^j (\text{mod } n),\ 0 < j < \phi(n)$$

The number of different periods is again denoted by e so that

(3) $$ef = \phi(n)$$

where $\phi(n)$ is the totient function. The index k is chosen so that kq^ν ranges over all the $\phi(n)$ totatives of n. He noted that if n is not squarefree, i.e. if $\mu(n) = 0$, then in some cases all $\eta_k = 0$. He dismissed these cases from further consideration without inquiring as to when this vanishing occurs.

In 1863 Fuchs [3] studied this question and gave a necessary and sufficient condition for the vanishing of η_k, namely

(4) $$f = t(n) = pt(n/p) \qquad \text{for some } p|n \quad ,$$

where $t(m) = t_q(m)$ is the order of q(mod m). He also showed that in case $\eta_k \neq 0$ it is an algebraic integer of degree e. These facts have been recently proved and generalized by Evans [2] in a more modern way.

In 1879, Sylvester [13] developed the theory of cyclotomy for any composite n with f=2. He envisaged a generalization for any value of f without referring to Kummer and gave examples of period polynomials for n=15,20,21,25,26,28 and 33. In the case of n=20 he noted that the periods vanish when q=11, but gave no explanation.

Although Sylvester predicted a great future for the theory of cyclotomy for composite n, it did not come to pass. Thus far only the cyclotomy modulo $n = p_1p_2$ has been considered by Whiteman [15] and Storer [12] in connection with difference sets. Beginning with Storer [11] all the work for a general n has been done by him and others in Galois fields and domains.

One reason why the case $\mu(n) = 0$ was avoided is the high probability that the periods may vanish for a given q. When Mahler [10] in 1977 studied the behavior of the function

$$f(z) = \sum_{i=0}^{\infty} z^{2^i}$$

near the unit circle he needed a Kummer period with q=2 and $n=p^2$. By (4) the necessary and sufficient condition for $\eta_k \neq 0$ becomes the Wieferich criterion

$$2^{p-1} \equiv 1 \pmod{p^2}$$

which holds for only two values of p below $6 \cdot 10^9$, namely p = 1093 and p = 3511 [8]. Figures 1 and 2 illustrate these two sums.

Another difficulty with composite n is that there may be more than one cyclotomy associated with a given n and e.

Finally we call attention to the fact that the set of subscripts k in Kummer's way of writing the periods may not have a single generator modulo n. For n = 56 for example the minumum set of k's for e=8,f=3 and q=9 is

$$k_i = 1,3,5,11,15,17,29,31 \quad .$$

In this case no simple algorithm can be given for computing the periods. For all these reasons a worker may prefer to go into Galois fields and domains.

In case $n = p^\alpha$, p odd, however, one has a primitive root to serve as a generator. In this case Klösgen [5] began to develop the theory of cyclotomic periods in his study of Fermat's Last Theorem.

We begin by finding all values of n with $\mu(n) = 0$ for which cylcotomy exists for the cases of e = 2,3,4,6 and by working out some of these cyclotomies. We next consider the case of p^α with particular emphasis on $\alpha = 2$. We conclude the paper with some remarks on difference sets.

1. Cyclotomy with small e. In case $f = 1$, $\eta_k = \zeta_n^k$ and the η's satisfy the cyclotomic polynomial

$$Q_n(X) = \prod_{\delta | n} (X^\delta - 1)^{\mu(n/\delta)}$$

This familiar cyclotomy is unique and exists for every n. It will not be considered in this section.

If $e = 1$, then $f = \phi(n)$ and since n is not squarefree

$$\eta = \sum_{(\nu,n) = 1} \zeta_n^\nu = \mu(n) = 0.$$

We shall assume from now on that $\eta_k \neq 0$, hence, we write n in the form

$$n = p^\alpha m \quad , \quad \alpha \geq 2, \ p \nmid m \ . \tag{5}$$

The letters p, α, n and m are reserved with these meanings. The case $p = 2$ is a little special as we shall see from

LEMMA 1. $\eta_k \neq 0$ if and only if for every p for which $p^2 | n$ either

(A) $$t(n) = t(pm)$$

or

(B) $p = 2$, $\alpha \geq 3$, $t(m)$ odd, $t(n) = 2t(m)$, $q \equiv 3 \pmod 4$.

This lemma is proved in the paper of Evans [2, Lemma 3].

If e is small, then Lemma 1 puts a heavy restriction on n in order that the periods do not vanish. If fact if (A) holds then

$$t(n) = t(mp) = \lambda(mp)/d \tag{6}$$

for some divisor d of $\lambda(mp)$, where $\lambda(h)$ is the least positive value of x for which

$$a^x \equiv 1 (\text{mod } h)$$

holds for all $(a,h) = 1$. Therefore

$$e = \phi(n)/t(n) = d\, p^{\alpha-1}\, \phi(mp)/\lambda(mp) \tag{7}$$

unless $\eta_k = 0$ for all values of k.

If (B) holds, then

$$t(n) = 2t(m) = 2\lambda(m)/d \tag{8}$$

and so

$$e = 2^{\alpha-2}\, d\, \phi(m)/\lambda(m). \tag{9}$$

We need the following

LEMMA 2. If $h>2$ is any integer, then $\phi(h)/\lambda(h)$ is divisible by 2^β, where

$$\beta = \begin{cases} \nu - 2 & \text{if } 2||h \\ \nu - 1 & \text{if } 4||h \text{ or if } h \text{ is odd} \\ \nu & \text{if } 8|h \end{cases} \tag{10}$$

where ν is the number of distinct prime factors of h.

For proof see for example Le Veque [9] for necessary details.

LEMMA 3. If n contains τ distinct prime factors then

$$e \text{ is divisible by } \begin{cases} p^{\alpha-1}2^{\tau-2} & \text{if } 2||n \\ p^{\alpha-1}2^{\tau-1} & \text{if } p \text{ is odd and } 4||n \text{ or } n \text{ odd} \\ p^{\alpha-1}2^{\tau} & \text{if } 8|n,\ p \text{ odd} \end{cases} \tag{11}$$

$$e \text{ is divisible by } 2^{\alpha+\tau-3} \qquad \text{if } p = 2,\ \alpha \geq 2,\ m > 1 \tag{12}$$

$$e = 2^{\alpha-2} \qquad \text{if } p = 2,\ \alpha > 2,\ m = 1,\ q\equiv 3 \pmod 4). \tag{13}$$

PROOF. This follows at once from (7) and (9) and lemma 2.

THEOREM 1. If $e = 2$, then $n = 8$ or $4m$, where m is a prime.

PROOF. Using (11) with $p = 2$, $\alpha = 2$ gives $\tau = 2$, $n = 4m$, m a prime. By (13) $n = 8$ is the only other case with $e = 2$.

If $n = 8$, then $q\equiv 3 \pmod 4$ and

$$\eta_1 = \zeta_8 + \zeta_8^3 = \zeta_8(1 + i) = \sqrt{2}\, i \qquad \text{if } q\equiv 3 \pmod 8$$

$$\eta_1'' = \zeta_8 + \zeta_8^7 = \bar{\zeta}_8(1 - i) = \sqrt{2} \qquad \text{if } q\equiv 7 \pmod 8.$$

This illustrates the fact that there are two cyclotomies with the same n and e.

The corresponding period polynomials are:

$$x^2 + 2 \qquad \text{if } q\equiv 3 \pmod 8$$

$$x^2 - 2 \qquad \text{if } q\equiv 7 \pmod 8) .$$

THEOREM 2. If $e = 2$, then $n = 4m$, m an odd prime and

$$\eta_1 = -\left(\frac{-1}{m}\right) i \qquad \text{if } q \equiv 1 \pmod 4$$

$$\eta_1 = \begin{cases} i\sqrt{m} & m\equiv 1 \pmod 4 \\ \sqrt{m} & m\equiv 3 \pmod 4 \end{cases} \qquad q\equiv 3 \pmod 4 .$$

PROOF. Since $f = \phi(4m)/2 = m-1$ we must choose q to be a primitive root g of m. If $q\equiv 1 \pmod 4$, then all the powers of q are congruent to one modulo 4 and letting $m' = \left(\frac{-1}{m}\right)m$ we have

$$\eta_1 = \sum_{\nu=0}^{m-2} \zeta_{4m}^{g^\nu} = \left(\frac{-1}{m}\right) \zeta_{4m}^{m} \sum_{\nu=0}^{m-2} \zeta_{4m}^{g^{\nu-m'}} = i\left(\frac{-1}{m}\right) \sum_{\nu=1}^{m-1} \zeta_m^{\nu} = -i\left(\frac{-1}{m}\right)$$

If $q = 3 \pmod 4$, then $q^\nu \equiv (-1)^\nu = \left(\frac{q^\nu}{m}\right) \pmod 4$, hence factoring out ζ^m as above we get

$$\eta_1 = \left(\frac{-1}{m}\right) i \sum_{j=1}^{m-1} \left(\frac{j}{m}\right) \zeta_m^j = \left(\frac{-1}{m}\right) i\, G_m = \begin{cases} i\sqrt{m}, & m\equiv 1 \\ \sqrt{m}, & m\equiv 3 \end{cases} \pmod 4,$$

since G_m is the Gauss sum.

COROLLARY. The period polynomial for $n = 4m$, m an odd prime and $e = 2$ is

$$x^2 + 1 \qquad \text{if } q = 1 \pmod 4$$

$$x^2 + m' \qquad \text{if } q = 3 \pmod 4$$

THEOREM 3. If $e = 4$, then $n = 16$, $4m$ or $4m_1 m_2$, where the m's are all odd primes.

PROOF. Since $p = 2$ by (11), $\alpha = 2$ and $\tau = 2$ or 3, so that $n = 4m$, m a prime or $n = 4m_1 m_2$. By (12) $\alpha = 3$ and $\tau = 2$, which gives $n = 8m$. By (13), $\alpha = 4$ and $n = 16$.

THEOREM 4. If $n = 4m$, m an odd prime and $e = 4$, then the period polynomial is

$$x^4 + (m' + 1/2)x^2 + ((m' - 1)/4)^2$$

where $m' = (\frac{-1}{m})m$.

PROOF. In this case $f = \phi(n)/4 = (m - 1)/2$, hence $q = g^2$, where g is a primitive root of m, so that the powers of q will all be quadratic residues of m, congruent to one modulo 4. If we choose $k \equiv 1 \pmod 4$ a quadratic non-residue of m, then obviously $\eta_1 + \eta_k$ will be a period modulo n for $e = 2$, $q \equiv 1 \pmod 4$ so by Theorem 2

(14) $$\eta_1 + \eta_k = -i(\tfrac{-1}{m}) .$$

We next show that

(15) $$\eta_1\eta_k = (m' - 1)/4 .$$

In fact

$$\eta_1\eta_k = \sum_{i,j=0}^{(m-3)/2} \zeta_n^{g^{2i} + kg^{2j}} .$$

The exponents

$$g^{2i} + kg^{2j} \equiv 2 \pmod 4) .$$

Since $(\frac{k}{m}) = -1$, we must have $(\frac{-1}{m}) = -1$ in order that

(16) $$g^{2i} + kg^{2j} \equiv 0 \pmod m$$

in which case there are $(m - 1)/2$ solutions (i,j) of (16). The total number of (i,j) is $((m - 1)/2)^2$. Hence each of the numbers prime to m must appear $(m-1)/4$ times if $(\frac{-1}{m}) = 1$ and $(m - 3)/4$ times if $(\frac{-1}{m}) = -1$. Therefore $\eta_1\eta_k$ becomes

$$\eta_1\eta_k = ((m - 3)/4) \sum_{\substack{\nu=0 \\ 2\nu+1\neq m}}^{m-1} \zeta_{2m}^{2\nu+1} + ((m - 1)/2)\, \zeta_{2m}^{m} = -(m+1)/4 \quad \text{if } m \equiv 3 \pmod 4)$$

and

$$\eta_1\eta_k = ((m-1)/4) \sum_{\substack{\nu=0 \\ 2\nu+1\neq m}}^{m-1} \zeta_{2m}^{2\nu+1} = (m-1)/4 \qquad \text{if } m\equiv 1 (\text{mod } 4).$$

This established (15), which together with (14) implies that

$$(17) \qquad (x-\eta_1)(x-\eta_k) = x^2 + (\tfrac{-1}{m})\, ix + (m'-1)/4$$

Multiplying (17) by its conjugate gives the period polynomial

$$x^4 + ((m'+1)/2)\, x^2 + ((m'-1/4))^2.$$

The discriminant of this quartic is

$$D = m^2(m - (\tfrac{-1}{m}))^2 \quad .$$

For $n = 20$ we have the Sylvester quartic $x^4 + 3x^2 + 1$ with $D = 400$.

<u>THEOREM 5</u>. If $e = 4$, $n = 8m$, m an odd prime and $q\equiv 3(\text{mod } 8)$, then the period polynomial is

$$x^4 - (m'-1)\, x^2 + (m'+1)^2/4, \qquad \text{where } m' = (\tfrac{-1}{m})\, m.$$

<u>PROOF</u>. In this case $f = \phi(n)/4 = m - 1$ and we let $q\equiv 3(\text{mod } 8)$ be a primitive root of m, then

$$\eta_1 = \sum_{\nu=0}^{\frac{m-3}{2}} (\zeta_n^{q^{2\nu}} + \zeta_n^{q^{2\nu+1}}) = \begin{matrix} \zeta_8(G_m-1)/2 - \zeta_8^3(G_m+1)/2, & m\equiv 1,3(\text{mod } 8) \\ \bar{\zeta}_8(G_m-1)/2 - \bar{\zeta}_8^3(G_m+1)/2, & m\equiv 5,7(\text{mod } 8) \end{matrix}$$

since $\zeta_n^m = \zeta_8 = (1+i)/\sqrt{2}$ and G_m is the Gauss sum. Therefore

$$(18) \qquad \eta_1 = \begin{cases} \sqrt{2}(\sqrt{m} - (\tfrac{2}{m})i)/2 & \text{if } m\equiv 1(\text{mod } 4) \\ \sqrt{2}i(\sqrt{m} + (\tfrac{2}{m}))/2 & \text{if } m\equiv 3(\text{mod } 4). \end{cases}$$

Since the other η's can be obtained by changing the sign of $\sqrt{m}$ and of i, the theorem follows.

The discriminant of this period polynomial is

$$(19) \qquad D = 2^6 \cdot m^2\, (m + (\tfrac{-1}{m}))^2$$

For $n = 40$, the equation is $x^4 - 4x^2 + 9$ with $D = 240^2$.

One could obtain similar polynomials for other values of q (mod 8) and for $n = 4m_1m_2$, but even more cases would have to be discussed depending on m_1, m_2 and q modulo 4. It is interesting to note that all these polynomials are simpler than the classical cyclotomic quartic for n a prime, which depends on the partition of $n = a^2 + 4b^2$.

For $n = 16$ we have $f = 2$ by (13) and the Sylvester quartic with $q = 5$

$$x^4 - 4x^2 + 2.$$

In fact, by (13), $f = 2$ for $n = 2^\alpha$ with $\alpha \geq 3$. These will be discussed in the next section.

It follows from (11) that $e = 3$ implies $n = 9$ or $n = 18$, and hence $f = 2$. These correspond to the two Sylvester polynomials

$$x^3 - 3x + 1 \qquad \text{and} \qquad x^3 - 3x - 1$$

respectively. However, Weber [14] constructed cyclotomic cubics for $n = 9m$ with $m \equiv 1 \pmod 6$. His cubics can be obtained from Kummer's cyclotomy with $e = 6$ as we shall see. First we state

THEOREM 6. If $e = 6$, then $n = 9m$ or $18m$, where m is a prime, or $n = 36$, or $n = 4m$, $m \equiv 1 \pmod 6$ is a prime.

This follows at once from conditions (11).

If $n = 36$, $e = 6$, then $f = 2$ and we have the Sylvester sextic

$$x^6 - 6x^4 + 9x^2 - 3\ .$$

All the Sylvester polynomials given above are as corrected in [7].

If $n = 9m$, $m \equiv 1 \pmod 6$, m a prime, $e = 6$, $f = m - 1$, $q \equiv 2$ or $5 \pmod 9$ the six η_k's can be added in pairs to give

$$\theta_k = \eta_k + \eta_{ks}\ , \qquad \text{where} \quad s^2 \equiv 1 \pmod n),\ s \not\equiv 1 \pmod n)$$

where the θ_k are the roots of Weber's cubic

(20) $$x^3 - 3mx - mL/3 \qquad (L/3 \equiv 1 (\mathrm{mod}\ 3))$$

with

(21) $$4n = L^2 + 27M^2.$$

There are two cyclotomies depending on whether q is a quadratic residue of m or not and two values of L in (21), but it is not clear how they are related. For example for $n = 63 = 9\cdot 7$ we have two sextics as follows:

$q = 2$: $\quad x^6 + 14x^3 - 63x^2 - 10x - 224 \quad$ with $D = 3^{10}\cdot 5^{10}\cdot 7^6$

$q = 5$: $\quad x^6 - 21x^4 + 14x^3 + 63x^2 - 21x - 35 \quad$ with $D = 2^{26}\cdot 3^8\cdot 7^6$

Since no power of 2 is congruent to $-1 (\mathrm{mod}\ n)$ we can choose $s = -1$ in the first case and $s = 8$ for $q = 5$. This leads to the two Weber cubics (20)

$$x^3 - 21x - 28 \qquad \text{and} \qquad x^3 - 21x + 35$$

respectively, so that the corresponding values of L in (21) are $L = 12$ and $L = 15$.

Finally if $e = p$ is an odd prime, then by (11), p^2 divides n, so that $n = p^2$ is the least value of n with $e = p$. We shall consider this case in the next section.

2. The case $n = p^\alpha$. For $p = 2$, the case $n = 2^\alpha$ is particularily simple. By (B) of Lemma 1, $\alpha \geq 3$, $f = 2$, $e = 2^{\alpha-2}$ and $q \equiv 3 (\mathrm{mod}\ 4)$. Therefore $q = 2^\alpha - 1$ and $q = 2^{\alpha-1} - 1$ are the only two values of the square root of one modulo n.

Taking the first cyclotomy we have

$$\eta_1 = \zeta_{2^\alpha} + \overline{\zeta}_{2^\alpha} = 2\cos(2\pi/2^\alpha) .$$

Thus in this case Sylvester's period polynomial is the well-known Chebyshev polynomial $C_n(x)$ for $n = 2^{\alpha-2}$. Its coefficients are known. In fact this polynomial is

(22) $$\sum_{\nu=0}^{2^{\alpha-3}} (-1)^\nu \frac{2^{\alpha-2}}{2^{\alpha-2}-\nu} \binom{2^{\alpha-2}-\nu}{\nu} x^{2^{\alpha-2}-2\nu}$$

Its discriminant is

$$D = 2^{3\cdot 2^{\alpha-3}} .$$

The odd prime factors of the numbers represented by this polynomial are all of the form $2^{\alpha}z \pm 1$. In fact this polynomial is also the Lucas function

$$V_s(P,Q) = a^s + b^s, \text{ where a and b are roots of } y^2 - Py + Q = 0$$

for $s = 2^{\alpha-2}$, $P = x$ and $Q = 1$.

The second cyclotomy with $q = 2^{\alpha-1} - 1$ can be obtained from the above by changing x to ix.

The case of $n = 3^{\alpha}$ is also a simple one, since by (A) of Lemma 1 this requires $f = 2$, $e = 3^{\alpha-1}$ and we have $q = 3^{\alpha}-1$. Thus

$$\eta_1 = \zeta_{3^{\alpha}} + \bar{\zeta}_{3^{\alpha}} = 2 \cos (2\pi/3^{\alpha}) .$$

We can therefore obtain the period polynomial for $n = 3^{\alpha}$ by transforming the cyclotomic polynomial for $n = 3^{\alpha}$, namely

$$Q_{3^{\alpha}}(y) = y^{2\cdot 3^{\alpha-1}} + y^{3^{\alpha-1}} + 1 = y^{3^{\alpha-1}} (y^{3^{\alpha-1}} + 1 + y^{-3^{\alpha-1}})$$

(with ζ_3 as a root) into Sylvester's $\psi_{3^{\alpha}}(x)$ with the root $\zeta_{3^{\alpha}} + \bar{\zeta}_{3^{\alpha}}$ via the substitution $y + 1/y = x$. This gives us the period polynomial

$$\psi_{3^{\alpha}}(x) = 1 + V_{3^{\alpha-1}}(x,1) = 1 + \sum_{\nu=0}^{(3^{\alpha-1}-1)/2} (-1)^{\nu} \frac{3^{\alpha-1}}{3^{\alpha-1} - \nu} \binom{3^{\alpha-1}-\nu}{\nu} x^{3^{\alpha-1} - 2\nu} .$$

The discriminant of the period polynomial is

$$D = 3^{((2\alpha-1)3^{\alpha-1} - 1)/2} .$$

The prime factors of $\psi_{3^{\alpha}}(h)$ are all of the form $3^{\alpha}m \pm 1$. For example for $n = 27$ we have the polynomial

$$x^9 - 9x^7 + 27x^5 - 30x^3 + 9x + 1 \qquad \text{with } D = 3^{22} .$$

From now on p will be a prime ≥ 5 and $n = p^{\alpha}$, $\alpha \geq 2$. By (A) of Lemma 1 we have

$$f = t(p^{\alpha}) = t(p) = (p-1)/d \quad \text{and} \quad e = dp^{\alpha-1} \quad .$$

To start with we confine ourselves to the case $\alpha = 2$, $d = 1$, $e = p$ and $f = p - 1$. Hence $q = g^p$, where g is a primitive root of p^2. This defines a unique cyclotomy with $k = g^j$, $j = 0(1)p-1$, where

$$\eta_{g^j} = \eta'_j = \sum_{\nu=0}^{p-2} \zeta_{p^2}^{g^{p\nu + j}} \quad .$$

Here η' is written in Gauss notation, similar to (1).

We now introduce a third notation which facilitates the proofs of the theorems that follow, namely

$$\eta_i^* = \sum_{\nu=0}^{p-2} \zeta_{p^2}^{q^{\nu}(q + pi)} \quad . \tag{23}$$

For example for $n = 25$ we have $g = 2$ and $q = 2^5 \equiv 7 (\text{mod } 25)$ the five η's are

$$\eta_0^* = \zeta^7 + \zeta^{24} + \zeta^{18} + \zeta = \eta_0' = \eta_1$$
$$\eta_1^* = \zeta^{12} + \zeta^9 + \zeta^{13} + \zeta^{16} = \eta_4' = \eta_{16}$$
$$\eta_2^* = \zeta^{17} + \zeta^{19} + \zeta^8 + \zeta^6 = \eta_3' = \eta_8$$
$$\eta_3^* = \zeta^{22} + \zeta^4 + \zeta^3 + \zeta^{21} = \eta_2' = \eta_4$$
$$\eta_4^* = \zeta^2 + \zeta^{14} + \zeta^{23} + \zeta^{11} = \eta_1' = \eta_2$$

and the period polynomial in this case is

$$x^5 - 10x^3 + 5x^2 + 10x + 1 .$$

In general, the connection between the η^*'s and the η's is given by the following lemma.

<u>LEMMA 4</u>. Let $\gamma \equiv \text{ind}_g(p + q) \pmod p$. Then for all integers i, $\eta_i^* = \eta'_{i\gamma}$.

<u>PROOF</u>. If $i = 0$ the proof is easy; both sums are equal to $\sum_{\nu=0}^{p=2} \zeta_{p^2}^{g^{p\nu}}$,

If $i > 0$ we have

$$(p + q)^i \equiv q^i + ipq^{i-1} \equiv q^{i-1}(q + ip) \pmod{p^2}.$$

Therefore

$$\eta_i^* = \sum_{\nu=0}^{p-2} \zeta_{p^2}^{(q+ip)q^\nu} = \sum_{\nu=0}^{p-2} \zeta_{p^2}^{(q+p)^i q^{\nu+1-i}}$$

$$= \sum_{\nu=0}^{p-2} \zeta_{p^2}^{g^{\gamma i} g^{p(\nu+1-i)}} = \sum_{\nu=0}^{p-2} \zeta_{p^2}^{g^{\gamma i + p\nu}} = \eta'_{\gamma i} .$$

We can use this lemma to establish

LEMMA 5. Let $H(\zeta_{p^2})$ denote any polynomial in ζ_{p^2}, and let $T_t(H(\zeta_{p^2}))$ denote the result of replacing ζ_{p^2} by $\zeta_{p^2}^t$ in the expression $H(\zeta_{p^2})$. Then, if $p \nmid t$, $T_t(\eta_i^*) = \eta^*_{i + \bar{\gamma}\,\mathrm{ind}\,t}$, where $\gamma\bar{\gamma} \equiv 1 \pmod p$ and γ is defined in Lemma 4.

PROOF.

$$T_t(\eta_i^*) = T_t(\eta'_{i\gamma}) = \sum_{\nu=0}^{p-2} \zeta_{p^2}^{t(g^{p\nu + i\gamma})} = \sum_{\nu=0}^{p-2} \zeta_{p^2}^{g^{\mathrm{ind}\,t}\, g^{p\nu + i\gamma}}$$

$$= \sum_{\nu=0}^{p-2} \zeta_{p^2}^{g^{p\nu + i\gamma + \mathrm{ind}\,t}} = \eta'_{i\gamma + \mathrm{ind}\,t} = \eta^*_{i + \bar{\gamma}\,\mathrm{ind}\,t} .$$

3. Properties of period polynomials. The period polynomial can be written

$$F(x) = \prod_{k=0}^{p-1} (x - \eta_k^*) = \sum_{\nu=0}^{p} a_\nu x^{p-\nu} .$$

We use σ_N to denote the sum of the N-th powers of all the η^*'s. We begin with

THEOREM 7.

$$\sigma_N = \sum_{k=0}^{p-1} (\eta_k^*)^N$$

is divisible by p for all integer values of $N \geq 0$.

PROOF. For each composition

(25) $$N_1 + N_2 + \dots + N_{p-1} = N$$

of N into p-1 parts ≥ 0 we define the multinomial coefficient by

$$C(N_1, N_2, \dots, N_{p-1}) = \frac{N!}{N_1!\, N_2! \dots N_{p-1}!}$$

Then we can write using (23)

(26) $$\sigma_N = \sum_{\kappa=0}^{p-1} \Big(\sum_{\nu=0}^{p-2} \zeta_{p^2}^{(q+\kappa p)q^\nu} \Big)^N =$$

$$\sum_{(N)} C(N_1, N_2, \dots, N_{p-1})\, \zeta_{p^2}^{\sum_{\nu=1}^{p-1} N_\nu q^\nu} \sum_{\kappa=0}^{p-1} \zeta_p^{\kappa \sum_{i=1}^{p-1} N_i q^{i-1}}$$

where the outer sum extends over all compositions (25) of N. Now the inner sum over κ has the value p or zero according as p divides $\sum_{i=1}^{p-1} N_i q^{i-1}$ or not. Hence p divides σ_N.

A less direct proof of this and the next theorem will be found in Klösgen [5].

<u>THEOREM 8.</u> $\sigma_1 = 0$, $\sigma_2 = p(p-1)$, $a_1 = 0$, $a_2 = -p(p-1)/2$.

<u>PROOF.</u>

$$\sigma_1 = \sum_{(j,p)=1} \zeta_{p^2}^{j} = \mu(p^2) = 0 = a_1$$

If N = 2 there are two kinds of compositions of N. These are permutations of 2,0,0,...,0 and of 1,1,0,...,0 . In the first case p, being odd, fails to divide

$$\sum_{i=1}^{p-1} N_i q^{i-1} = 2q^j$$

and so this case fails to contribute to σ_2. In the second case

$$\sum_{i=1}^{p-1} N_i q^{i-1} = q^{s-1} + q^{t-1}$$

This is divisible by p when $q^{s-t} + 1$ is, that is when

$$s - t \equiv (p - 1)/2 \quad (\text{mod } p\text{-}1) \quad .$$

This happens precisely (p - 1)/2 times. Hence

$$\sigma_2 = p(p - 1) \quad .$$

From $a_2 = (\sigma_1^2 - \sigma_2)/2$ we have

$$a_2 = -p(p - 1)/2 \quad .$$

THEOREM 9. If $1 \leq j < p$, then

$$a_j \equiv 0(\text{mod } p)$$

PROOF. If $j < p$, Newton's formula gives

$$\sigma_j + \sigma_{j-1} a_1 + \sigma_{j-2} a_2 + \ldots + \sigma_1 a_{j-1} + ja_j = 0$$

From this by Theorem 7 the present theroem follows by induction on j.

THEOREM 10. $\sigma_p \equiv -p(\text{mod } p^2)$.

PROOF. If we put N = p in (25) we see that

$$C(N_1, N_2, \ldots, N_{p-1})$$

is divisible by p, except when one of the parts N_i is equal to p and the rest are zero. Since the sum on κ in (26) is p or zero, each term of (26) is a multiple of p^2, except as noted above. Hence

$$\sigma_p = \sum_{i=1}^{p-1} \zeta_{p^2}^{pq^i} \sum_{\kappa=0}^{p-1} \zeta_p^{\kappa pq^{i-1}} \equiv p \sum_{i=1}^{p-1} \zeta_p^{q^{i-1}} \equiv -p \ (\text{mod } p^2)$$

COROLLARY. $a_p \equiv 1(\text{mod } p)$.

This follows from Newton's formulas

$$\sigma_p + a_1\sigma_{p-1} + a_2 \sigma_{p-2} + \ldots + a_{p-1} \sigma_1 + a_p p = 0$$

and the preceding theorems.

An examination of the following table of Period Polynomials reveals that

$$a_p \equiv 1 \pmod{p^2} \qquad \text{for } p \leq 19.$$

In a private communication R.J. Evans has shown by Hilbert theory that this congruence holds for all primes p.

THEOREM 11. $F(-1) \equiv 0 \pmod{p}$.

PROOF. By Theorem 9 and the above corollary

$$F(-1) \equiv (-1)^p + 1 \equiv 0 \pmod{p}.$$

Table of Period Polynomials

p	$F_p(x)$
3	$x^3 - 3x + 1$
5	$x^5 - 10x^3 + 5x^2 + 10x+1$
7	$x^7 - 21x^5 - 21x^4 + 91x^3 +112x^2 - 84x - 97$
11	$x^{11} - 55x^9 + 33x^8 + 825x^7 - 396x^6 - 4972x^5 + 1287x^4 + 12760x^3 - 924x^2 - 10989x + 243$
13	$x^{13} - 78x^{11} - 65x^{10} + 2080x^9 + 2457x^8 - 24128x^7 - 27027x^6 + 137683x^5 + 110214x^4 - 376064x^3 - 128206x^2 + 363883x - 12167$
17	$x^{17} - 136x^{15} + 85x^{14} + 6154x^{13} - 6545x^{12} - 119680x^{11} + 168555x^{10} + 998835x^9 - 1749300x^8 - 2783546x^7 + 6581040x^6 - 678725x^5 - 3813882x^4 + 770593x^3 + 616267x^2 - 82620x - 577$
19	$x^{19} - 171x^{17} - 133x^{16} + 11476x^{15} + 15580x^{14} - 385833x^{13} - 673436x^{12} + 6916190x^{11} + 13391960x^{10} - 66283229x^9 - 126730380x^8 + 339213156x^7 + 582575340x^6 - 861915924x^5 - 1264657480x^4 + 868638105x^3 + 1138104275x^2 - 137550709x - 221874931$

4. Factors of the Discriminant. We shall need the following

LEMMA 6. Let $\lambda = 1 - \zeta_p$, then

$$p = \lambda^{p-1} R(\zeta_p)$$

where $R(x)$ is a polynomial with rational integer coefficients and λ^{p-1} is the highest power of λ dividing p.

Proof. This well known lemma is easily proved as follows:

$$p = \prod_{\nu=1}^{p-1} (1 - \zeta_p^{\nu}) = \prod_{\nu=1}^{p-1} (1 - \zeta_p)(1 + \zeta_p + \ldots + \zeta_p^{\nu-1}) = \lambda^{p-1} R(\zeta_p)$$

where

$$R(x) = \prod_{\nu=1}^{p-1} (1 + x + \ldots + x^{\nu-1}).$$

Since $\zeta_p^m \equiv 1 (\mathrm{mod}\ \lambda)$ for every m,

$$1 + \zeta_p + \ldots + \zeta_p^{\nu-1} \equiv \nu (\mathrm{mod}\ \lambda) \quad .$$

Therefore

$$R(\zeta_p) \equiv (p - 1)! \equiv -1 (\mathrm{mod}\ \lambda) \quad .$$

We now introduce the product

(27) $$P_k = \prod_{i=0}^{p-1} (\eta_i^* - \eta_{i+k}^*), \quad (1 \leq k < p) \quad .$$

Theorem 12. P_k defined above is a rational integer.

Proof. By Lemma 4, P_k is the norm of $\eta_0' - \eta_{k\gamma}'$ in the cyclotomic field, and hence is an integer.

Theorem 13. P_k is divisible by p^2.

Proof. We have

$$\eta_i^* - \eta_{i+k}^* = \sum_{\nu=0}^{p-2} (\zeta_{p^2}^{q^\nu(q + ip)} - \zeta_{p^2}^{q^\nu(q + (i + k)p)},$$

$$= \sum_{\nu=0}^{p-2} \zeta_{p^2}^{q^\nu(q + ip)} (1 - \zeta_p^{kq^\nu}) = \lambda \sum_{\nu=0}^{p-2} \zeta_{p^2}^{q^\nu(q + ip)} R_\kappa(\zeta_p)$$

where $R_k(x)$ is a polynomial with integer coefficients. Taking the product over

$i = 0(1)p - 1$ we find using Lemma 6 that

$$P_k = \lambda^p S_1(\zeta_{p^2}) = p\,\lambda\, S_2(\zeta_{p^2})$$

where S_1 and S_2 are polynomials with integer coefficients. Since λ divides p we have p^2 divides P_k.

<u>COROLLARY</u>. The discriminant D of F(x) is divisible by p^{2p-2}.

<u>PROOF</u>. This follows at once because

$$D = \prod_{0 \le i < j < p} (\eta_i^* - \eta_j^*)^2 = \prod_{k=1}^{p-1} P_k$$

and each factor of this last product is divisible by p^2.

We now return to the Gauss notation

$$\eta_i' = \sum_{\nu=0}^{p-2} \zeta_{p^2}^{g^{\nu p + i}}$$

<u>LEMMA 7</u>. Let $r \neq p$ be a prime and let $g^s \equiv r \pmod{p^2}$, then for any integer β

(28) $$(\eta_0' - \eta_k')^{r^\beta} \equiv \eta_{\beta s}' - \eta_{\beta s + k}' \pmod{r}.$$

<u>PROOF</u>. The proof is by induction on β.

<u>THEOREM 14</u>. Let r be a prime factor of the discriminant D, then if $r \neq p$, r is a p-th power residue of p^2.

<u>PROOF</u>. Suppose that r is not a p-th power residue, then

$$r \equiv g^s \pmod{p^2} \quad \text{and} \quad s \not\equiv 0 \pmod{p}.$$

Since r divides D we can suppose that it divides some P_k. Applying Lemmas 4 and 7 for $\beta = 0(1)p-1$ and multiplying together the congruences (28) gives, with $h = k\gamma$

$$(\eta_0' - \eta_h')(r^p-1)/(r-1)^{(r^p-1)/(r-1)} \equiv P_k \equiv 0 \pmod{r}$$

since s and p are coprime. From this it follows that

$$0 \equiv (\eta_0' - \eta_h')^{r^p} \equiv \eta_0' - \eta_h' (\mathrm{mod}\ r) \quad .$$

But this is impossible since $h \neq 0$, and the η's are incongruent (mod r).

Theorem 15. Let r be a prime factor of a value F(N) of the period polynomial F(x). Then either r = p or r is a p-th power residue modulo p^2.

Proof. Suppose $r \neq p$ and r is not a p-th power residue mod p^2 so that $r = g^s (\mathrm{mod}\ p^2)$, where $s \not\equiv 0 (\mathrm{mod}\ p)$; then

$$\eta_i'^{\,r} \equiv \eta_{i+s}' \qquad (\mathrm{mod}\ r).$$

By Lagrange's identical congruence

$$\eta_i' - \eta_{i+s}' \equiv (-1)^r \eta_i' (1 - \eta_i') \ldots (r - 1 - \eta_i') \quad (\mathrm{mod}\ r).$$

Letting i = 0(1)p-1 and multiplying these congruences together we get

$$P_{s\bar{\gamma}} \equiv F(0)F(1) \ldots F(r-1) \qquad (\mathrm{mod}\ r).$$

But then r divides F(N) and therefore $P_{s\bar{\gamma}}$ and hence D. By Theorem 14 this is impossible since r is not a p-th power residue modulo p^2.

Linear forms of primes dividing D_p and F_p

p	
3	18z ± 1
5	50z ± 1, 7
7	98z ± 1, 19, 31
11	242z ± 1, 3, 9, 27, 81
13	338z ± 1, 19, 23, 89, 99, 147
17	578z ± 1, 65, 75, 131, 155, 179, 249, 251
19	722z ± 1, 69, 99, 127, 245, 293, 299, 307, 333

D_p

p	
3	3^4
5	$5^4 \cdot 7$
7	$7^6 \cdot 19 \cdot 31$
11	$11^{10} \cdot 3^{22} \cdot 457$
13	$13^{12} \cdot 19^3 \cdot 23^5 \cdot 337 \cdot 823 \cdot 7121 \cdot 21317$
17	$17^{16} \cdot 131^2 \cdot 179 \cdot 827 \cdot 8669 \cdot 32237 \cdot 58313 \cdot 106417 \cdot 122611 \cdot 544631 \cdot 1005971 \cdot 1746007$
19	$19^{18} \cdot 307 \cdot 389 \cdot 1571 \cdot 251501 \cdot 1596341 \cdot 1694603 \cdot 5649949 \cdot 7131623 \cdot 34404091 \cdot 239214961 \cdot 1342190653 \cdot 15613677091$

Factorization of $F_p(N)$

p	N = -1	N = 0	N = 1	N = 2
3	3	1	-1	3
5	5	1	7	-7
7	7	-97	-19	31
11	$3^5 \cdot 11$	3^5	-3^7	3^6
13	$-13 \cdot 23^3$	-23^3	$19 \cdot 2557$	$89 \cdot 239$
17	$17 \cdot 216751$	-577	-93481	2366267
19	$19 \cdot 1499593$	-221874931	$1867 \cdot 143971$	$127 \cdot 567881$

The reader cannot have failed to notice that all the above numbers are odd. Before concluding that there is a theorem let the reader consider the following.

THEOREM 16. For D or $F_p(N)$ to be even, the prime p must be 1093 or 3511, or else $p > 6 \cdot 10^9$.

PROOF. For either D or $F_p(N)$ to be even the prime 2 must be a p-th power residue modulo p^2 by Theorems 14 and 15, but for $x^p \equiv 2 \pmod{p^2}$ to hold, x must be congruent to 2(mod p). This implies

$$2^p \equiv 2 \pmod{p^2}$$

and the conclusion of the theorem follows from [8].

It is an open question whether the discriminant or any value of $F_p(N)$ can be even for p = 1093 or 3511.

Most of the above results for $n = p^2$ are easily generalized to $n = p^\alpha$, $\alpha > 2$ and can be proved by the same methods. The simplest case is again that of f=p-1 and so $e = p^{\alpha-1}$. The products P_k are now divisible by higher powers of p than p^2, whenever k is divisible by p. In fact D in now divisible by p^T, where

$$T = (\alpha - 1)p^{\alpha-1} + (p - 2)(p^{\alpha-1} - 1)/(p - 1) .$$

Thus when n = 81, $D = 3^{94}$. The other prime divisors r of D and of F(N) are $p^{\alpha-1}$ - st power residues of p^α.

A more complicated cyclotomy arises for $n = p^\alpha$, $\alpha \geq 2$, when f = (p - 1)/d for d > 1. Here $e = dp^{\alpha-1}$ and has other prime factors besides p. In this case we have

$$T = (\alpha d - 1)p^{\alpha-1} + (p - 1 - d)(p^{\alpha-1} - 1)/(p - 1) .$$

The prime factors r of D and of F(N) are now such that r^d is a p-th power residue of p^α.

The above results are generalizations of Kummer's work for n = p and were envisaged by both Kummer and Sylvester.

5. Application to difference sets. It is well known that for p a prime the theory of cyclotomy led to the discovery of difference sets consisting of quadratic and quartic and octic residues and of mixtures of various cosets of sextic and other residues [1].

However the Singer difference sets, which arose from finite [1, Ch.V] projective geometry, and were later studied in Galois fields have not been previously connected with cyclotomy.

These sets have parameters (v,k,λ) given by

$$v = (q^{N+1} - 1)/(q - 1), \quad k = (q^N - 1)/(q - 1), \quad \lambda = (q^{N-1} - 1)/(q - 1).$$

For $q = N = 3$ we have the set (40,13,4) with elements

$$d_i = 1,2,3,5,6,9,14,15,18,20,25,27,35$$

which can be grouped as follows

$$d_i = 3^\nu,\ 2 \cdot 3^\nu\ (\nu = 0,1,2,3) \text{ and } 5\nu\ (\nu = 1,3,4,5,7)$$

The Hall-Ryser polynomial [1,p.8] is

$$\theta = \sum_{i=1}^{13} \zeta_{40}^{d_i} = \eta_1 + \eta_1'' - 1 \tag{29}$$

where η_1 is the Kummer period for $n = 40$, $e = 4$, $q = 3$

and η_1'' is the Kummer period for $n = 20$, $e = 2$, $q = 3$

By Theorem 5

$$\eta_1 = (\sqrt{10} + i\sqrt{2})/2$$

and by Theorem 2

$$\eta_1'' = i\sqrt{5}$$

so that by (29)

$$\theta = (\sqrt{10} - 2 + i\ (\sqrt{2} + 2\sqrt{5}))/2$$

and hence

$$\theta\bar{\theta} = 9 = k - \lambda\ .$$

Another interesting example is provided by the two non-isomorphic difference sets with parameters (63,31,15) with $N = 5$, $q = 2$. They can be written:

A: $d_i = 0, 9 \cdot 2^\nu, 27 \cdot 2^\nu (\nu = 0,1,2);\ 2^\nu, 3 \cdot 2^\nu, 7 \cdot 2^\nu, 13 \cdot 2^\nu (\nu = 0,1,\ldots,5)$

B: $d_i = 0, 9 \cdot 2^\nu, 27 \cdot 2^\nu (\nu = 0,1,2);\ 2^\nu, 3 \cdot 2^\nu, 5 \cdot 2^\nu, 23 \cdot 2^\nu (\nu = 0,1,\ldots,5)$

They are both connected with the cyclotomy for $n = 63$, $e = 6$ and $q = 2$, discussed in §1 under $n = 9m$ and $e = 6$. In fact we can write

$$\theta = \eta_1 + \eta_{13} + \eta_1'' - 1 \qquad \text{for set A}$$

and

$$\theta = \eta_1 + \eta_5 + \bar{\eta}_5 + \eta_1'' - 1 \qquad \text{for set B}$$

where $\eta'' = (1 - \sqrt{7}\, i)/2$ is the period for $n = 21$, $e = 2$. One can verify that $\theta\bar{\theta} = 16$ in both cases.

Similarily the four non-isomorphic sets with parameters (121,40,13) for $N = 4$ and $q = 3$ can be related to the cyclotomy for $n = 121$, $e = 11$, $q = 3$ discussed in §2. In all four cases the elements of the difference set are made up of powers of 3 and of multiples of powers of 3 so that θ is again a sum of the η's. Thus for the set D in [1, p.153] we have

$$\theta = \eta_1 + \bar{\eta}_2 + \eta_4 + \eta_5 + \bar{\eta}_5 + \eta_7 + \bar{\eta}_7 + \eta_{17}.$$

It is hoped that this connection may bring to light some new difference sets, just as the development of cyclotomy for $n = pq$ in Whiteman [15] and Storer [12] led to new difference sets.

REFERENCES

1. L.D. Baumert, Cyclic Difference Sets, Lecture Notes in Math. 182, Springer - Verlag, 1971.
2. R.J. Evans, Generalized Cyclotomic Periods, to Proc. Amer. Math. Soc., 81 (1981) 207-212.
3. L. Fuchs, Ueber die Perioden, welche aus den Wurzeln der Gleichung $w^n = 1$ gebildet sind, wenn n eine zusammengesetzte Zahl ist, J. reine angew. Math. 61 (1863) 374-386.
4. C.F. Gauss, Disquisitiones Arithmeticae, Yale Univ. Press. 1966.
5. W. Klösgen, Untersuchungen Uber Fermatische Kongruenzen, Gesell. Math. und Datenverarbeitung, no. 36, 124pp, 1970, Bonn.
6. E. Kummer, Theorie der idealen Primfaktoren der complexen Zahlen, welche aus den Wurzeln der Gleichung $w^n = 1$ gebildet sind, wenn n eine zusammengesetzte

Zahl ist, Math. Abh. Akad. Wiss. Berlin (1856) 1 - 47; Collected Papers v.1, 583 - 629, Springer-Verlag, 1975.

7. D.H. Lehmer, Guide to Tables in the Theory of Numbers, Nat. Res. Council, Bull. 105, 1941, 169p.

8. D.H. Lehmer, On Fermat's Quotient, base 2, Math. Comp. 36 (1981) 289-290.

9. W.J. LeVeque, Topics in Number Theory, v.1, p. 53, Addison-Wesley, 1956.

10. K. Mahler, On a special function, Jn. Number Theory, 12 (1980) 20-26.

11. T. Storer, Cyclotomy and Difference Sets, Lectures in Advanced Math. V.2, Markham, Chicago, 1967.

12. T. Storer, Cyclotomies and difference sets modulo a product of two distinct odd primes, Mich. Math. Jn. 14 (1967) 117-127.

13. J.J. Sylvester, Collected papers, v.3, pp 325-339, Cambridge 1909.

14. H. Weber, Lehrbuch der Algebra, v.2, § 23, 3rd ed. Chelsea, N.Y. 1961.

15. A.L. Whiteman, A family of difference sets, Ill. Jn. Math. 4(1962) 107-112.

WARING'S PROBLEM FOR SETS OF DENSITY ZERO

Melvyn B. Nathanson
Department of Mathematics
Southern Illinois University
Carbondale, Illinois 62901

Dedicated to Emil Grosswald

1. Introduction

Lagrange proved in 1770 that every positive integer is the sum of four squares. Waring conjectured in 1770 and Hilbert [7] proved in 1909 that, for every $k \geq 2$, every positive integer is the sum of a bounded number of nonnegative k-th powers. Denote by $g(k)$ the smallest number h such that every positive integer is the sum of h nonnegative k-th powers. Denote by $G(k)$ the smallest number h such that every sufficiently large positive integer is the sum of h nonnegative k-th powers.

Let $r_{k,s}(n)$ denote the number of solutions of the equation

$$a_1^k + a_2^k + \ldots + a_s^k = n$$

in positive integers $a_1, a_2, \ldots, a_s$. Hardy and Littlewood [6] proved that if $s \geq s_0(k)$, then

$$0 < \liminf_{n \to \infty} \frac{r_{k,s}(n)}{n^{(s/k)-1}} \leq \limsup_{n \to \infty} \frac{r_{k,s}(n)}{n^{(s/k)-1}} < \infty.$$

This implies that for every $s \geq s_0(k)$ there exist positive constants $c_1 = c_1(k,s)$ and $c_2 = c_2(k,s)$ such that

$$(1) \qquad c_1 n^{(s/k)-1} \leq r_{k,s}(n) \leq c_2 n^{(s/k)-1}$$

where the inequality on the right holds for all $n \geq 1$, and the inequality on the left holds for all $n \geq n_0(k,s)$.

In general, a sequence A of nonnegative integers is a basis of order h (resp. asymptotic basis of order h) if every (resp. every sufficiently large) integer is the sum of h elements of A. The sequence A is a basis (resp. asymptotic basis) if A is a basis (resp. asymptotic basis) of order h for some h. Waring's problem asserts that the sequence of k-th powers of the nonnegative integers is a basis. Let A(x) denote the number of elements of the sequence A that do not exceed x. Then A has positive density if $A(x) \geq \alpha x$ for some $\alpha > 0$ and all $x \geq x_0$, and A has density zero if $\lim_{x \to \infty} A(x)/x = 0$. Shnirel'man [8] proved in 1930 that if A is any sequence of positive density and $0,1 \in A$, then the sequence of k-th powers of the elements of A forms a basis. Of course, if A has density zero, then the k-th powers of the elements of A do not necessarily form a basis. I prove in Theorem 1 that for any $s > s_0(k)$ and $\varepsilon > 0$ there exists a probability measure on the space of all strictly increasing sequences of positive integers such that, with probability 1, a random sequence A satisfies $A(x) = O(x^{1-(1/s)+\varepsilon})$ and the k-th powers of the elements of A form an asymptotic basis of order s. Thus, A has density zero.

Erdős and Nathanson [3] obtained a similar result in the case k = 2. For every $\varepsilon > 0$, they proved the existence of a sequence A whose squares form a basis of order 4 and which satisfies $A(x) = O(x^{(3/4)+\varepsilon})$. I show in Theorem 2 that there exists a sequence A whose squares form a basis of order 4 and which satisfies $A(x) = O(x^{(2/3)+\varepsilon})$. Of course, $A(x) = O(x^{1/2})$ would be best possible.

Choi, Erdős, and Nathanson [2] constructed, for every n > 1, a finite set A of squares such that

$$\text{card}\,(A) < \left(\frac{4}{\log 2}\right) n^{1/3} \log n$$

and each of the integers 0,1,2,...,n is the sum of four squares belonging to A. I construct in Theorem 3 an analogue of this result for k-th powers.

The proofs of Theorems 1 and 2 use the probabilistic method of Erdős and Rényi [4]. (Halberstam and Roth [5, Chapter 3] contains an excellent exposition of this method.) Let p(1), p(2), p(3),... be an arbitrary sequence of real numbers in [0,1]. Erdős and Rényi constructed a probability measure μ on the space Ω of all strictly increasing sequences of positive integers such that, for every $n \geq 1$, the event

$$E_n = \{A \in \Omega | n \in A\}$$

is measurable, and $\mu(E_n) = p(n)$. Moreover, the events $E_1, E_2, \ldots$ are independent. In particular, if $0 < \delta < 1$ and $p(n) = n^{-\delta}$ for all $n \geq 1$, then the law of large numbers implies that, with probability 1, a random sequence $A \in \Omega$ satisfies

$$A(x) \sim \left(\frac{1}{1-\delta}\right) x^{1-\delta}$$

as $x \to \infty$ (Halberstam and Roth [5, p. 145]).

2. Results.

THEOREM 1. For any $s > s_0(k)$ and $0 < \varepsilon < 1/s$, there exists a probability measure on the space Ω of all strictly increasing sequences of positive integers such that, with probability 1, a random sequence A has the following properties:

(i) $A(x) \sim cx^{1-(1/s)+\varepsilon}$ for some $c > 0$, and

(ii) Every sufficiently large integer n can be represented in the form

$$n = a_1^k + a_2^k + \ldots + a_s^k \tag{2}$$

with $a_1, a_2, \ldots, a_s \in A$.

PROOF. Consider two representations of n as a sum of s k-th powers,

$$n = a_1^k + a_2^k + \ldots + a_s^k = b_1^k + b_2^k + \ldots + b_s^k.$$

These representations intersect if $a_i = b_j$ for some i and j. If $a_i \neq b_j$ for all i and j, then the representations are disjoint. Let $d_{k,s}(n)$ denote the number of representations in some maximal collection of pairwise disjoint representations of n in the form (2). Recall that $r_{k,s}(n)$ denotes the number of representations of n as a sum of s positive k-th powers, and that the Hardy-Littlewood asymptotic formula (1) holds for $s \geq s_0(k)$. Consider a fixed representation

$n = u_1^k + \ldots + u_s^k$. For each i, the number of representations $n = a_1^k + \ldots + a_s^k$ with $a_j = u_i$ is exactly $r_{k,s-1}(n-u_i^k)$, and so the number of representations with $a_j = u_i$ for some j is at most $s r_{k,s-1}(n-u_i^k)$. Since $s-1 \geq s_0(k)$, the number of representations that intersect the fixed representation is at most

$$(3) \quad \sum_{i=1}^{s} s\, r_{k,s-1}(n-u_i^k) \leq \sum_{i=1}^{s} s\, c_2(k,s-1)(n-u_i^k)^{((s-1)/k)-1} \leq c_3 n^{((s-1)/k)-1}$$

Because every representation (2) intersects at least one representation in the maximal collection of $d_{k,s}(n)$ pairwise disjoint representations, it follows from (1) and (3) that for $n \geq n_0(k,s)$

$$c_1 n^{(s/k)-1} \leq r_{k,s}(n) \leq c_3 n^{((s-1)/k)-1} d_{k,s}(n)$$

and so

$$d_{k,s}(n) \geq c_4 n^{1/k}.$$

Construct an Erdös-Rényi probability measure on Ω by setting $p(n) = n^{-\delta}$, where $\delta = (1/s)-\varepsilon$. The probability that a given representation $n = a_1^k + \ldots + a_s^k$ is possible with a random sequence A is precisely the probability that A contains each of the numbers $a_1, a_2, \ldots, a_s$, that is, $\Pi' a_i^{-\delta}$, where the product is taken over the distinct integers among $a_1, a_2, \ldots, a_s$. Since (2) implies that $a_i \leq n^{1/k}$, it follows that

$$\Pi' a_i^{-\delta} \geq \prod_{i=1}^{s} a_i^{-\delta} \geq \prod_{i=1}^{s} n^{-\delta/k} = n^{-\delta s/k}.$$

Therefore, the probability that the representation $n = a_1^k + \ldots + a_s^k$ is <u>not</u> possible with a random sequence A is

$$1 - \Pi' a_i^{-\delta} \leq 1 - n^{-\delta s/k}.$$

Because of the independence of the events E_n, the probability that a random sequence A does not permit any one of the $d_{k,s}(n)$ representations in the maximal collections of pairwise disjoint representations of n is at most

$$(1-n^{-\delta s/k})^{d_{k,s}(n)} \leq (1-n^{-\delta s/k})^{c_4 n^{1/k}}$$

for $n \geq n_0(k,s)$. Since $\delta = (1/s) - \varepsilon$ and $(\delta s/k) < 1/k$, it follows that the series

$$\sum_{n=n_0}^{\infty} (1-n^{-\delta s/k})^{c_4 n^{1/k}}$$

converges. The Borel-Cantelli lemma implies that, with probability 1, a random sequence A will permit a representation $n = a_1^k + \ldots + a_s^k$ for all n sufficiently large.

Finally, since $p(n) = n^{-\delta}$, the law of large numbers implies that

$$A(x) \sim c\, x^{1-\delta} = c\, x^{1-(1/s)+\varepsilon}$$

for $c = (1-(1/s)+\varepsilon)^{-1} > 0$ and almost all sequences A. This completes the proof.

THEOREM 2. For any $\varepsilon > 0$ there exists a sequence A of nonnegative integers such that $A(x) = O(x^{(2/3)+\varepsilon})$ and every positive integer n can be represented in the form

$$n = a_1^2 + a_2^2 + a_3^2 + a_4^2$$

with $a_1, a_2, a_3, a_4 \in A$.

PROOF. Fix $0 < \varepsilon < 2/3$. Let $r_k(n)$ denote the number of solutions of the equation $n = a_1^2 + a_2^2 + \ldots + a_k^2$ in positive integers. Then $r_2(n) = O(n^{\varepsilon/4})$. Let

$$S = \{n > 0 \mid n \not\equiv 0, 4, 7 \pmod 8\}.$$

Bateman [1] and Siegel [9] proved that $r_3(n) \geq n^{(1/2)-(\varepsilon/4)}$ for all $n \in S$, $n \geq n_1(\varepsilon)$. Let $d_3(n)$ denote the number of representations in a maximal collection of pairwise disjoint representations of n in the form $n = a_1^2 + a_2^2 + a_3^2$. As in the proof of Theorem 1, it follows that

$$n^{(1/2)-(\varepsilon/4)} \leq r_3(n) \leq c_5 n^{(\varepsilon/4)} d_3(n)$$

and so

$$d_3(n) \geq c_6 n^{(1/2)-(\varepsilon/2)}$$

for some $c_6 > 0$ and all $n \geq n_1(\varepsilon)$.

Construct an Erdös-Rényi probability measure on the space Ω of all strictly increasing sequences of positive integers by setting $p(n) = n^{-\delta}$, where $\delta = (1/3) - (\varepsilon/2)$. Let $n = a_1^2 + a_2^2 + a_3^2$. Then $a_1, a_2, a_3 \leq n^{1/2}$, and the probability that a random sequence A contains a_1, a_2, a_3 is at least

$$a_1^{-\delta} a_2^{-\delta} a_3^{-\delta} \geq n^{-3\delta/2}.$$

The probability that A does not permit the representation $n = a_1^2 + a_2^2 + a_3^2$ is thus at most $1-n^{-3\delta/2}$. For $n \geq n_1(\varepsilon)$, the probability that a random sequence A does not permit any one of the $d_3(n)$ representations in the maximal collection of pairwise disjoint representations of n is at most

$$(1-n^{-3\delta/2})^{d_3(n)} \leq (1-n^{-3\delta/2})^{c_6 n^{(1/2)-(\varepsilon/2)}}.$$

Since $\delta = (1/3) - (\varepsilon/2)$ and

$$3\delta/2 = (1/2) - (3\varepsilon/4) < (1/2) - (\varepsilon/2),$$

it follows that the series

$$\sum_{\substack{n \geq n_1(\varepsilon) \\ n \in S}}^{\infty} (1-n^{-3\delta/2})^{c_6 n^{(1/2)-(\varepsilon/2)}}.$$

converges. The Borel-Cantelli lemma implies that, with probability 1, every sufficiently large number $n \in S$ can be written in the form $n = a_1^2 + a_2^2 + a_3^2$ with $a_1, a_2, a_3 \in A$.

By the law of large numbers, with probability 1, a random sequence A also satisfies $A(x) \sim c\, x^{(2/3)+(\varepsilon/2)}$.

Let A_1 be a fixed sequence of positive integers such that $A_1(x) = O(x^{(2/3)+(\varepsilon/2)})$ and all but finitely many $n \in S$ can be written in the form $n = a_1^2 + a_2^2 + a_3^2$ with $a_1, a_2, a_3 \in A_1$. Let F be a finite set of integers such that $0,1 \in F$ and <u>every</u> $n \in S$ is of the form $n = a_1^2 + a_2^2 + a_3^2$ with $a_1, a_2, a_3 \in A_2 = F \cup A_1$. If $n > 0$ and $n \equiv 7 \pmod 8$, then $n-1 \equiv 6 \pmod 8$ and so $n-1 \in S$. Then $n-1$ is a sum of three squares, hence n is a sum of four squares of elements of A_2.

The sequence A_2 satisfies $A_2(x) = O(x^{(2/3)+(\varepsilon/2)})$ and every integer $n \geq 0$ with $n \not\equiv 0 \pmod 4$ is of the form $n = a_1^2 + a_2^2 + a_3^2 + a_4^2$ with $a_1, a_2, a_3, a_4 \in A_2$.

Let the sequence A consist of all numbers of the form $2^k a$, where $a \in A_2$ and $k \geq 0$. Every integer $n \geq 0$ is of the form $n = 4^k m$, where $k \geq 0$ and $m \not\equiv 0 \pmod 4$. Then $m = a_1^2 + a_2^2 + a_3^2 + a_4^2$ with $a_1, a_2, a_3, a_4 \in A_2$, and so

$$n = 4^k m = (2^k a_1)^2 + (2^k a_2)^2 + (2^k a_3)^2 + (2^k a_4)^2$$

with $2^k a_1, 2^k a_2, 2^k a_3, 2^k a_4 \in A$. Thus, every positive integer is the sum of four squares of elements of A.

Finally, if $2^k a \leq x$, then $k \leq \log x/\log 2$ and so $A(x) \leq (1 + (\log x/\log 2))A_2(x)$. Therefore,

$$A(x) = O(x^{(2/3)+(\varepsilon/2)} \log x) = O(x^{(2/3)+\varepsilon}).$$

This completes the proof.

<u>THEOREM 3</u>. Let $k \geq 2$ and $s = g(k) + 1$. For any $\varepsilon > 0$ and $n \geq n_2(\varepsilon)$ there exists a finite set A of k-th powers such that

$$\text{card }(A) \leq (2 + \varepsilon)n^{1/(k+1)}$$

and each of the integers $0,1,2,\ldots,n$ is the sum of s k-th powers belonging to A.

. Let $\varepsilon > 0$. Denote by $[x]$ the integer part of x. Let $A = A_1 \cup A_2$, where

$$A_1 = \{a^k | 0 \leq a \leq (1 + \varepsilon)n^{1/(k+1)}\}$$

$$A_2 = \{[q^{1/k} n^{1/(k+1)}]^k | 1 \leq q \leq n^{1/(k+1)}\}$$

Since $[n^{1/(k+1)}] \in A_1 \cap A_2$, it follows that

$$\text{card }(A) \leq \text{card }(A_1) + \text{card }(A_2) - 1 \leq (2+\varepsilon)n^{1/(k+1)}.$$

By definition of $g(k)$, each integer $m \in [0,(1+\varepsilon)^k n^{k/(k+1)}]$ is a sum of $g(k)$, hence also of $s = g(k) + 1$, elements of $A_1 \subseteq A$.

Suppose $n^{k/(k+1)} < m \leq n$. Define

$$q = [\frac{m}{n^{k/(k+1)}}].$$

Then $1 \leq q \leq n^{1/(k+1)}$. Define

$$b = [q^{1/k}n^{1/(k+1)}].$$

Then $b^k \in A_2 \subseteq A$. Let $r = m - b^k$. Then

$$r = m-b^k \geq m-qn^{k/(k+1)} \geq 0$$

and

$$\begin{aligned} r = m-b^k &< m-(q^{1/k}n^{1/(k+1)}-1)^k \\ &= m-qn^{k/(k+1)} - \sum_{i=1}^{k-1} \binom{k}{i}(-1)^{k-i}q^{i/k}n^{i/(k+1)} \\ &< n^{k/(k+1)} + \sum_{i=0}^{k-1} \binom{k}{i} q^{i/k}n^{i/(k+1)} \\ &\leq n^{k/(k+1)} + 2^k(q^{1/k}n^{1/(k+1)})^{k-1} \\ &\leq n^{k/(k+1)} + 2^k(n^{1/k(k+1)}n^{1/(k+1)})^{k-1} \\ &= n^{k/(k+1)} + 2^k n^{(k-1)/k} \\ &\leq (1+\varepsilon)n^{k/(k+1)} \end{aligned}$$

if $n \geq n_2(\varepsilon)$. Therefore, r is a sum of $g(k)$ elements of $A_1 \subseteq A$, and so $m = b^k + r$ is a sum of $g(k) + 1 = s$ elements of A. This completes the proof.

3. Open Problems

The preceding results suggest new problems in additive number theory.

1. Let $h(k)$ (resp. $H(k)$) denote the smallest integer s such that there exists a sequence A of density zero such that the sequence of k-th powers of elements of A is a basis (resp. asymptotic basis) of order s. Theorem 2 implies

that $H(2) = G(2) = 4$ and $h(2) = g(2) = 4$. Do $H(k) = G(k)$ and $h(k) = g(k)$ for $k \geq 3$?

2. This is perhaps an easier question. Let $k \geq 3$. Does there exist an infinite set X of integers such that $\{a^k | a \notin X\}$ is a basis (resp. asymptotic basis) of order $g(k)$ (resp. $G(k)$)?

3. Let A be a strictly increasing sequence of nonnegative integers. Define

$$\alpha(A) = \limsup_{x \to \infty} \frac{\log A(x)}{\log x} .$$

If $A(x) = O(x^{\alpha})$, then $\alpha(A) \leq \alpha$. Define

$$\alpha(k,s) = \inf_{A} \alpha(A)$$

where the infimum is taken over all sequences A whose k-th powers form an asymptotic basis of order s. Theorem 1 implies that $\alpha(k,s) \leq 1-(1/s)$ for all $s > s_0(k)$. Estimate $\alpha(k,s)$. Clearly, $\alpha(k,s) \geq k/s$.

4. Theorem 2 implies that $1/2 \leq \alpha(2,4) \leq 2/3$. Are these inequalities strict? Erdös and Nathanson [3] conjectured that for every $\varepsilon > 0$ there exists a sequence A with $A(x) = O(x^{(1/2)+\varepsilon})$ such that every sufficiently large integer is the sum of four squares of elements of A. If $A(x) = O(x^{1/2})$, can the squares of the elements of A form an asymptotic basis of order 4?

5. Choi, Erdös, and Nathanson [2] conjectured that for every $\varepsilon > 0$ and for all n sufficiently large there exists a finite set A of squares such that

$$\text{card } (A) \leq n^{(1/4)+\varepsilon}$$

and each integer $m = 0,1,2,\ldots,n$ is a sum of four squares in A. Let $s \geq g(k)$. Denote by $f(n,k,s)$ the cardinality of the smallest set A of k-th powers such that each $m = 0,1,2,\ldots,n$ is the sum of s elements of A. Determine

$$\beta(k,s) = \limsup_{n \to \infty} \frac{\log f(n,k,s)}{\log n} .$$

Clearly, $\beta(k,s) \geq 1/s$. Does $\liminf_{s \to \infty} s\beta(k,s) = 1$? Does

$$\lim_{n \to \infty} \frac{\log f(n,k,s)}{\log n}$$

exist?

REFERENCES

1. P.T. Bateman, On the representations of a number as the sum of three squares, Trans. Amer. Math. Soc. 71 (1951), 70-101.

2. S.L.G. Choi, P. Erdös, and M.B. Nathanson, Lagrange's theorem with $N^{1/3}$ squares, Proc. Amer. Math. Soc. 79 (1980), 203-205.

3. P. Erdös and M.B. Nathanson, Lagrange's theorem and thin subsequences of squares, in J. Gani and V.K. Rohatgi, editors, Contributions to Probability: A Collection of Papers Dedicated to Eugene Lukacs, Academic Press, New York, 1981.

4. P. Erdös and A. Rényi, Additive properties of random sequences of positive integers, Acta. Arithm. 6 (1960), 83-110.

5. H. Halberstam and K.F. Roth, Sequences Vol. I, Clarendon Press, Oxford, 1966.

6. G.H. Hardy and J.E. Littlewood, Some problems of 'Partitio Numerorum': IV. The singular series in Waring's Problem and the value of the number G(k), Math. Zeit, 12 (1922), 161-188.

7. D. Hilbert, Beweis für die Darstellbarkeit der ganzen Zahlen durch eine feste Anzahl n-ter Potenzen (Waringsche Problem), Math. Ann. 67 (1909), 281-300.

8. L.G. Shnirel'man, On additive properties of integers, Izvestiyakh Donskogo Instituta v Novocherkasske 14 (1930), 3-27.

9. C.L. Siegel, Über die Klassenzahl quadratische Zahlkörper, Acta Arith. 1 (1935), 83-86.

SEQUENCES WITHOUT ARITHMETIC PROGRESSIONS

D.J. Newman

Dedicated to my friend and colleague, Emil Grosswald

Our purpose is to give a simplified version of Roth's proof of the theorem that a sequence of integers without any 3 terms in arithmetic progression must have 0-density. The amusing thing about this proof is that it contains all the standard ingredients of the analytic method - trigonometric sum estimates - major and minor arcs, contour integrals, etc. - but all in small, palatable doses!

Affine Properties and Permissivity Constants

A property (collection), P, of finite sets of non-negative integers is to be called <u>affine</u> if it satisfies the following two conditions:

1. $\{a_n\}$ has P if and only if $\{\alpha a_n+\beta\}$ has P. Here $\alpha(\neq 0)$ and β are any integers.

2. Every subset of a set with P also has P.

Thus, for example, the property, P_A, of not containing any arithmetic progression (i.e. three elements $a < b < c$ such that $a+c = 2b$) is an affine property. Again the trivial property, P_0, of just being a set, is affine.

Now let us fix any affine property P and consider a largest subset, $S = S(n,P)$, of the interval $[0,n]$ which has P. We then denote $f(n) = f(n;P) = |S|$ and we observe, from 1. and 2. that $f(m)+f(n) \geq f(m+n)$. That is to say the function $f(n)$ is subadditive and it is a well known consequence that therefore $\frac{f(n)}{n} \to$ limit as $n \to \infty$ (in fact $\lim \frac{f(n)}{n} = \inf \frac{f(n)}{n}$). So we define the "permissivity constant", C_P, as this limit, i.e. $C_P = \lim_{n \to \infty} \frac{f(n;P)}{n}$.

Clearly for the trivial property, P_0, we have $C_{P_0} = 1$. The remarkable fact, which was proved by Szemeredi and then later by Furstenberg, is that, for <u>all other affine properties</u>, $C_P = 0$. Before this great achievement Roth proved the special case of P_A. The proofs of Szemeredi and Furstenberg are very complicated and Roth's proof is not really simple. Our purpose here is to give a fairly direct analytic proof of Roth's theorem that $C_{P_A} = 0$.

The Basic Approximation Lemma

It turns out that the extremal sets, $S(n,P)$, no matter what the affine property, all behave very much as though their elements were chosen at random (with probability C_P, of course) and this is our Theorem 1 below. The utility of this result for P_A stems from the fact that a sequence chosen randomly with positive probability would have to have lots of arithmetic progressions and so we will be forced to the conclusion that $C_{P_A} = 0$.

THEOREM 1: $\sum_{a \in S(n;P)} z^a = C_P \sum_{k < n} z^k + o(n)$, uniformly on $|z| = 1$.

PROOF: The basic strategy is to estimate $q(z) = \sum_{a \in S} z^a - C_P \sum_{k < n} z^k$, together with all its partial sums, at all the roots of unity of orders up to some N. These points cut the unit circle into arcs, the so-called Farey arcs, whose lengths are then easily estimated. Thereby we will be able to obtain our required bounds on the whole circle by using the following

LEMMA: Let $p(z)$ be any polynomial of degree n and A be any arc (on the unit circle) of length ℓ. If there is some $\zeta \in A$ at which all the partial sums $p_m(\zeta)$ are bounded by 1 then, throughout A, $p(z)$ is bounded by $1+n\ell$.

PROOF: By direct division we have

$$\frac{p(z)}{1-\frac{z}{\zeta}} = \sum_{v < n} p_v(\zeta)\left(\frac{z}{\zeta}\right)^v + \frac{p(\zeta)}{1-\frac{z}{\zeta}}\left(\frac{z}{\zeta}\right)^n, \text{ or}$$

$p(z) = (1-\frac{z}{\zeta}) \sum_{v < n} p_v(\zeta)(\frac{z}{\zeta})^v + p(\zeta)(\frac{z}{\zeta})^n$ so that $p(z) \ll |1-\frac{z}{\zeta}| \cdot n+1$ and the proof is completed by the observation that $|1-\frac{z}{\zeta}| \leq \ell$ all along A.

In order to estimate $q_m(\omega)$, where $\omega^\alpha = 1$, let us write it as $\sum_{\beta=1}^{\alpha} \omega^\beta \Big(\sum_{\substack{a \in S \\ a < m \\ a \equiv \beta(\alpha)}} 1 - C_P \sum_{\substack{k < m \\ k \equiv \beta(\alpha)}} 1 \Big)$ and let us note that the first inner sum,

$T = \sum_{\substack{a \in S \\ a < m \\ a \equiv \beta(\alpha)}} 1$, counts the size of a subset of S, which therefore has the property P, and which is affine to a subset of $[0, \frac{m}{\alpha})$ so that it has at most $f(\frac{m}{\alpha})$ elements (where we write $f(x)$ for $f(\lceil x \rceil)$). Thus we have

$$q_m(\omega) = -\sum_{\beta=1}^{\alpha} \omega^{\beta}(f(\tfrac{m}{\alpha}) - T) + \sum_{\beta=1}^{\alpha} \omega^{\beta}(f(\tfrac{m}{\alpha}) - C_P \sum_{\substack{k < m \\ k \equiv \beta(\alpha)}} 1)$$

$$<< \sum_{\beta=1}^{\alpha} | f(\tfrac{m}{\alpha}) - T| + \sum_{\beta=1}^{\alpha} |f(\tfrac{m}{\alpha}) - C_P \tfrac{m}{\alpha} |$$

$$= \sum_{\beta=1}^{\alpha} (f(\tfrac{m}{\alpha}) - T) + \sum_{\beta=1}^{\alpha} (f(\tfrac{m}{\alpha}) - C_P \tfrac{m}{\alpha}) \quad .$$

$$= 2\alpha f(\tfrac{m}{\alpha}) - f(m) - C_P\, m \le 2\alpha\, (f(\tfrac{m}{\alpha}) - C_P \tfrac{m}{\alpha}).$$

Now from the fact that $f(x)-C_P x = o(x)$ it follows that $\max_{t \le x} (f(t)-C_P t) = o(x)$. Thus for every $\varepsilon > 0$ there is a K such that $\frac{n}{\alpha} > K$ implies (for $m \le n$) $f(\frac{m}{\alpha})-C_P \frac{m}{\alpha} < \varepsilon \frac{n}{\alpha}$. Furthermore if $\alpha < K$ then $\frac{n}{\alpha}$ is much larger and we can conclude, for large n, that e.g. $f(\frac{m}{\alpha})-C_P \frac{m}{\alpha} < \frac{\varepsilon}{K} \frac{n}{\alpha}$.

To summarize, then, we have shown that

$$q_m(\omega) << 2\, \varepsilon\, n \text{ for all } \alpha < \tfrac{n}{K} \text{ (minor arc)}$$

$$q_m(\delta) << \tfrac{2\varepsilon}{K}\, n \text{ for all } \alpha < K \text{ (major arcs).}$$

If we next observe the elementary fact that the Farey arc contiguous to an αth root of unity in the Farey disection with $N = \frac{n}{K}$ has length $\le \frac{2\pi}{\alpha \cdot \frac{n}{K}}$ then our lemma gives

$$q(z) << 2\, \varepsilon\, n(1+ \tfrac{2\pi K}{\alpha}),\ K \le \alpha < \tfrac{n}{K} \text{ (minor arcs)}$$

$$q(z) << 2\tfrac{\varepsilon}{K}\, n(1+ \tfrac{2\pi K}{\alpha}),\ 1 \le \alpha < K \text{ (major arcs).}$$

In either case we obtain $q(z) << 2(1+2\pi)\,\epsilon\, n$ and so this holds over the entire unit disc. Theorem 1 is proved.

THEOREM 2 (Roth). $C_{P_A} = 0$.

PROOF: With $g(z) = \sum_{a \in S} z^a$ (as above) we form $I = \frac{1}{2\pi i} \int_{|z|=1} g(z)\cdot g(z)\cdot g(z^{-2})\frac{dz}{z}$

and note that multiplying out shows that I is exactly equal to the number of triples from S which form an arithmetic progression (counting trivial ones where all three terms are equal). Similarly with $G(z) = \sum_{k<n} z^k$ we find that

$J = \frac{1}{2\pi i} \int_{|z|=1} G(z)\cdot G(z)\cdot G(z^{-2})\,\frac{dz}{z}$ is exactly equal to the number of triples below n which form an arithmetic progression. This is clearly equal to $\left\lceil \frac{1}{2} n^2 \right\rceil$ by elementary counting.

The heart of the matter is Theorem 1 from which we can conclude that I is well approximated by $C_P^3 \cdot J$. Indeed writing $g(z) = C_P G(z)+q(z)$ and inserting into I we find that $I = C_P^3 J$ + Error terms. Said error terms are integrals of products of three G or q functions with at least one factor being a q. By Theorem 1, $q = o(n)$ and so these error terms may be estimated by $o(n)\cdot$ [integral of products of 2 functions] where each of these functions is a $|q|$ or a $|G|$. Any such integral, however, is estimated using Schwarz' inequality, by $\sqrt{n\cdot n} = n$ and so, altogether, the error terms are $o(n^2)$.

Combining these remarks, we see that the number of triples from S which are in arithmetic progression is $C_P^3 \frac{n^2}{2} + o(n^2)$. Now specialize $P = P_A$ and note that then there are no nontrivial such triples in S.

CONCLUSION

$$0(n) = C_{P_A}^3 \frac{n^2}{2} + o(n^2)$$

and of course it follows that $C_{P_A} = 0$, just as we claimed.

ON POLYGON GROUPS

by

Joseph Lehner

Dedicated to My Good Friend and Former Colleague Emil Grosswald

1. Let SL(2,R) be the group of real 2×2 matrices of determinant 1. It acts on the upper half-plane $H = \{z = x+iy: y > 0\}$ by $z \mapsto A(z) = (az+b)/(cz+d)$ for $A \in SL(2,R)$. We write the transformation $z \mapsto A(z)$ as Az, for brevity. The corresponding group of Möbius transformations {Az} is written PSL(2,R) and we have the homomorphism ψ: SL(2,R) $\to$ PSL(2,R) with kernel (-I,I), where I = (1,0:0,1). Let $\sigma(A)$ be the trace of A.

If $G \subset SL(2,R)$ we denote the image $\psi(G) \subset PSL(2,R)$ by $\bar{G}$. Then G (as well as $\bar{G}$) is called fuchsian if G is discrete, i.e., if there is no infinite sequence of different elements in G that converges to I or to -I.

A polygon group is an abstract group Γ with presentation

$$(1.1) \qquad \Gamma = < a_1,\ldots,a_n: a_i^{m_i} = a_n a_{n-1}\ldots a_1 = 1,\ i = 1,\ldots,n \geq 3 >,$$

where $m_i \geq 2$ are integers subject to

$$(1.2) \qquad \sum_{i=1}^{n} m_i^{-1} < n-2.$$

(This is restrictive only for n = 3,4.) If G is a faithful representation of Γ as a matrix group in SL(2,R), G (as well as $\bar{G}$) is also called a polygon group; its presentation is

$$(1.3) \qquad G = < A_1,\ldots,A_n: A_i^{m_i} = \pm I,\ A_n A_{n-1}\ldots A_1 = \pm I > ,$$

where the A_i are elliptic elements ($|\sigma| < 2$). The case n = 3 comprises the familiar triangle groups.

The problem considered in this paper is whether a matrix polygon group G is discrete. That is, we wish to find necessary and sufficient conditions on the matrices of G to insure discreteness. Every element of finite order in G is conjugate to a power of a generator A_i; this is true whether G is discrete or not ([2], Cor. 2). There may be elliptic elements of infinite order ("infinitesimal elements"), and in fact G will contain such elements precisely when G is not discrete ([6], p. 96, ℓℓ. 3-5). If G is discrete, it will not contain parabolic elements ($\sigma = \pm 2$) but this is not necessarily true if G is not discrete. All groups considered here have hyperbolic elements ($|\sigma| > 2$).

2. We regard H as a model of hyperbolic geometry, which we use throughout. First, let G be a discrete polygon group and $\bar{G} = \psi(G)$ its image in PSL(2,R). $\bar{G}$ has a fundamental region bounded by a hyperbolic polygon ("fundamental polygon") with a finite number of sides. The minimum number of sides is 2n-2. There are many fundamental polygons, but any one that attains the minimum number of sides and is strictly convex is here called <u>canonical</u>. Fricke ([1], 285-310) and L. Keen ([3]) have discussed the existence and construction of canonical polygons for discrete groups, in particular, for discrete polygon groups.

Using Keen's method one shows that every discrete polygon group admits a canonical polygon P_n of the form shown in Figure 1 for n = 4. (The hyperbolic lines are drawn straight for convenience.) The condition (1.2) is necessary in order that the polygon be geometrically possible. P_4 has the following properties:

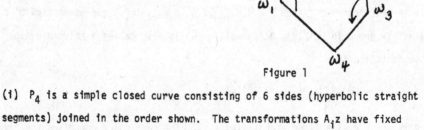

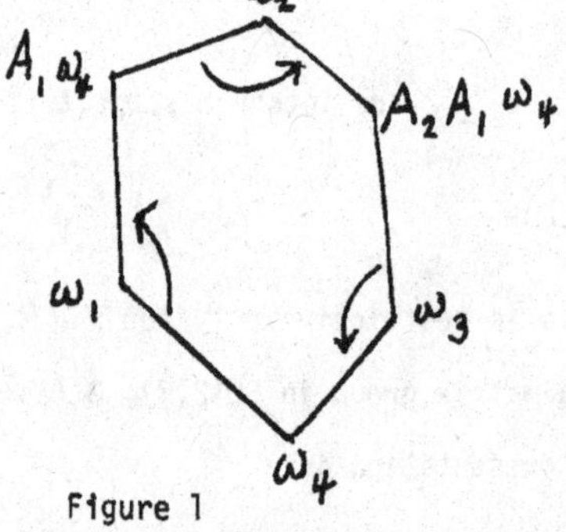

Figure 1

(i) P_4 is a simple closed curve consisting of 6 sides (hyperbolic straight line segments) joined in the order shown. The transformations A_iz have fixed points ω_i (i = 1,2,3) and A_iz maps one side of the angle at ω_i on the other. A_iz is a rotation in the positive sense.

(ii) The interior of P_4 is a fundamental region for G.

(iii) There are 3 one-point cycles $\{\omega_1\}$, $\{\omega_2\}$, $\{\omega_3\}$ and one three-point cycle $C = \{\omega_4, A_1\omega_4, A_2A_1\omega_4\}$. Since A_i is of order m_i $(i = 1,2,3,4)$ and (ii) holds, the angle at ω_i is $2\pi/m_i$ $(i = 1,2,3)$, while the sum of the angles at C is $2\pi/m_3$. ([6], p. 126).

(iv) By following a small vector around P_4 we see that $A_4^{-1}z$ has negative rotation, so A_4z has positive rotation.

(v) P_4 is strictly convex; this means that all angles of P_4 are less than π when $m_i > 2$ and equal to π when $m_i = 2$.

In general P_n will be a simple closed strictly convex polygon of 2n-2 sides that are mated in pairs by the A_i: it will have n-1 one-point cycles and a single (n-1)-point cycle. The existence of a P_n of this form is a necessary condition that G be discrete.

Consider the converse situation. Let G be a polygon group, not assumed to be discrete. G is given by (1.3) with

$$\sigma(A_i) = 2 \cos \frac{\pi}{m_i}, \quad i = 1,\dots,n \tag{2.1}$$

$$\sum_{1}^{n} m_i^{-1} < n-2, \quad n = 3,4,\dots$$

The A_iz are taken with positive rotation; the necessary and sufficient condition for this is

$$c_i\sigma(A_i) < 0, \quad A_i = (a_i,b_i:c_i,d_i). \tag{2.2}$$

We construct the polygon $P_n = \{\omega_n \equiv \tau_0, \omega_1, A_1\omega_n \equiv \tau_1, \omega_2, A_2A_1\omega_n \equiv \tau_2, \dots, \omega_{n-1}, A_{n-1}A_{n-2}\dots A_1\omega_n \equiv \tau_{n-1}\}$. This is a closed polygon by (1.3), since $\tau_{n-1} = A_n^{-1}A_nA_{n-1}\dots A_1\omega_n = A_n^{-1}\omega_n = \omega_n = \tau_0$. Suppose P_n is strictly convex. The $\{\omega_i\}$, $i = 1,\dots,n-1$, constitute one-point cycles. Since $A_i\omega_i = \omega_i$ and A_iz maps the side (τ_{i-1},ω_i) on (τ_i,ω_i), $i = 1,\dots,n$, the angle of P_n at ω_i is $2\pi/m_i$. The set $(\tau_1,\tau_2,\dots,\tau_{n-1})$ is an (n-1)-point cycle. Let θ_i be the angle of P_n at τ_i. Under the mapping $A_iA_{i-1}\dots A_1z$, the angular region at τ_i, say R_i, is mapped into a region with vertex τ_0, enclosing the angle θ_i. The regions appear in clockwise order beginning with $A_{n-1}\ldots A_1(R_{n-1})$ and ending with R_0. The union of these regions is an "elliptic sector" at τ_0 of angle $2\pi/m_i$, since the sides of the sector are mapped into each other by $A_n z$, as is easily verified. Hence

$$\theta_1 + \theta_2 + \ldots + \theta_{n-1} = 2\pi/m_n.$$

We are now in a position to make use of Poincaré's Theorem ([8]), which, for the present application, may be stated as follows. Let P_n be a hyperbolic polygon of 2n-2 sides, the sides being mated in pairs by transformations $A_i z$, $i = 1,\ldots,n$. Let there be n-1 one-point cycles with angle $2\pi/m_i$, $i = 1,\ldots,n-1$, $m_i \geq 2$, and one (n-1)-point cycle with angle sum $2\pi/m_n$, $\sum_1^n m_i^{-1} < n-2$. Then the group $< A_1,\ldots,A_n >$ is discrete. The hypotheses of Poincaré's Theorem hold in the present case and so G is discrete. Hence

<u>THEOREM 1.</u> If G is a polygon group satisfying (1.3), (2.1), then G is discrete if and only if the polygon P_n is strictly convex.

3. As we wish to use the matrix group G rather than $\bar{G}$, we must determine the $\pm$ sign in the presentation (1.3). We assume all A_i have nonnegative trace. Here we can rely on the results of J. Milnor[1] ([7], Sect. 3).

Let G be discrete. Let $\hat{P}$ be the universal covering group of PSL(2,R) with projection map π. Then $\pi^{-1}(G) = \hat{G}$ is called the centrally extended group G. We have $\hat{G}/C \cong G$ with C the center of $\hat{G}$. Milnor proves that C is cyclic: $C = < c >$, and that the presentation of $\hat{G}$ is

$$(3.1) \qquad < \hat{A}_1,\ldots,\hat{A}_n : \hat{A}_i^{m_i} = c,\ \hat{A}_n\ldots\hat{A}_1 = c^{n-2} > .$$

Now SL(2,R) is a 2-sheeted covering of PSL(2,R), so $\hat{P}$ is its universal covering group. The center of SL(2,R) is $\{I,-I\}$. Since c generates the center of $\hat{P}$, it follows that c projects to $-I$ in SL(2,R). We therefore have from (3.1),

$$(3.2) \qquad G = < A_1,\ldots, A_n : A_i^{m_i} = -I,\ A_n\ldots A_1 = (-I)^n > .$$

---
[1] This reference was kindly supplied by A.M. Macbeath.

(We remark that the equation $A_i^{m_i} = -I$ can easily be proved directly.) A discrete matrix polygon group, then, has the presentation (3.2). In future we shall use this presentation instead of (1.3) for a matrix polygon group.

4. Let us consider the case n = 3 (triangle group) in more detail. We have

$$(4.1) \qquad G = \langle A_1, A_2, A_3 : A_1^{m_1} = A_2^{m_2} = A_3^{m_3} = A_3A_2A_1 = -I \rangle, \ m_1^{-1} + m_2^{-1} + m_3^{-1} < 1.$$

Since the A_i may be interchanged, we can insist on

$$m_1 \le m_2 \le m_3.$$

Now assume G is discrete. Then, as we have seen,

$$(4.2) \qquad \sigma(A_i) = 2 \cos \pi/m_i.$$

Also $\sigma(A_2A_1) = \sigma(-A_3^{-1}) = -\sigma(A_3)$, so

$$(4.3) \qquad \sigma(A_2A_1) = -2 \cos \pi/m_3.$$

Moreover, we may assume

$$(4.4) \qquad A_i z \text{ has positive rotation.}$$

The conditions (4.1) - (4.4) are therefore necessary for G to be discrete.

Conversely, let us assume G is a group satisfying (4.1) - (4.4). By a transformation of PSL(2,R) we may map the line $\omega_1\omega_2$ into the imaginary axis so that $\omega_1 = i$, $\omega_2 = i\rho$ for some ρ in $0 < \rho < 1$. We use the same notation for the transformed group. Remembering (4.4) we can write

$$(4.5) \quad A_1 = \begin{pmatrix} \cos \frac{\pi}{m_1} & \sin \frac{\pi}{m_1} \\ -\sin \frac{\pi}{m_1} & \cos \frac{\pi}{m_1} \end{pmatrix}, \ A_2 = \begin{pmatrix} \cos \frac{\pi}{m_2} & \rho \sin \frac{\pi}{m_2} \\ -\rho^{-1} \sin \frac{\pi}{m_2} & \cos \frac{\pi}{m_2} \end{pmatrix}$$

Then (4.3) becomes

$$(4.6) \quad 2 \cos \frac{\pi}{m_1} \cos \frac{\pi}{m_2} - (\rho+\rho^{-1}) \sin \frac{\pi}{m_1} \sin \frac{\pi}{m_2} = -2 \cos \frac{\pi}{m_3},$$

which has a unique solution ρ in $0 < \rho < 1$. This is clear when $m_1, m_2 \geq 3$, $m_3 \geq 4$; the remaining cases ($m_1 = 2$, $m_2 \geq 3$, $m_3 \geq 4$) have to be handled individually.

Let ω_{12} be the fixed point of A_1A_2, etc. By calculation from (4.5), using $0 < \rho < 1$, we find

$$\text{(4.7)} \qquad \operatorname{Re} \omega_{12} > 0, \quad \operatorname{Re} \omega_{21} < 0.$$

The polygon P_3 has the vertices $\{\omega_3, \omega_1, A_1\omega_3, \omega_2, A_2A_1\omega_3\}$. Now $\omega_3 = \omega_{21}$ by $A_3A_2A_1 = -I$, so $A_2A_1\omega_3 = A_2A_1\omega_{21} = \omega_{21}$. Also $A_1\omega_3 = A_1\omega_{21} = \omega_{12}$. So P_3 becomes $\{\omega_{21}, \omega_1, \omega_{12}, \omega_2, \omega_{21}\}$. Clearly P_3 is strictly convex and by Theorem 1 G is discrete. Hence

<u>THEOREM 2</u>. The triangle group G, given by (4.1), (4.2), is discrete if and only if (4.3) is satisfied.

Theorem 2 is well known. See, for example, [4], [5].

We observe that G has no free parameters. This is in accordance with Teichmüller theory, which states that the dimensionality of the space of conjugacy classes of sets of generators for G is 2n-6.

5. In this final section we consider the case n = 4. Our object is to show that a polygon 4-group depends on two real parameters, these being subject to certain inequalities.

Let G be a polygon 4-group with the presentation

$$\text{(5.1)} \qquad G = \langle A_1, \ldots, A_4 : A_i^{m_i} = -I,\ A_4 \cdots A_1 = I;\ i = 1,2,3,4 \rangle,\ m_1^{-1} + \ldots + m_4^{-1} < 2,$$

and we may assume

$$\text{(5.2)} \qquad \sigma(A_i) \geq 0, \quad m_4 \geq 3.$$

We first assume G is discrete. Then we must have

$$\text{(5.3)} \qquad \sigma(A_i) = 2 \cos \pi/m_i, \quad i = 1, \ldots, 4$$

and we can require

$$\text{(5.4)} \qquad c_i < 0, \quad i = 1, \ldots, 4$$

where $A_i = (a_i, b_i : c_i, d_i)$, so that the transformations $A_i z$ have positive rotation. As we have seen G has a strictly convex canonical polygon P_4 with vertices $\{\omega_4, \omega_1, A_1\omega_4, \omega_2, A_2A_1\omega_4, \omega_3\}$ in clockwise order. Since this is convex, the line joining ω_4 and ω_2 lies within P_4. By a Möbius transformation we map this line onto the segment of the imaginary axis joining $i\rho$ and i, where $0 < \rho < 1$. In the transformed group, then, ω_1 and $A_1\omega_4$ are in the left half-plane, $A_2A_1\omega_4$ and ω_3 are in the right half-plane. In particular

(5.5) $$a_1 > d_1, \qquad a_3 < d_3.$$

This implies

(5.6) $$b_1 > 0, \qquad b_3 > 0$$

(5.7) $$a_1 > \cos \pi/m_1, \qquad d_3 > \cos \pi/m_3.$$

From $A_3 = A_4^{-1}A_1^{-1}A_2^{-1}$ we get

$$a_3 = (d_1 \cos \frac{\pi}{m_2} - b_1 \sin \frac{\pi}{m_2}) \cos \frac{\pi}{m_4} + (c_1 \cos \frac{\pi}{m_2} - a_1 \sin \frac{\pi}{m_2})\rho \sin \frac{\pi}{m_4}$$

(5.8) $$d_3 = -(d_1 \sin \frac{\pi}{m_2} + b_1 \cos \frac{\pi}{m_2})\rho^{-1} \sin \frac{\pi}{m_4} + (c_1 \sin \frac{\pi}{m_2} + a_1 \cos \frac{\pi}{m_2}) \cos \frac{\pi}{m_4}$$

$$c_3 = (d_1 \cos \frac{\pi}{m_2} - b_1 \sin \frac{\pi}{m_2})\rho^{-1} \sin \frac{\pi}{m_4} + (-c_1 \cos \frac{\pi}{m_2} + a_1 \sin \frac{\pi}{m_2}) \cos \frac{\pi}{m_4}$$

and

$$c_3 < 0, \qquad a_3 < d_3$$

by (5.4), (5.5). Thus

(5.9) $$a_3 + d_3 = 2 \cos \frac{\pi}{m_3} = \{2 \cos \frac{\pi}{m_1} \cos \frac{\pi}{m_2} + (c_1 - b_1) \sin \frac{\pi}{m_2}\} \cos \frac{\pi}{m_4}$$
$$+ \{(\rho c_1 - \rho^{-1} b_1) \cos \frac{\pi}{m_2} - (\rho a_1 + \rho^{-1} d_1) \sin \frac{\pi}{m_2}\} \sin \frac{\pi}{m_4}$$
$$\equiv \phi(\rho),$$

ϕ being defined by the last expression. We know there is a solution for ρ in $(0,1)$.

Fricke proved ([1], p. 365, (3)) that $\sigma(A_2A_1) < -2$, i.e.,

$$(5.10)\qquad \sigma(A_2A_1) = 2\cos\frac{\pi}{m_1}\cos\frac{\pi}{m_2} + (c_1-b_1)\sin\frac{\pi}{m_2} < -2,$$

$$(5.11)\qquad \phi(\rho) = \sigma(A_2A_1)\cos\frac{\pi}{m_4} + \{(\rho c_1-\rho^{-1}b_1)\cos\frac{\pi}{m_2} - (\rho a_1+\rho^{-1}d_1)\sin\frac{\pi}{m_2}\}\sin\frac{\pi}{m_4}.$$

It follows from this that $\phi(1) \leq 0$. In fact,

$$(5.12)\qquad \phi(1) = \sigma(A_2A_1)\cos\frac{\pi}{m_4} + (c_1-b_1)\cos\frac{\pi}{m_2}\sin\frac{\pi}{m_4} - 2\cos\frac{\pi}{m_1}\sin\frac{\pi}{m_2}\sin\frac{\pi}{m_4} \leq 0,$$

since $c_1 < 0$, $b_1 > 0$ by (5.4), (5.6). We now investigate

$$\phi(0^+) = -\infty \cdot \operatorname{sgn}(d_1\sin\frac{\pi}{m_2} + b_1\cos\frac{\pi}{m_2}).$$

Note that the expression under sgn is the 12 entry of A_2A_1. There are now 3 possibilities.

I. $d_1\sin\frac{\pi}{m_2} + b_1\cos\frac{\pi}{m_2} = 0.$

Here by (5.11)

$$\phi(0^+) = \sigma(A_2A_1)\cos\frac{\pi}{m_4} \leq 0.$$

Now

$$(5.13)\qquad \begin{aligned}\phi'(\rho) &= \{(-a_1\sin\frac{\pi}{m_2} + c_1\cos\frac{\pi}{m_2}) + \rho^{-2}(d_1\sin\frac{\pi}{m_2} + b_1\cos\frac{\pi}{m_2})\}\sin\frac{\pi}{m_4}\\ &= (-a_1\sin\frac{\pi}{m_2} + c_1\cos\frac{\pi}{m_2})\sin\frac{\pi}{m_4} < 0\end{aligned}$$

by (5.7). Hence $\phi(\rho)$ is never positive and (5.9) is violated; this case is impossible.

II. $d_1\sin\frac{\pi}{m_2} + b_1\cos\frac{\pi}{m_2} > 0.$

Then $\phi(0^+) = -\infty$. Either ϕ is monotone in (0,1), in which case there is clearly no solution of (5.9), or ϕ has a maximum. There cannot be more than one maximum, since ϕ is quadratic in ρ. Let $\rho = \rho_0$ be the maximum point; by (5.13)

$$\rho_0 = \left(\frac{d_1 \sin \pi/m_2 + b_1 \cos \pi/m_2}{a_1 \sin \pi/m_2 - c_1 \cos \pi/m_2}\right)^{1/2}, \quad 0 < \rho < 1.$$

Because of the present assumption we have $a_1 \sin \pi/m_2 - c_1 \cos \pi/m_2 > 0$, and

$$\max_{0 < \rho < 1} \phi(\rho) = \phi(\rho_0) = \sigma(A_2A_1)\cos \pi/m_4$$

$$-2\left(d_1 \sin \frac{\pi}{m_2} + b_1 \cos \frac{\pi}{m_2}\right)^{1/2}\left(a_1 \sin \frac{\pi}{m_2} - c_1 \cos \frac{\pi}{m_2}\right)^{1/2} < 0.$$

Hence this case is also excluded and we can say that

(5.14) $$d_1 \sin \pi/m_2 + b_1 \cos \pi/m_2 < 0,$$

with the consequence that

(5.15) $$d_1 < 0.$$

Equation (5.14) means that $\phi(0^+) = +\infty$, so that (5.9) has indeed one, and only one, solution ρ in $0 < \rho < 1$.

Finally, the convexity of P_4 imposes conditions. The point $A_1\omega_4$ lies outside the line joining ω_1 to ω_2. Similarly, $A_2A_1\omega_4$ lies outside $\omega_2\omega_3$, and ω_4 lies outside $\omega_1\omega_3$. Let θ_1 be the angle of P_4 at $A_1\omega_4$, θ_2 the angle at $A_2A_1\omega_4$, and θ_3 the angle at ω_4. Then the above condition is

(5.16) $$\theta_i \leqq \pi, \quad i = 1,2,3.$$

Now as we saw in the proof of Theorem 1,

(5.17) $$\theta_1 + \theta_2 + \theta_3 = 2\pi/m_3 < \pi,$$

in view of $m_4 > 2$. But (5.17) implies (5.16).

Suppose G is a discrete polygon 4-group satisfying (5.1) - (5.3). A conjugate group, still called G, has ω_2, ω_4 lying on the imaginary axis, as described above. We have

$$\omega_1 = u_1 + iv_1 = \frac{a_1 - d_1}{2c_1} - \frac{\sin \pi/m_1}{c_1}.$$

If ω_1 is given, we calculate c_1 from v_1, then a_1, d_1 from u_1 and $a_1 + d_1 = 2 \cos \pi/m_1$. Then $a_3, \ldots, d_3$ are found from (5.8) and ρ from (5.9). So G is determined and the conditions (5.4) - (5.10), (5.14), (5.17) are fulfilled. G depends on 2 real parameters, the coordinates of ω_1, these parameters being subject to the conditions listed above.

Conversely, suppose $A_1,\ldots,A_4$ are matrices in SL(2,R) generating a group G satisfying (5.1) - (5.3). Let ω_1 be assigned and calculate A_3, A_4 from (5.8), (5.9). Let the conditions (5.4) - (5.10), (5.14), (5.17) be fulfilled. Note that these are conditions on the coordinates of ω_1. We can now draw the polygon P_4; it will be strictly convex. By Theorem 1 G is discrete. This justifies the statement made at the beginning of this section.

REFERENCES

1. R. Fricke - F. Klein, Vorlesungen über die Theorie der automorphen Funktionen, vol. 1. Teubner, Leipzig, 1897.
2. A. Hoare - A. Karrass - D. Solitar, Subgroups of infinite index in fuchsian groups. Math. Z. 125 (1972), 59-69.
3. L. Keen, Canonical polygons for finitely generated fuchsian groups. Acta Mathematica 115 (1966), 1-16.
4. A.W. Knapp, Doubly generated fuchsian groups. Michigan Math. Jour. 15 (1968), 289-304.
5. J. Lehner, On pqr-groups. Bull. Inst. Math. Academia Sinica 6 (1978), 419-422.
6. J. Lehner, Discontinuous groups and automorphic functions. Surveys no. 8, American Mathematical Society, Providence, 1964.
7. J. Milnor, On the 3-dimensional Brieskorn manifolds M(p,q,r). Annals of Math. Studies 84 (1975), 175-225.
8. G. de Rham, Sur les polygones générateurs de groupe Fuchsiens. Enseignement Mathématique 17 (1971), 49-61.

THETA FUNCTION IDENTITIES
AND
ORTHOGONAL POLYNOMIALS

D.M. Bressoud*

Pennsylvania State University

University Park, Pennsylvania

This paper deals with the interaction of two areas of mathematics to which I was first introduced by Professor Grosswald. It is an exposition of some very important but long neglected work done by L.J. Rogers in the early 1890's on the use of certain families of orthogonal polynomials to obtain identities related to theta functions. Emil Grosswald has always stressed the importance of recognizing and crediting the work of earlier mathematicians, and it is appropriate that I dedicate this article to him.

1. Introduction

The work to which I refer is Rogers' three Memoirs on the Expansion of Certain Infinite Products, [8], [9] and [10]. These are best known for containing a proof of the Rogers-Ramanujan identities: $(|q| < 1)$

(1.1) $$\sum_{m=0}^{\infty} \frac{q^{m^2}}{(1-q)(1-q^2)\dots(1-q^m)} = \sum_{m=-\infty}^{\infty} \frac{(-1)^m q^{\frac{5}{2}m^2+\frac{1}{2}m}}{(1-q)(1-q^2)\dots} = \prod_{i=1}^{\infty} \frac{1}{(1-q^{5i-4})(1-q^{5i-1})},$$

(1.2) $$\sum_{m=0}^{\infty} \frac{q^{m^2+m}}{(1-q)(1-q^2)\dots(1-q^m)} = \sum_{m=-\infty}^{\infty} \frac{(-1)^m q^{\frac{5}{2}m^2+\frac{3}{2}m}}{(1-q)(1-q^2)\dots} = \prod_{i=1}^{\infty} \frac{1}{(1-q^{5i-3})(1-q^{5i-2})},$$

* Partially supported by National Science Foundation Grants MCS 77-18723 (02) and MCS 77-22992.

which would later be conjectured by Ramanujan, and not proved by him until he had seen Rogers' proof. The second equalities of (1.1) and (1.2) follow from Jacobi's triple product identity:

$$(1.3) \qquad \sum_{m=-\infty}^{\infty} (-1)^m x^{m^2} z^m = \prod_{i=1}^{\infty} (1-z^{-1}x^{2i-1})(1-zx^{2i-1})(1-x^{2i}).$$

What I shall concentrate on are not Rogers' results but one of the methods which is still capable of producing new and striking identities. The following sequence of polynomial identities, which become (1.1) as N approaches ∞, are new, but arise simple and naturally out of Rogers' approach:

$$1 = 1$$

$$1+q = \frac{(1-q^2)}{(1-q)}$$

$$1+q\,\frac{(1-q^2)}{(1-q)} + q^4 = -q^2 + \frac{(1-q^3)(1-q^4)}{(1-q)(1-q^2)} - q^3$$

$$1+q\,\frac{(1-q^3)}{(1-q)} + q^4\,\frac{(1-q^3)}{(1-q)} + q^9 = -q^2\,\frac{(1-q^6)}{(1-q)} + \frac{(1-q^4)(1-q^5)(1-q^6)}{(1-q)(1-q^2)(1-q^3)} - q^3\,\frac{(1-q^6)}{(1-q)}$$

$$1+q\,\frac{(1-q^4)}{(1-q)} + q^4\,\frac{(1-q^3)(1-q^4)}{(1-q)(1-q^2)}$$

$$+q^9\,\frac{(1-q^4)}{(1-q)} + q^{16} = q^9 - q^2\,\frac{(1-q^7)(1-q^8)}{(1-q)(1-q^2)} + \frac{(1-q^5)(1-q^6)(1-q^7)(1-q^8)}{(1-q)(1-q^2)(1-q^3)(1-q^4)}$$

$$-q^3\,\frac{(1-q^7)(1-q^8)}{(1-q)(1-q^2)} + q^{11}$$

$$\vdots$$

$$(1.4) \qquad \sum_{m=0}^{N} q^{m^2} \left[{N \atop m};q\right] = \sum_{m=-\infty}^{\infty} (-1)^m q^{\frac{5}{2}m^2+\frac{1}{2}m} \left[{2N \atop N+2m};q\right].$$

$$\vdots$$

$\left[{N \atop m};q\right]$ is the Gaussian polynomial defined to be zero for $m < 0$ or $m > N$, one for $m = 0$ or N and

$$\left[{N \atop m};q\right] = \prod_{i=1}^{m} \frac{(1-q^{N-m+i})}{(1-q^i)} \quad \text{for } 0 < m < N.$$

2. Rogers' Difference Operator

We begin with the binomial theorem in the following form:

$$(2.1)\qquad (1-z)^{-\lambda} = \sum_{n=0}^{\infty} \binom{-\lambda}{n}(-z)^n$$

$$= \sum_{n=0}^{\infty} \frac{\lambda(\lambda+1)\ldots(\lambda+n-1)}{1\cdot 2\cdot\ldots\cdot n} z^n$$

$$= \sum_{n=0}^{\infty} \frac{(\lambda)_n}{(1)_n} z^n, \quad (|z| < 1)$$

where $(\lambda)_n$ is the rising factorial defined by

$$(\lambda)_n = \frac{\Gamma(\lambda+n)}{\Gamma(\lambda)}.$$

For positive integral values of λ, (2.1) can be proved by applying the differential operator, $\frac{d}{dz}$, $\lambda-1$ times to each side of the equality

$$(2.2)\qquad (1-z)^{-1} = \sum_{n=0}^{\infty} z^n.$$

Rogers begins his first memoir by exploring the consequences of replacing the differential operator by the difference operator, δ_q, defined by

$$\delta_q f(z) = \frac{f(z)-f(zq)}{z-zq}.$$

δ_q becomes $\frac{d}{dz}$ in the limit as q approaches 1. When δ_q is applied $\lambda-1$ times to each side of (2.2) we obtain

$$(2.3)\qquad \prod_{i=0}^{\lambda-1} \frac{1}{1-zq^i} = \sum_{n=0}^{\infty} \frac{(1-q^{\lambda})(1-q^{\lambda+1})\ldots(1-q^{\lambda+n-1})}{(1-q)(1-q^2)\ldots(1-q^n)} z^n.$$

For $|q| < 1$, $|z| < 1$, we may rewrite this as

$$(2.4)\qquad \frac{(az;q)_\infty}{(z;q)_\infty} = \sum_{n=0}^{\infty} \frac{(a;q)_n}{(q;q)_n} z^n,$$

where $a = q^{\lambda}$ and $(a;q)_n = \prod_{j=0}^{\infty} \frac{1-aq^j}{1-aq^{n+j}}$, the rising q-factorial defined for all real n. So far, (2.4) has only been proved for $a = q^{\lambda}$, $\lambda \in \mathbb{Z}^+$. However, each side of (2.4) is actually an entire function of a. Since $|q| < 1$, these two functions agree on an infinite sequence of points with limit in the region of analyticity, and so agree for all values of a.

Equation (2.4) is known as the q-binomial theorem and is considerably older than Rogers, going back to H.A. Rothe in 1811. The binomial theorem in its full generality is the limit as q approaches 1 of the case $z = q^\lambda$ for arbitrary real λ.

3. Rogers' Orthogonal Polynomials

There is a family of orthogonal polynomials whose definition is intimately tied to the binomial theorem, namely the ultraspherical polynomials $C_n^\lambda(x)$ defined by

$$(3.1)\qquad (1-re^{i\theta})^{-\lambda}(1-re^{-i\theta})^{-\lambda} = \sum_{n=0}^{\infty} r^n C_n^\lambda(\cos\theta).$$

The fact that these are polynomials in cos θ follows from the binomial theorem. Rogers defined a more general family of functions which are polynomials in cos θ by virtue of the q-binomial theorem, the q-ultraspherical polynomials which I shall denote $C_n^{\lambda,q}(x)$, defined by

$$(3.2)\qquad \frac{(re^{i\theta}q^\lambda;q)_\infty(re^{-i\theta}q^\lambda;q)_\infty}{(re^{i\theta};q)_\infty(re^{-i\theta};q)_\infty} = \sum_{n=0}^{\infty} r^n\, C_n^{\lambda,q}(\cos\theta).$$

From (2.4), it follows that

$$(3.3)\qquad C_n^{\lambda,q}(\cos\theta) = \sum_{m=0}^{n} \frac{(q^\lambda;q)_m(q^\lambda;q)_{n-m}}{(q;q)_m(q;q)_{n-m}} \cos(n-2m)\theta\ .$$

A natural and conceivably useful question to ask at this point is what is the expansion of cos mθ in terms of $C_n^{\lambda,q}(\cos\theta)$? Rogers gives this expansion in his third memoir. He used (3.3) to work out a sufficient number of cases so that he could guess the general formula which is provable by induction. More recently, Askey and Wilson [3] found the weight functions for the q-ultraspherical polynomials which reduces the expansion of cos mθ to a question of computation. The result is

(3.4) $$\cos m\theta = \sum_{k=0}^{[m/2]} \frac{(1-q^m)q^{\lambda k}(q^{-\lambda};q)_k(q;q)_{m-k-1}}{2(q;q)_k(q^{1+\lambda};q)_{m-k}}$$

$$\times \frac{(1-q^{\lambda+m-2k})}{(1-q^{\lambda})} C_{m-2k}^{\lambda,q}(\cos\theta), \quad (m \geq 1).$$

This implies the following relationship between families of the q-ultraspherical polynomials:

(3.5) $$C_m^{\lambda,q}(x) = \sum_{k=0}^{[m/2]} \frac{(q^{\lambda-\mu};q)_k(q^{\lambda};q)_{m-k}(1-q^{\mu+n-2k})}{(q;q)_k(q^{1+\mu};q)_{m-k}(1-q^{\mu})} q^{\mu k}C_{m-2k}^{\mu,q}(x).$$

4. Rogers' Idea

While Rogers did not have equation (3.4) until his third memoir, certain special cases were proved in the second memoir. These were all he needed for the Rogers-Ramanujan identities. Throughout the third memoir, Rogers states that his results should be applied to the techniques of the second memoir in order to generalize those identities, but he never does this for us. What I shall do here is to follow the reasoning of the second memoir which led him to the Rogers-Ramanujan identities, but with some of the generality made possible by the third memoir.

We begin with the infinite product of (3.2) which, when expanded as a power series in r, provided the definition of $C_n^{\lambda,q}(\cos\theta)$. Instead, we use the q-binomial theorem to expand it as a Fourier series in θ:

(4.1) $$\frac{(re^{i\theta}q^{\lambda};q)_\infty(re^{-i\theta}q^{\lambda};q)_\infty}{(re^{i\theta};q)_\infty(re^{-i\theta};q)_\infty}$$

$$= \sum_{t=-\infty}^{\infty} r^t \cos t\,\theta \frac{(q^{\lambda};q)_t}{(q;q)_t} \sum_{m=0}^{\infty} \frac{(q^{\lambda};q)_m(q^{\lambda+t};q)}{(q;q)_m(q^{t+1};q)_m} r^{2m}.$$

It is clearly desirable to know when the inner sum of the right side can be expressed in a simple, closed form. On examination, one realizes that in the limit as q approaches 1, this inner sum becomes the hypergeometric function: $F(\lambda,\lambda+t;t+1;r^2)$. Thus, if we want a closed form, we shall probably need to have r depend on q in such a way that $\lim_{q\to 1^-} r = 1$.

In fact, this inner sum was extensively studied by Heine [7] in the mid 1800's, and is known as the basic hypergeometric function. While there are several choices of parameters which will permit its expression in closed form, the simplest is the choice $r^2 = q^{1-2\lambda}$ or $r = q^{\frac{1}{2}-\lambda}$. The summation is accomplished by appropriate use of the q-binomial theorem (see [2], corollaries 2.3 and 2.4) to yield:

$$(4.2)\quad \frac{(e^{i\theta}q^{\frac{1}{2}};q)_\infty(e^{-i\theta}q^{\frac{1}{2}};q)_\infty}{(e^{i\theta}q^{\frac{1}{2}-\lambda};q)(e^{-i\theta}q^{\frac{1}{2}-\lambda};q)_\infty}$$

$$= \frac{(q^{1-\lambda};q)_\infty(q^{1-\lambda};q)_\infty}{(q;q)_\infty(q^{1-2\lambda};q)_\infty} \sum_{t=-\infty}^{\infty} (-1)^t q^{\frac{1}{2}t^2} \cos t\theta \, \frac{(q^{1-t-\lambda};q)_t}{(q^{1-\lambda};q)_t} .$$

In the limit as λ approaches $-\infty$, this becomes Jacobi's triple product identity. If $-\lambda = N \in Z^+$, we get the finite form due to Cauchy ([6], eq. (6)):

$$(4.3)\quad (e^{i\theta}q^{\frac{1}{2}};q)_N(e^{-i\theta}q^{\frac{1}{2}};q)_N(q;q)_N$$

$$= \frac{(q;q)_{2N}}{(q;q)_N} \sum_{t=-\infty}^{\infty} (-1)^t q^{\frac{1}{2}t^2} \cos t\theta \, \frac{(q^{1+n-t};q)_\infty(q^{1+N+t};q)_\infty}{(q^{1+N};q)_\infty(q^{1+N};q)_\infty} .$$

Returning to equation (4.2), Rogers realized that the left side is easily expanded, by (3.2), in terms of the q-ultraspherical polynomials, $C_n^{\lambda,q}(\cos\theta)$, yielding

$$(4.4)\quad \sum_{n=0}^{\infty} q^{n(\frac{1}{2}-\lambda)} C_n^{\lambda,q}(\cos\theta)$$

$$= \frac{(q^{1-\lambda};q)_\infty (q^{1-\lambda};q)_\infty}{(q;q)_\infty (q^{1-2\lambda};q)_\infty} \sum_{t=-\infty}^{\infty} (-1)^t q^{\frac{1}{2}t^2} \cos t\theta \frac{(q^{1-t-\lambda};q)_t}{(q^{1-\lambda};q)_t} .$$

Both sides can be expanded in terms of another family of q-ultraspherical polynomials, say $C_n^{\mu,q}$, by using (3.4) and (3.5). In both expansions, the constant terms (relative to $C_n^{\mu,q}(\cos\theta)$) must be equal, and this equality is precisely

$$(4.5)\quad \sum_{n=0}^{\infty} q^{n(1+\mu-2\lambda)} \frac{(q^{\lambda-\mu};q)_n (q^{\lambda};q)_n}{(q;q)_n (q^{1+\mu};q)_n}$$

$$= \frac{(q^{1-\lambda};q)_\infty (q^{1-\lambda};q)_\infty}{(q;q)_\infty (q^{1-2\lambda};q)_\infty} \left(1+ \sum_{t=1}^{\infty} q^{2t^2+\mu t}(1+q^t) \frac{(q^{1-2t-\lambda};q)_{2t}(q^{-\mu};q)_t}{(q^{1-\lambda};q)_{2t}(q^{1+\mu};q)_t}\right).$$

In the limit as μ approaches ∞ and with $-\lambda = N \in \mathbb{Z}^+$, this is the finite form of (1.1) given by equation (1.4). If, instead of comparing constant terms we compare the coefficients of $C_1^{\mu,q}$, then the same choices for λ and μ yield a finite form of (1.2), namely

$$(4.6)\quad \sum_{n=0}^{N} q^{n^2+n} \begin{bmatrix} N \\ n \end{bmatrix}_q = \frac{1}{(1-q^{N+1})} \sum_{m=-\infty}^{\infty} (-1)^m q^{\frac{5}{2}m^2+\frac{3}{2}m} \begin{bmatrix} 2N+2 \\ N+2m+2 \end{bmatrix}_q .$$

5. <u>Final Remarks</u>

Other choices for λ and μ are possible, and a more detailed treatment of what can be done along these lines has been given in [4] and [5]. Also, (1.4) and (4.6) are not the only known finite forms of the Rogers-Ramanujan identities. Andrews has given one pair in [2], p. 50 and another in [1].

Finally, the proofs outlined in this paper are, in general, not the most efficient. The trappings of orthogonal polygonals can be done away with entirely. But I hope that I have demonstrated that they are useful indicators for what is the right question to ask at each step.

REFERENCES

1. G.E. Andrews, The Rogers-Ramanujan identities, problem 74-12, SIAM Review, 16 (1974), 390.
2. __________, The Theory of Partitions, vol. 2 in Encyclopedia of Mathematics, ed. G.-C. Rota, Addison Wesley, Reading, Mass., 1976.
3. R. Askey and J. Wilson, Some basic hypergeometric orthogonal polynomials that generalize Jacobi polynomials, to appear.
4. D.M. Bressoud, On partitions, orthogonal polynomials and the expansion of certain infinite products, Proc. London Math. Soc., to appear.
5. __________, Some identities for terminating q-series, Math. Proc. Cambridge Philos. Soc., to appear.
6. A. Cauchy, Second Mémoire sur les fonctions dont plusiers valeurs sont liées entre elles par une equation lineaire, C.R. 17 (1843), 567-572. Reprinted in Oeuvres, Ser. 1, vol. 8, pp. 50-55, Gauthier-Villars, Paris.
7. E. Heine, Untersuchungen uber die Reihe ..., J. Reine Angew. Math., 34 (1847), 285-328.
8. L.J. Rogers, On the expansion of some infinite products, Proc. London Math. Soc., 24 (1893), 337-352.
9. __________, Second memoir on the expansion of certain infinite products, Proc. London Math. Soc., 25 (1894), 318-343.
10. __________, Third memoir on the expansion of certain infinite products, Proc. London Math. Soc., 26 (1895), 15-32.

RAMANUJAN CONGRUENCES FOR q(n)

B. Gordon and K. Hughes

Dedicated to Professor Emil Grosswald

1. <u>Introduction</u>. Let p(n) denote the number of partitions of n, and q(n) the number of partitions of n into distinct parts, with the usual convention that $p(0) = q(0) = 1$. In 1919, Ramanujan [8] conjectured that

(1) $p(n) \equiv 0 \pmod{5^\alpha}$ if $24n \equiv 1 \pmod{5^\alpha}$,

(2) $p(n) \equiv 0 \pmod{7^\beta}$ if $24n \equiv 1 \pmod{7^\beta}$,

(3) $p(n) \equiv 0 \pmod{11^\gamma}$ if $24n \equiv 1 \pmod{11^\gamma}$,

for all $\alpha, \beta, \gamma \geq 0$. In 1938, Watson [9] proved (1). By that time Gupta had noted that (2) is false when $\beta = 3$; Watson formulated and proved a modification of it, namely

(2') $$p(n) \equiv 0 \pmod{7^\beta} \text{ if } 24n \equiv 1 \pmod{7^{2\beta-2}} \text{ for } \beta \geq 2.$$

Finally, in 1967 Atkin [2] proved (3). In the same year, he and O'Brien [4] proved that

(4) $$p(169n-7) \equiv \kappa_\delta p(n) \pmod{13^\delta} \text{ if } 24n \equiv 1 \pmod{13^\delta},$$

where κ_δ is an integer depending only on δ.

In this paper we will obtain the analogous congruences for q(n) modulo powers of 5 and 7. The results are as follows:

<u>THEOREM 1</u>. <u>For all</u> $\alpha, \beta \geq 0$,

(5) $$q(n) \equiv 0 \pmod{5^\alpha} \text{ if } 24n \equiv -1 \pmod{5^{2\alpha+1}},$$

(6) $$q(49n+2) \equiv \lambda_\beta q(n) \pmod{7^\beta} \text{ if } 24n \equiv -1 \pmod{7^\beta},$$

<u>where</u> λ_β <u>is an integer depending only on</u> β.

Thus the behavior of q(n) (mod 5^α) resembles that of p(n) (mod 7^β), while the behavior of q(n) (mod 7^β) resembles that of p(n) (mod 13^δ). It is unlikely that q(n) satisfies any such congruences modulo powers of 3, but in contrast to p(n) it has remarkable properties modulo powers of 2. These are discussed by Allatt and Slater in [1] and a forthcoming sequel. As far as congruences for q(n) (mod 11^{γ}) are concerned, the problem is greatly complicated by the fact that the group $\Gamma_0(22)$ has genus 2. The computational difficulties seem too formidable to justify pursuing this problem here.

2. Function-theoretic preliminaries. Let $\Gamma = PSL(2,Z)$ be the modular group, and $\Gamma_0(N) = \{\begin{pmatrix} a & b \\ c & d \end{pmatrix} \in \Gamma : c \equiv 0 \pmod N\}$. Let $H = \{\tau : \text{Im}\,\tau > 0\}$ be the upper half-plane, and put $H^* = H \cup Q \cup \{i\infty\}$. We let $\Gamma_0(N)$ act on H^*, and give $R_0(N) = H^* \bmod \Gamma_0(N)$ the structure of a compact Riemann surface in the usual way; we denote by $K_0(N)$ the field of meromorphic functions on $R_0(N)$. The orbits of $Q \cup \{i\infty\}$ mod $\Gamma_0(N)$ are the cusps of $R_0(N)$; if $r_1,\ldots,r_\kappa$ form a complete system of representatives of these orbits, we will also refer to them as the cusps of $R_0(N)$. It is known that when $N = p^{\alpha}$ is a prime power, such a set of representatives is given by the numbers ν/p^{β}, where $0 \le \beta < \alpha$, $0 \le \nu < \min(p^{\beta}, p^{\alpha-\beta})$, and $(\nu,p^{\beta}) = 1$. From this fact, a set of representatives for arbitrary N can be obtained using the Chinese remainder theorem. Without going into details, we mention that the resulting system consists of numbers ν/δ, where $\delta|N$, $0 \le \nu < \delta$, and $(\nu,\delta) = 1$. Moreover, for a given divisor δ of N, the number of distinct cusps ν/δ of $R_0(n)$ is $\phi((\delta,\delta'))$, where $\delta' = N/\delta$; here (δ,δ') denotes g.c.d., and ϕ is the Euler function. In this system $i\infty$ is represented by $1/N$.

Now let $\eta(\tau) = e^{\pi i\tau/12} \prod_{n=1}^{\infty} (1-e^{2\pi i n\tau})$ be the Dedekind η-function. For a fixed integer $N > 0$, a function of the form

$$f(\tau) = \prod_{\substack{\varepsilon|N \\ \varepsilon > 0}} \eta(\varepsilon\tau)^{m_\varepsilon}, \quad m_\varepsilon \in Z \tag{7}$$

will be called an η-product. The following theorem of Newman [7] gives necessary and sufficient conditions for an η-product to belong to $K_0(N)$.

THEOREM 2. The function (7) is in $K_0(N)$ if and only if

(a) $\sum_{\varepsilon|N} m_\varepsilon = 0$,

(b) $\sum_{\varepsilon|N} \varepsilon m_\varepsilon \equiv 0 \pmod{24}$,

(c) $\sum_{\varepsilon|N} \varepsilon' m_\varepsilon \equiv 0 \pmod{24}$, where $\varepsilon' = N/\varepsilon$,

(d) $\prod_{\varepsilon|N} \varepsilon^{m_\varepsilon} \in Q^2$.

Ligozat [6] has determined the order of the η-product (7) at the cusps ν/δ of $R_0(N)$ in the case where the conditions of Theorem 2 are fulfilled. The order depends only on the denominator δ of the cusp, and is given by the following theorem.

THEOREM 3. If (7) is in $K_0(N)$, then its order at the cusp ν/δ of $R_0(N)$ is

$$\frac{1}{24} \sum_{\varepsilon|N} \frac{N}{(\delta,\delta')} \frac{(\delta,\varepsilon)^2}{\delta\varepsilon} m_\varepsilon .$$

Now let p be a prime, and

$$f(x) = \sum_{n \geq n_0} a_n x^n$$

a meromorphic Laurent series. We define

$$U_p f(x) = \sum_{pn \geq n_0} a_{pn} x^n = \frac{1}{p} \sum_{\lambda=0}^{p-1} f(\omega^\lambda x^{1/p}), \text{ where } \omega = e^{2\pi i/p}.$$

We write U instead of U_p when p is clear from the context. We note that U is a linear operator, and that $U[f(x^p)g(x)] = f(x)Ug(x)$.

If $f(\tau) \in K_0(N)$ for some N, then $f(\tau)$ has an expansion about the point $i\infty$ of $R_0(N)$ of the form $f(\tau) = \sum_{n \geq n_0} a_n x^n$, where $x = e^{2\pi i\tau}$. We call this expansion the Fourier series of $f(\tau)$, and sometimes commit the abuse of language of denoting it by $f(x)$; this seems to cause no confusion in practice. We define $U_p f(\tau)$ to be the result of applying U_p to the Fourier series $f(x)$; thus

$$U_p f(\tau) = \frac{1}{p} \sum_{\lambda=0}^{p-1} f\left(\frac{\tau+\lambda}{p}\right). \tag{8}$$

It is known that if $f \in K_0(pN)$, where $p|N$, then $U_p f \in K_0(N)$. The next theorem gives bounds on ord $U_p f$ at cusps of $R_0(N)$ in terms of ord f at cusps of $R_0(pN)$. We use the following notation. The p-adic order of an integer n, i.e. the highest power of p dividing n, is denoted by $\pi(n)$. The order of $U_p f$ at a cusp r of $R_0(N)$ is denoted by $\mathrm{ord}_r\, U_p f$, and the order of f at a cusp s of $R_0(pN)$ is denoted by $\mathrm{ord}_s f$.

<u>THEOREM 4</u>. <u>Suppose</u> $f \in K_0(pN)$, where $p|N$, and let $r = \nu/\delta$ be a cusp of $R_0(N)$ (where as usual $\delta|N$ and $(\nu,\delta) = 1$). Then

$$\mathrm{ord}_r U_p f \geq \begin{cases} \frac{1}{p}\, \mathrm{ord}_{r/p} f & \text{if } \pi(\delta) \geq \frac{1}{2}\pi(N) \\ \mathrm{ord}_{r/p} f & \text{if } 0 < \pi(\delta) < \frac{1}{2}\pi(N) \\ \min\limits_{0 \leq \lambda \leq p-1} \mathrm{ord}_{(r+\lambda)/p} f & \text{if } \pi(\delta) = 0. \end{cases}$$

For a proof of this, see [5]. For convenience of application, we specialize Theorem 4 to the case where $N = 2p$, p odd, and f is an η-product. In this case the Riemann surface $R_0(2p)$ has 4 cusps, represented by 0, 1/2, 1/p and 1/(2p). The surface $R_0(2p^2)$ has 2p+2 cusps, represented by 0, 1/2, λ/p, $(2\lambda-1)/(2p)$, $1/p^2$, and $1/(2p^2)$, where $1 \leq \lambda \leq p-1$. If f is an η-product in $K_0(2p^2)$, its orders at all the cusps λ/p are equal, and similarly those at $(2\lambda-1)/(2p)$ are all equal. Hence from Theorem 4 we obtain the following result.

THEOREM 5. <u>Suppose</u> f <u>is</u> <u>an</u> η-<u>product</u> <u>in</u> $K_0(2p^2)$, <u>where</u> p <u>is</u> <u>an</u> <u>odd</u> <u>prime</u>. <u>Then</u> $U_p f = Uf$ <u>is</u> <u>in</u> $K_0(2p)$, <u>and</u>

$$\mathrm{ord}_0 Uf \geq \min(\mathrm{ord}_0 f,\ \mathrm{ord}_{1/p} f),$$

$$\mathrm{ord}_{1/2} Uf \geq \min(\mathrm{ord}_{1/2} f,\ \mathrm{ord}_{1/(2p)} f),$$

$$\mathrm{ord}_{1/p}Uf \geq (1/p)\ \mathrm{ord}_{1/p^2}f,$$

$$\mathrm{ord}_{1/(2p)}Uf \geq (1/p)\ \mathrm{ord}_{1/(2p^2)}f.$$

<u>Moreover</u> Uf <u>has no poles on</u> $R_0(p)$ <u>except at the cusps</u>.

3. <u>Congruences for $q(n)$ (mod 5^α)</u>. The generating function of $q(n)$ is

$$Q(x) = \sum_{n=0}^{\infty} q(n)x^n = \prod_{n=1}^{\infty} (1+x^n)$$

$$= e^{-\pi i\tau/12}\eta(2\tau)/\eta(\tau),$$

where $x = e^{2\pi i\tau}$. For each $\nu \geq 0$, let ℓ_ν be the least non-negative solution of the congruence $24\ell \equiv -1 \pmod{5^\nu}$. The actual value is $\ell_\nu = (25^{[(\nu+1)/2]}-1)/24$; thus $\ell_0 = 0$, $\ell_1 = \ell_2 = 1$, $\ell_3 = \ell_4 = 26$, etc. In general

(9) $$\ell_{2\alpha+1} = \ell_{2\alpha+2} = 25\ell_{2\alpha} + 1.$$

Congruence (5) asserts that $q(\ell_{2\alpha+1} + m\cdot 5^{2\alpha+1}) \equiv 0 \pmod{5^\alpha}$ for all $m \geq 0$. If $D_\nu(x) = \sum_{m=0}^{\infty} q(\ell_\nu + m\cdot 5^\nu)x^m$, this can be written as

(10) $$D_{2\alpha+1}(x) \equiv 0 \pmod{5^\alpha},$$

where as usual $\sum a_n x^n \equiv \sum b_n x^n \pmod{m}$ means that $a_n \equiv b_n \pmod{m}$ for all n. In terms of the operator $U = U_5$, we have

$$D_0(x) = Q(x)$$

(11) $$D_{2\alpha+1}(x) = U(x^{-1}D_{2\alpha}(x))$$

$$D_{2\alpha+2}(x) = U(D_{2\alpha+1}(x)) \text{ for } \alpha \geq 0;$$

this follows readily from (9). Now the key idea is to "approximate" the functions $D_\nu(x)$ and x^{-1} appearing in (11) by suitable functions $L_\nu(x)$ and $F(x)$ in $K_0(50)$; we can then exploit the fact that U maps $K_0(50)$ into $K_0(10)$. To this end we set

(12) $$F(x) = x^{-1} \frac{Q(x)}{Q(x^{25})} = \frac{\eta(2\tau)}{\eta(\tau)} \frac{\eta(25\tau)}{\eta(50\tau)},$$

and we define a sequence of functions $L_\nu(x)$ $(\nu \geq 0)$ inductively by putting

(13) $$\begin{cases} L_0 = 1 \\ L_{2\alpha+1} = U(FL_{2\alpha}) \\ L_{2\alpha+2} = U(L_{2\alpha+1}) \text{ for } \alpha \geq 0. \end{cases}$$

We then have $D_{2\alpha}(x) = Q(x)L_{2\alpha}(x)$ and $D_{2\alpha+1}(x) = Q(x^5)L_{2\alpha+1}(x)$ for all $\alpha \geq 0$. Since $Q(x^5)$ has leading coefficient 1, congruence (10) is equivalent to

(14) $$L_{2\alpha+1}(x) \equiv 0 \pmod{5^\alpha}.$$

We will prove (14) by induction on α, using the formation rules (13) for the functions L_ν.

By Theorems 2 and 3, applied to the right side of (12), we find that F is in $K_0(50)$, and that its orders at the cusps of $R_0(50)$ are as shown below.

$R_0(50)$

δ	1	2	5	10	25	50
$\text{ord}_{\nu/\delta}F =$	-1	1	0	0	1	-1

Hence by Theorem 5, we have the following lower bounds for the orders of UF at the cusps of $R_0(10)$ (where we have rounded up to the nearest integer, as is clearly permissible).

$R_0(10)$

δ	1	2	5	10
$\text{ord}_{\nu/\delta}UF \geq$	-1	0	1	0

Moreover, UF is holomorphic at all other points of $R_0(10)$.

Now define $\theta(x) = \sum_{-\infty}^{\infty} x^{n^2}$; by Jacobi's product formula we have

$$\theta(-x) = \prod_{n=1}^{\infty} (1-x^{2n-1})^2(1-x^{2n})$$

$$= \eta(\tau)^2/\eta(2\tau).$$

Put

$$G(x) = [\frac{\theta(-x^5)}{\theta(-x)}]^2 = \frac{\eta(2\tau)^2\eta(5\tau)^4}{\eta(\tau)^4\eta(10\tau)^2}.$$

Applying Theorems 2 and 3 we find that G is in $K_0(10)$, and that its orders at the cusps of $R_0(10)$ are as follows.

$R_0(10)$

δ	1	2	5	10
$\text{ord}_{\nu/\delta}G =$	-1	0	1	0

Elsewhere on $R_0(10)$, G is holomorphic and non-zero. A comparison of the last two tables shows that UF = cG, where c is a constant. Since the Fourier series of UF and G both have constant term 1, we have UF = G.

The same reasoning shows that if $f \in K_0(10)$ is holomorphic on $R_0(10)$ except for a pole at $\tau = 0$ of multiplicity $\leq m$, then f is a polynomial in G of degree $\leq m$. Now if G is regarded as a function of $K_0(50)$, rather than of the subfield $K_0(10)$, Theorem 3 shows that its orders at the cusps of $R_0(50)$ are as shown below.

$R_0(50)$

δ	1	2	5	10	25	50
$\text{ord}_{\nu/\delta}G$	-5	0	1	0	1	0

Hence by Theorem 5 applied to G^i and FG^i, we obtain the following lower bounds for the orders of $U(G^i)$ and $U(FG^i)$ at the cusps of $R_0(10)$.

$R_0(10)$

δ	1	2	5	10
$\text{ord}_{\nu/\delta}U(G^i) \geq$	$-5i$	0	$i/5$	0
$\text{ord}_{\nu/\delta}U(FG^i) \geq$	$-5i-1$	0	$(i+1)/5$	0

If $i \geq 0$, this implies that $U(G^i)$ and $U(FG^i)$ are polynomials in G of degrees at most $5i$ and $5i+1$ respectively. Thus

$$(15)\qquad \begin{aligned} U(G^i) &= \sum_{j \geq 0} a_{ij}G^j \\ U(FG^i) &= \sum_{j \geq 0} b_{ij}G^j, \end{aligned}$$

with complex coefficients a_{ij}, b_{ij}. By considering our lower bounds for the orders of $U(G^i)$ and $U(FG^i)$ at $\tau = 1/5$ as well as at $\tau = 0$, we see that $a_{ij} = 0$ unless $i/5 \leq j \leq 5i$, and $b_{ij} = 0$ unless $(i+1)/5 \leq j \leq 5i+1$. In particular $a_{i0} = 0$ for all $i \geq 1$ and $b_{i0} = 0$ for all $i \geq 0$.

Let S be the vector space of all polynomials $P = \sum_{j \geq 0} c_jG^j$, and T the subspace of such polynomials with $c_0 = 0$. Equations (15) show that S is mapped into itself by U and also by the linear transformation $V:P \to U(FP)$. Moreover, by the above remarks about a_{ij} and b_{ij}, we see that V maps S into T, while U maps T into itself. With respect to the basis $G, G^2, G^3,\ldots$ of T, the matrices of U and V restricted to T are respectively $A = (a_{ij})$ and $B = (b_{ij})$, $1 \leq i, j < \infty$.

We can now prove that the functions L_ν are in T for all $\nu \geq 1$. Indeed $L_1 = UF = G \in T$, and by equations (13) we have $L_{2\alpha+1} = V(L_{2\alpha})$ and $L_{2\alpha+2} = U(L_{2\alpha+1})$ for $\alpha > 0$. By induction this yields $L_\nu \in T$ for all $\nu \geq 1$. In terms of the basis $G, G^2,\ldots$, we have

$$\begin{aligned} L_1 &= G = (1,0,0,\ldots) \\ L_{2\alpha+1} &= (1,0,0,\ldots)(AB)^\alpha \\ L_{2\alpha+2} &= (1,0,0,\ldots)(AB)^\alpha A. \end{aligned}$$

To conclude the proof of (14), we will show that $L_{2\alpha+1} = (\ell_1^{(2\alpha+1)}, \ell_2^{(2\alpha+1)},\ldots)$, where the components $\ell_i^{(2\alpha+1)}$ are integers divisible by 5^α. But to do this we need to know more about the matrices A and B, and this requires a digression.

Define

$$\phi(x) = \frac{\theta(-x^{25})}{\theta(-x)} = \frac{\eta(2\tau)}{\eta(\tau)^2} \; \frac{\eta(25\tau)^2}{\eta(50\tau)}.$$

Application of Theorems 2 and 3 shows that ϕ is in $K_0(50)$, and that its only singularity on $R_0(50)$ is a triple pole at $\tau = 0$. Hence for any $\mu \geq 0$, $U(\phi^\mu) = \frac{1}{5} \sum_{\lambda=0}^{4} \phi(\frac{\tau+\lambda}{5})^\mu$ is a polynomial in G of degree $\leq 3\mu$. Thus the power sums $5U(\phi\mu)$ of the quantities $\phi(\frac{\tau+\lambda}{5})^\mu$ $(0 \leq \lambda \leq 4)$ are in C[G], so by Newton's identities their elementary symmetric functions σ_i are also in C[G]. Therefore $\phi(\frac{\tau+\lambda}{5})$ $(0 \leq \lambda \leq 4)$ are the roots of an equation

(16) $$t^5-\sigma_1 t^4+\sigma_2 t^3-\sigma_3 t^2+\sigma_4 t-\sigma_5 = 0, \quad \sigma_i \in C[G].$$

We call (16) the modular equation between ϕ and G. To compute the coefficients σ_i explicitly, it is convenient to consider the reciprocal equation

$$u^5 - \frac{\sigma_4}{\sigma_5} u^4 + \frac{\sigma_3}{\sigma_5} u^3 - \frac{\sigma_2}{\sigma_5} u^2 + \frac{\sigma_1}{\sigma_5} u - \frac{1}{\sigma_5} = 0.$$

The roots of this are the functions $\phi(\frac{\tau+\lambda}{5})^{-1}$ $(0 \leq \lambda \leq 4)$. If $k \geq 0$, the function $\phi(\tau)^{-k}$ is holomorphic on $R_0(50)$ except for a 3k-fold pole at $\tau = 1/25$. Hence by Theorem 5, $U(\phi^{-k})$ is holomorphic on $R_0(10)$ except for a pole of multiplicity $\leq [3k/5]$ at $\tau = 1/5$. Since G^{-1} has a simple pole at $\tau = 1/5$ and is elsewhere holomorphic on $R_0(10)$, it follows that $U(\phi^{-k})$ is a polynomial in G^{-1} of degree $\leq [3k/5]$. By examining the Fourier series of $U(\phi^{-k})$ we can easily determine these polynomials for $k = 0,1,2,3,4$. We can then use Newton's identities to compute the elementary symmetric functions σ_i/σ_5 of the quantities $\phi(\frac{\tau+\lambda}{5})^{-1}$ $(0 \leq \lambda \leq 4)$; finally these clearly determine the σ_i. The result is the following:

(17)
$$U(\phi^{-4}) = 6G^{-2}-5$$
$$U(\phi^{-3}) = 6G^{-1}-5$$
$$U(\phi^{-2}) = 2G^{-1}-1$$
$$U(\phi^{-1}) = U(1) = 1,$$

$$\sigma_1 = 5G-25G^2+25G^3$$

$$\sigma_2 = -15G^2+25G^3$$

(18)
$$\sigma_3 = -5G^2+15G^3$$

$$\sigma_4 = 5G^3$$

$$\sigma_5 = G^3.$$

By the Newton recurrence for power sums, we have for all $\mu \in Z$

(19)
$$U(\phi^\mu) = \sigma_1 U(\phi^{\mu-1}) - \sigma_2 U(\phi^{\mu-2}) + \sigma_3 U(\phi^{\mu-3}) - \sigma_4 U(\phi^{\mu-4}) + \sigma_5 U(\phi^{\mu-5}).$$

Since the coefficients σ_i and the initial values (18) are in Z [G], it follows from (19) that for all $\mu \in Z$,

(20)
$$U(\phi^\mu) = \sum_{\nu=-\infty}^{\infty} c_{\mu\nu} G^\nu, \ c_{\mu\nu} \in Z \ .$$

(Actually the series in (20) is of course finite.) Substituting (18) and (20) into (19) and equating coefficients of like powers of G on both sides, we obtain the recurrence

(21)
$$c_{\mu\nu} = 5c_{\mu-1,\nu-1} - 25c_{\mu-1,\nu-2} + 25_{\mu+1,\nu-3} + 15c_{\mu-2,\nu-2} - 25_{\mu-2,\nu-3} - 5c_{\mu-3,\nu-2} + 15c_{\mu-3,\nu-3} - 5c_{\mu-4,\nu-3} + c_{\mu-5,\nu-3}.$$

From (21) we see that if $\pi(n)$ denotes the 5-adic order of n, then

(22)
$$\pi(c_{\mu\nu}) \geq \min[\pi(c_{\mu-1,\nu-1}) + 1, \pi(c_{\mu-1,\nu-2}) + 2, \pi(c_{\mu-1,\nu-3}) + 2, \pi(c_{\mu-2,\nu-2}) + 1, \pi(c_{\mu-2,\nu-3}) + 2, \pi(c_{\mu-3,\nu-2}) + 1, \pi(c_{\mu-3,\nu-3}) + 1, \pi(c_{\mu-4,\nu-3}) + 1, \pi(c_{\mu-5,\nu-3})].$$

LEMMA 6. For all $\mu \in Z$, we have

$$\pi(c_{\mu\nu}) \geq [(5\nu-3\mu-1)/6].$$

PROOF. By checking (17) we find that the inequality holds for $-4 \leq \mu \leq 0$. For $\mu > 0$ it follows by induction on μ, using (22). To handle the case $\mu < -4$, we solve (21) for $c_{\mu-5,\nu-3}$ and apply backwards induction on μ.

LEMMA 7. For all i,j we have $a_{ij} \in Z$ and

$$\pi(a_{ij}) \geq [(5j-i-1)/6].$$

PROOF. We have

$$U(G^i) = U\left(\frac{\theta(-x^5)^{2i}}{\theta(-x)^{2i}}\right)$$

$$= U\left(\frac{\theta(-x^{25})^{2i}}{\theta(-x)^{2i}} \frac{\theta(-x^5)^{2i}}{\theta(-x^{25})^{2i}}\right) = U(\phi^{2i})G^{-i}$$

$$= \sum_k c_{2i,k}G^kG^{-i} = \sum_j c_{2i,i+j}G^j.$$

Thus $a_{ij} = c_{2i,i+j}$, and hence by Lemma 6,

$$\pi(a_{ij}) \geq \left[\frac{5(i+j)-3(2i)-1}{6}\right] = \left[\frac{5j-i-1}{6}\right].$$

In order to obtain an analogous lower bound for $\pi(b_{ij})$, where $U(FG^i) = \sum_j b_{ij}G^j$, we note that as in the proof of Lemma 7,

(23) $$U(FG^i) = U(F\phi^{2i})G^{-i}.$$

The functions $U(F\phi^\mu)$ satisfy the same recurrence (19) as $U(\phi^\mu)$. The initial values $U(F\phi^\mu)$, $-4 \leq \mu \leq 0$, can be found in essentially the same way as $U(\phi^\mu)$. The result is:

$$U(F\phi^{-4}) = 3G^{-2} - 5G^{-1} - 5$$

$$U(F\phi^{-3}) = -5$$

(24) $$U(F\phi^{-2}) = G^{-1} - 4$$

$$U(F\phi^{-1}) = -1$$

$$U(F) = G$$

It follows from this that for all $\mu \in Z$,

$$U(F\phi^{\mu}) = \sum_{\nu} d_{\mu\nu} G^{\nu},$$

where the $d_{\mu\nu}$ are integers satisfying the same recurrence (21) as $c_{\mu\nu}$. Moreover, inspection of (24) shows that

$$\pi(d_{\nu\mu}) \geq [(5\nu-3\mu-1)/6]$$

for $-4 \leq \mu \leq 0$. In addition we have the very crucial fact that $\pi(d_{-3,\nu}) \geq 1$ for all ν. By the method of proof of Lemma 6, we obtain

LEMMA 8. For all $\mu,\nu \in Z$,

$$\pi(d_{\mu\nu}) \geq [(5\nu-3\mu-1)/6]$$

$$\pi(d_{\mu\nu}) \geq 1 \text{ if } \mu \equiv 2 \pmod 5.$$

From (23) we see that $b_{ij} = d_{2i,i+j}$. Hence:

LEMMA 9. For all $i,j \in Z$, we have $b_{ij} \in Z$ and

$$\pi(b_{ij}) \geq [(5j-i-1)/6]$$

$$\pi(d_{\mu\nu}) \geq 1 \text{ if } i \equiv 1 \pmod 5.$$

We are now ready to show that if $L_{\nu} = (\ell_1^{(\nu)},\ell_2^{(\nu)},\ell_3^{(\nu)} ,\ldots)$, then $\ell_j^{(2\alpha+1)} \equiv 0 \pmod{5^{\alpha}}$ for all j. To facilitate the induction, we prove more, namely:

THEOREM 10. For all $\alpha \geq 0$ and $j \geq 1$, we have

$$(25)_\alpha \qquad \pi(\ell_j^{(2\alpha+1)}) \geq \alpha + [(j-1)/2]$$

$$(26)_\alpha \qquad \pi(\ell_j^{(2\alpha+2)}) \geq \alpha + [j/2].$$

PROOF. Since $L_1 = (1,0,0,\ldots)$, $(25)_0$ holds. Suppose $(25)_\alpha$ holds for some $\alpha \geq 0$. Since $L_{2\alpha+2} = L_{2\alpha+1}A$, we have

$$\ell_j^{(2\alpha+2)} = \sum_{i \geq 1} \ell_i^{(2\alpha+1)} a_{ij}.$$

Hence $\pi(\ell_j^{(2\alpha+2)}) \geq \min_{i \geq 1} (\alpha + [(i-1)/2] + \pi(a_{ij}))$. To deduce $(26)_\alpha$ from this we show that

$$(27) \qquad [(i-1)/2] + \pi(a_{ij}) \geq [j/2]$$

for all $i,j \geq 1$. Clearly (27) holds if $i > j$. If $i \leq j$, we have by Lemma 7

$$\pi(a_{ij}) \geq [(5j-i-1)/6] \geq [(4j-1)/6]$$

$$\geq [j/2].$$

Thus $(25)_\alpha$ implies $(26)_\alpha$. Next suppose $(26)_\alpha$ holds for some $\alpha \geq 0$. Since $L_{2\alpha+3} = L_{2\alpha+2}B$, we have

$$\ell_j^{(2\alpha+3)} = \sum_{i \geq 1} \ell_i^{(2\alpha+2)} b_{ij}.$$

Hence

$$\pi(\ell_j^{(2\alpha+3)}) \geq \min_{i \geq 1} (\alpha + [i/2] + \pi(b_{ij})).$$

From this we wish to derive $(25)_{\alpha+1}$, so must show that

$$(28) \qquad [i/2] + \pi(b_{ij}) \geq 1 + [(j-1)/2] = [(j+1)/2]$$

for all $i, j \geq 1$. Clearly (28) holds if $i > j$. If $i \leq j$, then Lemma 9 gives

$$\pi(b_{ij}) \geq [(5j-i-1)/6] \geq [(4j-1)/6]$$

$$\geq [(j+1)/2] \text{ for } j \geq 2.$$

This leaves only the case $i = j = 1$. But since $\pi(b_{11}) \geq 1$ by Lemma 9, (28) holds in that case, too. This completes the proof of Theorem 10 by induction on α.

We have now shown that $L_{2\alpha+1}(x) = \sum_{j \geq 1} \ell_j^{(2\alpha+1)} G(x)^j$, where $\ell_j^{(2\alpha+1)} \equiv 0 \pmod{5^\alpha}$. Since $G(x)$ has integer coefficients (being an η-product), this proves (14) and hence (5).

REMARK. The above proof of (5) is obviously based on Atkin's paper [3]. Nevertheless there are some significant differences between the generating functions

$$P(x) = \sum_{n=0}^{\infty} p(n)x^n$$

and

$$Q(x) = \sum_{n=0}^{\infty} q(n)x^n,$$

and between the groups $\Gamma_0(5)$ and $\Gamma_0(10)$. We therefore found it desirable to give a fairly self-contained account here.

4. Congruences for $q(n) \pmod{7^\beta}$. The discussion here is parallel to that of the previous section, except for the difficulties caused by the fact that $\Gamma_0(14)$ has genus 1. To begin with, we let ℓ_β be the least non-negative solution of the congruence $24\ell \equiv -1 \pmod{7^\beta}$. Then $\ell_\beta = (49^{[(\beta+1)/2]}-1)/24$, so that $\ell_0 = 0$, $\ell_1 = \ell_2 = 2$, $\ell_3 = \ell_4 = 100$, and in general $\ell_{\beta+2} = 49\ell_\beta+2$. Congruence (6) asserts that

$$q(49(\ell_\beta+m\cdot 7^\beta)+2) \equiv \lambda_\beta\, q(\ell_\beta+m\cdot 7^\beta) \pmod{7^\beta}$$

for all $m \geq 0$, in other words that

$$q(\ell_{\beta+2}+m\cdot 7^{\beta+2}) \equiv \lambda_\beta\, q(\ell_\beta+m\cdot 7^\beta) \pmod{7^\beta}.$$

Thus the object is to show that if

$$D_\beta(x) = \sum_{m=0}^{\infty} q(\ell_\beta+m\cdot 7^\beta)x^m \qquad (\beta \geq 0),$$

then

(29) $$D_{\beta+2}(x) \equiv \lambda_\beta D_\beta(x) \pmod{7^\beta}$$

for all $\beta \geq 1$. Put

$$F(x) = x^{-2}\,\frac{Q(x)}{Q(x^{49})} = \frac{\eta(2\tau)}{\eta(\tau)}\,\frac{\eta(49\tau)}{\eta(98\tau)},$$

and define functions $L_\beta(x)$, $\beta \geq 0$ inductively as follows:

$L_0 = 1$, $L_{2\alpha+1} = U(FL_{2\alpha})$, $L_{2\alpha+2} = U(L_{2\alpha+1})$ for $\alpha \geq 0$; here $U = U_7$. As in §3 find that $D_{2\alpha}(x) = Q(x)L_{2\alpha}(x)$ and $D_{2\alpha+1}(x) = Q(x^7)L_{2\alpha+1}(x)$. Since $Q(x)$ and $Q(x^7)$ have leading coefficient 1, congruence (29) is equivalent to

(30) $$L_{\beta+2}(x) \equiv \lambda_\beta\, L_\beta(x) \pmod{7^\beta}.$$

Using Theorems 2 and 3, we find that F is in $K_0(98)$, and that its orders at the cusps of $R_0(98)$ are as shown below:

$R_0(98)$

δ	1	2	7	14	49	98
$\mathrm{ord}_{\nu/\delta} F =$	-2	2	0	0	2	-2

Hence Theorem 5 shows that UF is in $K_0(14)$, and that its orders at the cusps of $R_0(14)$ satisfy the following inequalities:

$R_0(14)$

δ	1	2	7	14
$\mathrm{ord}_{\nu/\delta} UF \geq$	-2	0	1	0

We now require a basis for the vector space S of functions in $K_0(14)$ which are holomorphic on $R_0(14)$ except at $\tau = 0$. Since $R_0(14)$ has genus 1, such a basis cannot consists of powers of a single function. Instead we construct functions G and H in S with poles at $\tau = 0$ of multiplicities 2 and 3 respectively. The desired basis then has the form $\{G^r H^s : 0 \leq r \leq 2,\ 0 \leq s < \infty\}$. Proceeding to the details, let

$$H(x) = [\frac{\theta(-x^7)}{\theta(-x)}]^4 = \frac{\eta(2\tau)^4}{\eta(\tau)^8} \frac{\eta(7\tau)^8}{\eta(14\tau)^4} .$$

By Theorems 2 and 3 we have H in $K_0(14)$, with the following orders at the cusps of $R_0(14)$.

$R_0(14)$

δ	1	2	7	14
$\mathrm{ord}_{\nu/\delta}H$ =	-3	0	3	0

Next, let

$$E(x) = \frac{Q(x)^7}{Q(x^7)} = \frac{\eta(2\tau)^7}{\eta(\tau)^7} \frac{\eta(7\tau)}{\eta(14\tau)} .$$

From Theorems 2 and 3 we see that E is in $K_0(14)$, and that its orders at the cusps of $R_0(14)$ are as shown below.

$R_0(14)$

δ	1	2	7	14
$\mathrm{ord}_{\nu/\delta}E$ =	-2	2	0	0

<u>LEMMA 11</u>. <u>The function</u> $G = (8E-1)/7$ <u>has a simple zero at the cusp</u> $\tau = 1/7$ <u>of</u> $R_0(14)$.

<u>PROOF</u>. From the last two tables we see that the seven functions $1,G,G^2,G^3,H,GH,H^2$ are in *S*, with poles at $\tau = 0$ of multiplicity ≤ 6. By the Riemann-Roch Theorem they are linearly dependent; examination of their Fourier series shows that

$$G^3 = 7H^2 - 5GH - H. \tag{31}$$

Since H has a triple zero at $\tau = 1/7$, it follows from (31) that G has a simple zero there.

By Lemma 11, G has the following orders at the cusps of $R_0(14)$.

$R_0(14)$

δ	1	2	7	14
$ord_{\nu/\delta}G$	-2	0	1	0

Since $Q(x)^7 \equiv Q(x^7)$ (mod 7), the power series $G(x) = [8Q(x)^7 - Q(x^7)]/7$ has integer coefficients. Of course H also has this property.

Now for all integers ν (positive, negative or zero), put

$$J_\nu = \begin{cases} H^{\nu/3} & \text{if } \nu \equiv 0 \pmod 3 \\ GH^{(\nu-1)/3} & \text{if } \nu \equiv 1 \pmod 3 \\ G^2H^{(\nu-2)/3} & \text{if } \nu \equiv 2 \pmod 3). \end{cases}$$

From the above computation of the orders of G and H at the cusps of $R_0(14)$, we find that $ord_{1/7}J_\nu = \nu$, while $ord_0 J_\nu = -\nu, -\nu-1$ or $-\nu-2$ according as $\nu \equiv 0, 1$ or 2 (mod 3). Elsewhere on $R_0(14)$, J_ν is holomorphic. Hence the functions $J_0 = 1$, $J_1 = G$, $J_2 = G^2$, $J_3 = H$, $J_4 = GH$, $J_5 = G^2H$, ... form a basis of S. Moreover $J_1, J_2, \ldots$ form a basis for the subspace T of all functions in S which vanish at $\tau = 1/7$.

An application of Theorem 5 shows that S is mapped into itself by the linear transformations U and V, where $V(P) = U(FP)$, $P \in S$. Moreover, U maps the subspace T into itself, while V maps S into T. Let $A = (a_{ij})$ and $B = (b_{ij})$, $1 \leq i, j < \infty$ be the matrices of the restrictions of U and V to T, using the basis $J_1, J_2, \ldots$. Thus for all $i \geq 1$, we have

$$U(J_i) = \sum_{j \geq 1} a_{ij} J_j$$

$$U(FJ_i) = \sum_{j \geq 1} b_{ij} J_j .$$

As already noted, $L_1 = UF$ is in T, with at most a double pole at $\tau = 0$. Hence $L_1 = cG$ where c is a constant; since the Fourier series of L_1 and G both have leading coefficient 1, we have $L_1 = G$. Therefore, in terms of the basis $J_1, J_2, \ldots$ we have

$$L_1 = (1,0,0,\dots)$$

$$\vdots$$

$$L_{2\alpha+1} = (1,0,0,\dots)\ (AB)^{\alpha}$$

$$L_{2\alpha+2} = (1,0,0,\dots)(AB)^{\alpha}A.$$

In order to show that the entires of A and B are integers, and to obtain lower bounds for their 7-adic orders, we now derive a modular equation. Let

$$\phi(x) = \frac{\theta(-x^{49})}{\theta(-x)} = \frac{\eta(2\tau)}{\eta(\tau)^2}\ \frac{\eta(49\tau)^2}{\eta(98\tau)} .$$

By Theorems 2 and 3, ϕ is in $K_0(98)$, its orders at the cusps of $R_0(98)$ being as follows:

$R_0(98)$	δ	1	2	7	14	49	98
	$\mathrm{ord}_{\nu/\delta}\ \phi =$	-6	0	0	0	0	0

As explained in §3, this implies that the functions $\phi(\frac{\tau+\lambda}{7})$ $(0 \leq \lambda \leq 6)$ are the roots of an equation

$$t^7-\sigma_1 t^6+\sigma_2 t^5-\sigma_3 t^4+\sigma_4 t^3-\sigma_5 t^2+\sigma_6 t-\sigma_7 = 0, \tag{32}$$

where the coefficients σ_i are in T. As in §3, we determine them by computing the power sums $\sum_{\lambda=0}^{6} \phi(\frac{\tau+\lambda}{7})^{-k}$ for $0 \leq k \leq 6$. From the preceding table we see that for $k \geq 0$, the function $\phi(\tau)^{-k}$ is holomorphic on $R_0(98)$ except for a 6k-fold pole at $\tau = 1/49$. By Theorem 5, $U(\phi^{-k})$ is in the vector space V of functions in $K_0(14)$ which are holomorphic on $R_0(14)$ except perhaps at $\tau = 1/7$. Moreover the multiplicity of the pole of $U(\phi^{-k})$ at $\tau = 1/7$ is at most $[6k/7]$.

From the properties of J_ν developed above, it follows that $J_0 = 1$, $J_{-2} = GH^{-1}$, $J_{-3} = H^{-1}$, $J_{-4} = G^2H^{-2}, \dots$ form a basis of V. Following along the lines of §3, we express $U(\phi^{-k})$ $(0 \leq k \leq 6)$ in terms of this basis, and then use Newton's identities to find the coefficients σ_i in (32). The results are as follows.

$$(33)\qquad \begin{cases} U(\phi^{-6}) = -6J_{-5}+18.7J_{-4}+18.7J_{-3}-6.7^2J_{-2}+7^2 \\ U(\phi^{-5}) = 10J_{-4}+40J_{-3}-7^2 \\ U(\phi^{-4}) = 8J_{-3}-7 \\ U(\phi^{-3}) = -6J_{-2}+7 \\ U(\phi^{-2}) = U(\phi^{-1}) = U(1) = 1. \end{cases}$$

$$(34)\qquad \begin{cases} \sigma_1 = 7G-7^2H+7^2G^2-7^3GH+7^3H^2 \\ \sigma_2 = -6\cdot 7H+2\cdot 7G^2-6\cdot 7^2GH+7^3H^2 \\ \sigma_3 = -2\cdot 7H-2\cdot 7^2GH+3\cdot 7^2H^2 \\ \sigma_4 = -2\cdot 7GH+7^2H^2 \\ \sigma_5 = 3\cdot 7H^2 \\ \sigma_6 = 7H^2 \\ \sigma_7 = H^2. \end{cases}$$

To deal uniformly with positive and negative powers of ϕ, we consider the space $\mathcal{W}$ of functions in $K_0(14)$ which are holomorphic on $R_0(14)$ except at $\tau = 0$ and $\tau = 1/7$. Our analysis (together with the fact that $J_{-1} = G^2H^{-1}$ has simple poles at 0 and 1/7) shows that the functions $J_\nu (-\infty < \nu < \infty)$ form a basis of $\mathcal{W}$. Moreover $U(\phi^\mu)$ is in $\mathcal{W}$ for all integers μ, so we can write

$$(35)\qquad U(\phi^\mu) = \sum_{\nu=-\infty}^{\infty} c_{\mu\nu} J_\nu ,$$

where the sum is actually finite. The functions $U(\phi^\mu)$ satisfy the recurrence

$$U(\phi^{\mu}) = \sigma_1 U(\phi^{\mu-1}) - \sigma_2 U(\phi^{\mu-2}) + \sigma_3 U(\phi^{\mu-3}) - \sigma_4 U(\phi^{\mu-4}) + \sigma_5 U(\phi^{\mu-5}) - \sigma_6 U(\phi^{\mu-6}) + \sigma_7 U(\phi^{\mu-7}). \tag{36}$$

We now substitute (34) and (35) into (36), and we use (31) to express each term of the form GJ_i, G^2J_i or GHJ_i on the right in terms of the basis $\{J_\nu\}$. We can then equate coefficients of J_ν on both sides, and obtain a recurrence for $c_{\mu\nu}$. The details are straightforward but tedious, so we will omit them here. The resulting recurrence for $c_{\mu\nu}$ depends on the residue of ν (mod 3). If $\nu \equiv 0$ (mod 3), then

$$\begin{aligned} c_{\mu\nu} = {} & -7c_{\mu-1,\nu-1} - 7^2c_{\mu-1,\nu-2} - 7^2c_{\mu-1,\nu-3} + 8\cdot 7^2c_{\mu-1,\nu-4} \\ & + 7^3c_{\mu-1,\nu-5} + 7^3c_{\mu-1,\nu-6} - 7^4c_{\mu-1,\nu-7} + 2\cdot 7c_{\mu-2,\nu-2} \\ & + 6\cdot 7c_{\mu-2,\nu-3} - 6\cdot 7^2c_{\mu-2,\nu-4} - 2\cdot 7^2c_{\mu-2,\nu-5} - 7^3c_{\mu-2,\nu-6} \\ & + 6\cdot 7^3c_{\mu-2,\nu-7} - 2\cdot 7c_{\mu-3,\nu-3} + 2\cdot 7^2c_{\mu-3,\nu-4} + 3\cdot 7^2c_{\mu-3,\nu-6} \\ & - 2\cdot 7^3c_{\mu-3,\nu-7} - 2\cdot 7c_{\mu-4,\nu-4} - 7^2c_{\mu-4,\nu-6} + 2\cdot 7^2c_{\mu-4,\nu-7} \\ & + 3\cdot 7c_{\mu-5,\nu-6} - 7c_{\mu-6,\nu-6} + c_{\mu-7,\nu-6}. \end{aligned} \tag{37}$$

If $\nu \equiv 1$ (mod 3), then

$$\begin{aligned} c_{\mu\nu} = {} & 7c_{\mu-1,\nu-1} - 12\cdot 7c_{\mu-1,\nu-2} - 6\cdot 7^2c_{\mu-1,\nu-3} - 7^3c_{\mu-1,\nu-4} \\ & + 6\cdot 7c_{\mu-1,\nu-5} + 7^3c_{\mu-1,\nu-6} + 2\cdot 7c_{\mu-2,\nu-2} + 16\cdot 7c_{\mu-2,\nu-3} \\ & + 6\cdot 7^2c_{\mu-2,\nu-4} - 32\cdot 7^3c_{\mu-2,\nu-5} - 7^3c_{\mu-2,\nu-6} - 2\cdot 7c_{\mu-3,\nu-3} \\ & - 2\cdot 7^2c_{\mu-3,\nu-4} + 10\cdot 7^2c_{\mu-3,\nu-5} + 3\cdot 7^2c_{\mu-3,\nu-6} + 2\cdot 7c_{\mu-4,\nu-4} \\ & - 10\cdot 7c_{\mu-4,\nu-5} - 7^2c_{\mu-4,\nu-6} + 3\cdot 7c_{\mu-5,\nu-6} - 7c_{\mu-6,\nu-6} \\ & + c_{\mu-7,\nu-6}. \end{aligned}$$

If $\mu \equiv 2 \pmod 3$, then

$$\begin{aligned} c_{\mu\nu} &= c_{\mu-1,\nu-1}+7^2c_{\mu-1,\nu-2}-6\cdot7^2c_{\mu-1,\nu-3}-7^3c_{\mu-1,\nu-4} \\ &+ 7^3c_{\mu-1,\nu-6}-2\cdot7c_{\mu-2,\nu-2}+16\cdot7c_{\mu-2,\nu-3}+6\cdot7^2c_{\mu-2,\nu-4} \\ &+ 3\cdot7^2c_{\mu-2,\nu-6}+2\cdot7c_{\mu-3,\nu-3}-2\cdot7^2c_{\mu-3,\nu-4}+3\cdot7^2c_{\mu-3,\nu-6} \\ &+ 2\cdot7c_{\mu-4,\nu-4}-7^2c_{\mu-4,\nu-6}+3\cdot7c_{\mu-5,\nu-6}-7c_{\mu-6,\nu-6} \\ &+ c_{\mu-7,\nu-6}. \end{aligned}$$

LEMMA 12. *If* $\pi(n)$ *denotes the 7-adic order of* n, *we have* $\pi(c_{\mu\nu}) \geq [(7\nu-6\mu-1)/12]$ *for all* $\mu, \nu \in Z$.

The proof is similar to that of Lemma 6.

Next we note that G, regarded as an element of $K_0(98)$ rather than the subfield $K_0(14)$, is holomorphic on $R_0(98)$ except for a 14-fold pole at $\tau = 0$. Moreover, G has a simple zero at $\tau = 1/49$. As already noted, for $k > 0$ the function ϕ^{-k} has a 6k-fold pole at $\tau = 1/49$, a 6k-fold zero at $\tau = 0$, and is holomorphic elsewhere on $R_0(98)$. Hence for $k \geq 3$, the function $G\phi^{-k}$ is holomorphic on $R_0(98)$ except for a (6k-1)-fold pole at $\tau = 1/49$. By Theorem 5, $U(G\phi^{-k})$ is then in V , with a pole of multiplicity at most $[(6k-1)/7]$ at $\tau = 1/7$. We can therefore compute $U(G\phi^{-k})$ for $3 \leq k \leq 9$ in a reasonable number of steps, and we find that

$$(38)\qquad \begin{aligned} U(G\phi^{-9}) &= -52J_{-7}-458J_{-6}-144\cdot7J_{-5}-12\cdot7^2J_{-4}+36\cdot7^2J_{-3}+7^3J_{-2} \\ U(G\phi^{-8}) &= -48J_{-6}+7^2J_{-2} \\ U(G\phi^{-7}) &= 36J_{-5}-12\cdot7J_{-3}+7^2J_{-2} \\ U(G\phi^{-6}) &= J_{-5}-3\cdot7J_{-4}-3\cdot7J_{-3}+6\cdot7J_{-2} \\ U(G\phi^{-5}) &= -2J_{-4}-4J_{-3}+7J_{-2} \\ U(G\phi^{-4}) &= U(G\phi^{-3}) = J_{-2}. \end{aligned}$$

Now the functions $U(G\phi^{\mu})$ satisfy the same recurrence (36) as $U(\phi^{\mu})$. Hence we have

$$U(G\phi^{\mu}) = \sum_{\nu=-\infty}^{\infty} d_{\mu\nu} J_{\nu},$$

where the coefficients $d_{\mu\nu}$ satisfy (37). From this recurrence, together with the initial conditions (38), we see that $d_{\mu\nu} \in Z$. Moreover, induction on μ shows the following.

LEMMA 13. $\pi(d_{\mu\nu}) \geq [(7\nu-6\mu-1)/12]$.

Similarly, if we put

$$U(G^2\phi^{\mu}) = \sum_{\nu=-\infty}^{\infty} e_{\mu\nu} J_{\nu},$$

then the coefficients $e_{\mu\nu}$ satisfy (37). This time the simplest range in which to compute the initial values of $e_{\mu\nu}$ is $-11 \leq \mu \leq -5$. The result is:

$e_{\mu\nu}$	$\nu = -9$	-8	-7	-6	-5	-4	-3
$\mu = -11$	5	-207	$11\cdot 7$	$109\cdot 7$	$15\cdot 7^2$	7^3	$-5\cdot 7^3$
-10		-6	$18\cdot 7$	$18\cdot 7$	$-6\cdot 7^2$	7^2	
- 9			6	72	$24\cdot 7$	7^2	$-6\cdot 7^2$
- 8			8			-7	
- 7					-6	-7	$2\cdot 7$
- 6						1	
- 5						1	

From this, together with (37), we get

LEMMA 14. $\pi(e_{\mu\nu}) \geq [(7\nu-6\mu-1)/12]$.

The next task is to apply Lemmas 12,13 and 14 to estimate $\pi(a_{ij})$. If $i \equiv 0 \pmod 3$, then

$$\begin{aligned} U(J_i) &= U(H^{i/3}) = U(\phi^{4i/3})H^{-i/3} \\ &= \sum_{\nu} c_{4i/3,\nu} J_{\nu} H^{-i/3} \\ &= \sum_{j} c_{4i/3,i+j} J_j, \end{aligned}$$

and hence $a_{ij} = c_{4i/3,i+j}$. Similarly we find that if $i \equiv 1 \pmod 3$, then $a_{ij} = d_{4(i-1)/3,i+j-1}$, and if $i \equiv 2 \pmod 3$, then $a_{ij} = e_{4(i-2)/3,i+j-2}$. Hence by Lemmas 12, 13 and 14, we find that $\pi(a_{ij}) \geq [(7j-i+\delta_i)/12]$, where $\delta_i = -1, 0,$ or 1 according as $i \equiv 0,1$ or $2 \pmod 3$. In particular we have the following estimate

<u>LEMMA 15.</u> $\pi(a_{ij}) \geq [(7j-i+\delta_i)/12]$.

A similar discussion is now required to estimate $\pi(b_{ij})$. This time we will be briefer. Let

$$U(F\phi^\mu) = \sum_\nu c'_{\mu\nu} J_\nu$$

$$U(FG\phi^\mu) = \sum_\nu d'_{\mu\nu} J_\nu$$

$$U(FG^2\phi^\mu) = \sum_\nu e'_{\mu\nu} J_\nu .$$

Then $c'_{\mu\nu}$, $d'_{\mu\nu}$ and $e'_{\mu\nu}$ all satisfy (37). The most convenient initial values are $c'_{\mu\nu}(-6 \leq \mu \leq 0)$, $d'_{\mu\nu}$ $(-9 \leq \mu \leq -3)$, and $e'_{\mu\nu}(-11 \leq \mu \leq -5)$. These are given by the following tables, where the blank entries are zero.

$c'_{\mu\nu}$	$\nu=-4$	-3	-2	-1	0	1
$\mu = -6$					7^2	
-5	1	2	$4\cdot 7$			
-4		3	$3\cdot 7$		-7	
-3					7	
-2					1	
-1					-1	
0						1

$d'_{\mu\nu}$	$\nu=-7$	-6	-5	-4	-3	-2
$\mu=-9$	-9	32	$2\cdot 7^2$		$-9\cdot 7^2$	7^3
-8		-5	$-5\cdot 7$			7^2
-7			6	$6\cdot 7$		-7^2
-6						-7
-5					-2	-7
-4					-1	-6
-3						-1

$e'_{\mu\nu}$	$\nu=-8$	-7	-6	-5	-4	-3	-2	-1	0
$\mu=-11$	-30	$-30\cdot 7$	$12\cdot 7$	$10\cdot 7^2$					
-10	-1	$2\cdot 7$	$-18\cdot 7$	$-13\cdot 7^2$	$-6\cdot 7^2$	$3\cdot 7^3$			
-9		2	-12	$-8\cdot 7$	-7^2		$2\cdot 7^2$		
-8			-1	-7	-7				
-7				-2	-7				
-6					1				
-5						-6	$-4\cdot 7$		7^2

<u>LEMMA 16.</u> $\pi(c'_{\mu\nu}), \pi(d'_{\mu\nu}), \pi(e'_{\mu\nu}) \geq [(7\nu-6\mu-1)/12]$.

Hence $\pi(b_{ij}) \geq [(7j-i+\delta_i)/12]$.

From this point on, the technique is essentially that of [4]. For $\beta \geq 1$, let L_β be represented in the basis $J_1, J_2, \ldots$ of T by the vector $L_\beta = (\ell_1^{(\beta)}, \ell_2^{(\beta)}, \ldots)$.

<u>LEMMA 17.</u> $\pi(\ell_j^{(\beta)}) \geq [j/2]$ <u>for all</u> β.

PROOF. The inequality holds for $\beta = 1$, since $L_1 = G = (1,0,0,\ldots)$. Assume it holds for some $\beta \geq 1$. Then

$$\ell_j^{(\beta+1)} = \sum_i \ell_i^{(\beta)} m_{ij},$$

where $m_{ij} = a_{ij}$ or b_{ij} according as β is odd or even. To complete the induction, it suffices to prove that

$$[i/2] + \pi(m_{ij}) \geq [j/2] \tag{39}$$

for all $i,j \geq 1$. Clearly (39) holds if $i \geq j$. If $i < j$, then by Lemmas 15 and 16,

$$\pi(m_{ij}) \geq [(7j-i-1)/12] \geq [6j/12],$$

so (39) holds in that case too.

It is trivial to check from (37) and the above initial conditions for $d_{\mu\nu}$, $d'_{\mu\nu}$ that $a_{11} \equiv -b_{11} \equiv 1 \pmod 7$. Together with Lemma 17, this shows that

$$L_\beta \equiv ((-1)^{\beta(\beta-1)/2}, 0,0,\ldots) \pmod 7.$$

Now let

$$D_{rs}^{(\beta)} = \ell_r^{(\beta)} \ell_s^{(\beta+2)} - \ell_s^{(\beta)} \ell_r^{(\beta+2)}.$$

LEMMA 18. $\pi(D_{rs}^{(\beta)}) \geq \beta + [(r+s-3)/2]$.

PROOF. Since $\ell_j^{(1)} = 0$ for all $j > 1$, we have

$$D_{1s}^{(1)} = -D_{s1}^{(1)} = \ell_s^{(3)} \text{ for } s > 1,$$

and $D_{rs}^{(1)} = 0$ otherwise. By Lemma 17, we have

$$\pi(D_{1s}^{(1)}) = \pi(\ell_s^{(3)}) \geq [s/2] = 1 + [(1+s-3)/2],$$

which proves the desired inequality for $\beta = 1$. Suppose now that it holds for some $\beta \geq 1$. Then

$$D_{rs}^{(\beta+1)} = \sum_{t \neq u} D_{tu}^{(\beta)} m_{tr} m_{us},$$

where $m_{ij} = a_{ij}$ or b_{ij} according as β is odd or even. By induction and Lemmas 15 and 16, we get

$$\pi(D_{rs}^{(\beta+1)}) \geq \min_{t \neq u} (\beta + [(t+u-3)/2] + [(7r-t+\delta_t)/12] + [(7s-u+\delta_u)/12]).$$

Thus it suffices to prove that

(40) $$[(t+u-3)/2] + [(7r-t+\delta_t)/12] + [(7s-u+\delta_u)/12] \geq 1 + [(r+s-3)/2]$$

for all positive integers r,s,t,u with $t \neq u$. By the inequality $[x] + [y] + [z] \geq [x+y+z] - 2$, the left side of (40) is at least

$$[\frac{7(r+s)+5(t+u)-37}{12}] \geq [\frac{7(r+s)-7}{12}]$$

if $t + u \geq 6$. This in turn is at least

$$[\frac{6(r+s)-6}{12}] = 1 + [(r+s-3)/2],$$

completing the induction. The case $t + u \leq 5$ is easily checked.

As noted earlier, we have $\ell_1^{(\beta)} \equiv \pm 1 \pmod 7$ for all $\beta \geq 1$; in particular $7 \nmid \ell_1^{(\beta)}$. Hence we can solve the congruence $\ell_1^{(\beta)} x \equiv \ell_1^{(\beta+2)} \pmod{7^\beta}$. Denote the least positive solution by λ_β. Then

$$\ell_1^{(\beta)} \ell_s^{(\beta)} \lambda_\beta \equiv \ell_s^{(\beta)} \ell_1^{(\beta+2)} \pmod{7^\beta}$$

$$\equiv \ell_1^{(\beta)} \ell_s^{(\beta+2)} \pmod{7^\beta}$$

by Lemma 18. Cancelling $\ell_1^{(\beta)}$ we obtain $\ell_s^{(\beta+2)} \equiv \ell_s^{(\beta)} \lambda_\beta \pmod{7^\beta}$ for all β. This proves (30) and hence (6).

In conclusion, we note than an almost immediate consequence of (6) is that q(n) assumes all residues (mod 7) infinitely often. This is because $\lambda_1 = 6$ (as shown in the course of the proof), and $q(n) \equiv 0,1,2,4 \pmod 7$ for n = 65, 2, 30 and 16 respectively.

REFERENCES

1. P. Allatt and J.B. Slater, "Congruences on some special modular forms," J. London Math. Soc. (2) 17 (1978),380-392.

2. A.O.L. Atkin, "Proof of a conjecture of Ramanujan," Glasgow Math. J. 8 (1967), 14-32.

3. __________, "Ramanujan congruences for $p_{-k}(n)$," Canadian Math. J. 20 (1968), 67-78.

4. __________, and J.N. O'Brien, "Some properties of p(n) and c(n) modulo powers of 13," Trans. Amer. Math. Soc. 126 (1967), 442-459.

5. K. Hughes, Arithmetic Properties of Modular Forms, Ph.D. Thesis, UCLA, 1980.

6. G. Ligozat, Courbes Modulaires de Genre 1, Bull. Soc. Math. France, Mémoire 43, 1975.

7. M. Newman, "Construction and application of a class of modular functions II," Proc. London Math. Soc. (3) 9 (1959), 373-387.

8. S. Ramanujan, "Some properties of p(n), the number of partitions of n," Proc. Camb. Phil. Soc. 19 (1919), 207-210.

9. G.N. Watson, "Beweis von Ramanujans Vermutungen über Zerfällungsanzahlen," J. Reine und Angew. Math. 179 (1938), 97-128.

University of California, Los Angeles

GAPS IN THE FOURIER SERIES OF AUTOMORPHIC FORMS

by

M.I. Knopp and J. Lehner

Dedicated to Emil Grosswald on the occasion of his retirement.

1. Let Γ be a fuchsian group acting on the upper half-plane and possessing translations, so that an automorphic form on Γ has a Fourier series. In this paper we consider properties of the Fourier coefficients, especially gaps in the sequence of exponents, treating both Hadamard and Fabry gaps. In particular, we show that for a large class of automorphic forms there can be no Hadamard gaps. Section 4 is devoted to Ω-results on the Fourier coefficients.

We make the following

DEFINITION. Let Γ be a fuchsian group acting on the upper half-plane $H = \{z = x+iy: \ y > 0\}$. Let F be a complex-valued function defined on H and k a real number. We say F is an automorphic form on Γ of exponent k with multiplier system (MS) v if

$$F(Az) = v(A)(cz+d)^k F(z), \ A = (a, b: \ c, d) \in \Gamma, \tag{1.1}$$

and F is holomorphic on H and at the parabolic cusps of Γ. Here v is defined on Γ and

$$|v(A)| = 1, \ A \in \Gamma. \tag{1.2}$$

Remark. Our "exponent" is the same as the term "weight" as used by Serre, for example, and others. (Cf. [16])

By iteration we find v must satisfy the "consistency condition",

$$v(A_1A_2)((A_1A_2)'(z))^{-k/2} = v(A_1)v(A_2)(A_1'(A_2z))^{-k/2}(A_2'(z))^{-k/2} \tag{1.3}$$

for all A_1, $A_2 \in \Gamma$, the powers of the derivatives being assigned their principal values.

By $\{\Gamma, k, v\}$, ($\{\Gamma, k, v\}_0$) we shall mean the space of forms (cusp forms) on Γ of exponent k with multiplier system v. If $v(A) \equiv 1$, $A \in \Gamma$, we write instead $\{\Gamma, k\}$, $\{\Gamma, k\}_0$, respectively.

2. Hadamard Gaps.

Let Γ possess the minimal translation $z \to z+1$. Under natural and rather weak assumptions we show that the Fourier series of a nontrivial automorphic form on Γ cannot have Hadamard gaps. The proof is not complicated in any way by the introduction of multiplier systems and thus we assume here and in Sections 3, 4 that F is an automorphic form with MS v on Γ of arbitrary real exponent k. The equation (1.1), together with the fact that F is holomorphic on H and at the parabolic cusps of Γ, implies the existence of a Fourier expansion of the form

(2.1) $$F(z) = \sum_{n=\mu}^{\infty} a_n e^{2\pi i(n+\kappa)z}, \quad 0 \leq \mu \in Z, \ 0 \leq \kappa < 1,$$

valid in H. Changing notation, we rewrite (2.1) in the form

(2.2) $$F(z) = \sum_{h=1}^{\infty} a_h e^{2\pi i(n_h+\kappa)z},$$

where $a_h \neq 0$ for all h and $\mu \leq n_1 < n_2 < \ldots$. Specifically, we shall prove

THEOREM 2.1. Suppose the fuchsian group Γ is not cyclic parabolic and F is an automorphic form on Γ with expansion (2.2). Suppose also that $n_h+\kappa > 0$ for all h in (2.2).

(i) Assume Γ has finite area. If $k > 0$ and there are Hadamard gaps in (2.2) (i.e., $n_{h+1}/n_h \geq \theta > 1$ for all h), then $F(z) \equiv 0$.

(ii) If Γ has infinite area and $k > 1$, we draw the same conclusion.

The proof of this theorem depends in an essential way upon two results, which we state as lemmas. The first of these is an adaptation to H of [17, Th. XI.10, p. 518].

<u>LEMMA 2.2.</u> For $z_0 \in H$ and $Y > 0$ define $N(Y, z_0)$ to be the number of Γ-images of z_0 in $S_Y = \{z = x+iy: 0 \leq x \leq 1, y > Y\}$. ($N(Y, z_0)$ is finite by the discontinuity of Γ.) If Γ has finite area, then $N(Y, z_0) > KY^{-1}$ for some $K > 0$ depending upon Γ and z_0.

The second result, due to Binmore [4], deals with gap series in the unit disc D.

<u>LEMMA 2.3.</u> Suppose $\phi(\zeta) = \sum_{n=0}^{\infty} a_n \zeta^n$ is holomorphic in D and bounded on some curve in D whose closure contains a point of ∂D. If the power series for $\phi(\zeta)$ has Hadamard gaps, then $|a_n| < M$ for all n, where $M > 0$.

When applied to automorphic forms, Lemma 2.3 can be recast in the stronger form we need for the proof of Theorem 2.1. This is

<u>LEMMA 2.4.</u> Let F be an automorphic form on Γ. Suppose that in (2.2) there are Hadamard gaps and that $n_h+\kappa > 0$ for all h. If in addition Γ is not parabolic cyclic, then

$$|a_h| < M(n_h+\kappa)^{-1} \tag{2.3}$$

for all h, with $M > 0$.

<u>PROOF</u>. Differentiate (1.1):

$$F'(Az) = v(A)\{(cz+d)^{k+2}F'(z) + kc(cz+d)^{k+1}F(z)\}. \tag{2.4}$$

The Fourier series of F and F' are, respectively,

$$F(z) = \sum_{n_h+\kappa > 0} a_h e^{2\pi i(n_h+\kappa)z},$$

(2.5)

$$F'(z) = \sum_{n_h+\kappa > 0} 2\pi i(n_h+\kappa)a_h e^{2\pi i(n_h+\kappa)z}.$$

By the assumption on Γ there exists an $A \in \Gamma$ that is not a translation. Then $p = A\infty$ is a finite cusp Γ-equivalent to ∞. Consider the vertical line L joining p to ∞. By (2.5) $F \to 0$ exponentially as $z \to \infty$ along any vertical line, in particular along the vertical line $A^{-1}(L)$. Then (1.1) shows that $F \to 0$ exponentially as $z \to p$ along L. By the same reasoning, using (2.4) and the second equation of (2.5), we conclude that $F'(z) \to 0$ exponentially as $z \to p$ along L.

We now transfer this information from H to D. For $|\zeta| < 1$ put

$$\phi(\zeta) = \sum_{n_h+\kappa > 0} 2\pi i(n_h+\kappa)a_h\zeta^{n_h} = F'\left(\frac{\log \zeta}{2\pi i}\right) \zeta^{-\kappa} ,$$

a gap series holomorphic in D. Let C be the image of L under the transformation $\zeta = e^{2\pi iz}$. Then C is a radius of D, joining 0 to $\zeta(p) = e^{2\pi ip}$, a point on ∂D. Since $F' \to 0$ as $z \to p$ along L, it follows that $\phi(\zeta) \to 0$ as $\zeta \to e^{2\pi ip}$ along C. In particular, then, $\phi(\zeta)$ remains bounded on C and (2.3) follows from Lemma 2.3.

3. Proof of Theorem 2.1. Suppose R is a fundamental region for Γ contained in $|\mathrm{Re}\, z| \le 1/2$, z_0 is an interior point of R, and $\Delta \subset R$ is a disk about z_0 of hyperbolic radius ρ. Fix $Y > 0$. Then for each Γ-image of z_0 in $S_Y: = \{z = x+iy:\ 0 \le x \le 1,\ y > Y\}$ the corresponding Γ-image of Δ lies in the slightly larger set $S_{Y'}$, where $Y' = Ye^{-2\rho}$.

The expansion (2.2) leads by termwise integration to the Parseval identity:

$$\int_0^1 |F(z)|^2 dx = \int_0^1 F(z)\overline{F}(z)dx = \sum_{h=1}^{\infty} |a_h|^2 e^{-4\pi(n_h+\kappa)y} .$$

Next, integrate on y to obtain

$$(3.1) \qquad \int_0^1 \int_{Y'}^{\infty} y^{k-2}|F(z)|^2 dxdy = \sum_{h=1}^{\infty} |a_h|^2 \int_{Y'}^{\infty} e^{-4\pi(n_h+\kappa)y} y^{k-2} dy.$$

By way of contradiction assume the occurrence of Hadamard gaps in the expansion (2.2) and assume that $F \not\equiv 0$. The latter assumption in particular insures that

$$\sigma = \iint_{\Delta} y^{k-2}|F(z)|^2\, dxdy > 0.$$

The functional equation (1.1) implies that $y^{k/2}|F(z)|$ is invariant under the mappings $z \to Az$, $A \in \Gamma$; from this follows

$$\iint_{A(\Delta)} y^{k-2}|F(z)|^2 dxdy = \iint_{\Delta} y^{k-2}|F(z)|^2 dxdy = \sigma,$$

for $A \in \Gamma$. Consequently, by the choice of Y',

$$\int_0^1 \int_{Y'}^{\infty} y^{k-2}|F(z)|^2 dxdy \geq \sigma N(Y, z_0),$$

so that (3.1) yields

$$\sigma N(Y, z_0) \leq \sum_{h=1}^{\infty} |a_h|^2 \int_{Y'}^{\infty} e^{-4\pi(n_h+\kappa)y}\, y^{k-2} dy. \tag{3.2}$$

The change of variable $t = 4\pi(n_h+\kappa)y$ leads to

$$\sigma N(Y, z_0) \leq (4\pi)^{-k+1} \sum_{h=1}^{\infty} \frac{|a_h|^2}{(n_h+\kappa)^{k-1}} \int_{4\pi(n_h+\kappa)Y'}^{\infty} e^{-t} t^{k-2} dt,$$

where $n_h+\kappa > 0$ for all h, by assumption. Applying Lemma 2.4, we have, finally,

$$\sigma N(Y, z_0) \leq \frac{M^2}{(4\pi)^{k-1}} \sum_{h=1}^{\infty} (n_h+\kappa)^{-k-1} \int_{4\pi(n_h+\kappa)Y'}^{\infty} e^{-t} t^{k-2} dt. \tag{3.3}$$

The inequality (3.3) will lead to a contradiction. In order to expedite the proof we consider three cases.

CASE 1. Suppose $k > 1$. In this instance we make no assumption about the area of Γ; it may be either finite or infinite. Since $k-2 > -1$, the improper integral $\int_0^{\infty} e^{-t} t^{k-2} dt = \Gamma(k-1)$ converges. Thus $\sigma N(Y, z_0) \leq M_1 \Gamma(k-1) \sum_{h=1}^{\infty} (n_h+\kappa)^{-k-1}$. But the assumption $n_{h+1}/n_h \geq \theta > 1$ implies $n_h \geq C\theta^{h-1}$, for $h \geq 2$ ($n_1 = 0$ is possible). Thus the series converges and we obtain $\sigma N(Y, z_0) \leq C(k)$ for $Y > 0$.

But since Γ is an infinite group and not cyclic, $N(Y, z_0) \to +\infty$ as $Y \to 0$. This contradiction proves the theorem when $k > 1$.

CASE 2. $0 < k < 1$. For $x > 0$ put

$$g_k(x) = \int_x^\infty e^{-t} t^{k-2} dt$$

and get

$$g_k(x) = \frac{1}{1-k} \{x^{k-1} e^{-x} - \int_x^\infty e^{-t} t^{k-1} dt\} \leq \frac{x^{k-1} e^{-x}}{1-k} .$$

This and (3.3) imply

$$\sigma N(Y, z_0) \leq \frac{M^2}{(1-k)(4\pi)^{k-1}} \sum_{h=1}^{\infty} \frac{1}{(n_h+\kappa)^{k+1}} \{4\pi(n_h+\kappa)\}^{k-1} Y'^{k-1} e^{-4\pi(n_h+\kappa)Y'}$$

$$= C_1(k) Y'^{k-1} \sum_{h=1}^{\infty} \frac{e^{-4\pi(n_h+\kappa)Y'}}{(n_h+\kappa)^2} \leq C_2(k) Y^{k-1} ,$$

where we have used the convergence of the series and $Y' = Ye^{-2\rho}$. $C_2(k) > 0$ depends upon k, but is independent of Y. Now, when Γ has finite area and $k > 0$, the last inequality contradicts Lemma 2.2 as $Y \to 0$.

CASE 3. $k = 1$. Integration by parts yields in this case

$$g_k(x) = \int_x^\infty e^{-t} t^{-1} dt = -e^{-x} \log x + \int_x^\infty e^{-t} \log t \, dt$$

$$\leq e^{-x} |\log x| + \int_0^\infty e^{-t} |\log t| dt = e^{-x} |\log x| + C,$$

with $C > 0$. Thus, by (3.3),

$$\sigma N(Y, z_0) \leq C_1(k) \sum_{h=1}^{\infty} \frac{1}{(n_h+\kappa)^2} \{e^{-4\pi(n_h+\kappa)Y'} |\log 4\pi(n_h+\kappa)Y'| + C\}$$

$$\leq C_1(k) \sum_{h=1}^{\infty} \{\frac{|\log (n_h+\kappa)|}{(n_h+\kappa)^2} + \frac{|\log 4\pi Y'|}{(n_h+\kappa)^2} + \frac{C}{(n_h+\kappa)^2}\}$$

$$\leq C_2(k) \log |4\pi Y| + C_2(k).$$

Again, for Γ of finite area this contradicts Lemma 2.2. The proof of Theorem 2.1 is complete.

4. <u>Ω-results on the Fourier coefficients</u>. The point of view of Section 3 may be adapted to give lower bounds on the Fourier coefficients of automorphic forms. We continue to assume that the fuchsian group Γ has as a <u>proper</u> subgroup the cyclic group generated by the translation $z \to z+1$. Assume once again that the automorphic form F of exponent $k > 0$ is holomorphic in H and has the Fourier expansion $F(z) = \sum_{n+\kappa > 0}^{\infty} a_n e^{2\pi i(n+\kappa)z}$, $0 \leq \kappa < 1$. The result is

<u>THEOREM 4.1</u>. (i) If Γ has finite area, then $a_n = \Omega(n^{\frac{k-1}{2} - \varepsilon})$ for all $\varepsilon > 0$.

(ii) If Γ is of infinite area and $k > 1$, then $a_n = (n^{\frac{k}{2} - 1 - \varepsilon})$ for all $\varepsilon > 0$.

<u>REMARK</u>. For the coefficients $\tau(n)$ of the cusp form Δ, which belongs to $\{\Gamma(1), 12\}_0$, $\Gamma(1) = SL(2, Z)$ the modular group, Rankin [12] has proved the much stronger result:

$$\limsup_{n \to \infty} \tau(n)/n^{\frac{11}{2}} = \infty.$$

<u>PROOF</u>. Assume first that $k > 1$. We show with no assumption on the area of Γ that the weaker estimate (ii) holds. In fact, by the inequality (3.2),

$$\sigma N(Y, z_0) \leq \sum_{n+\kappa > 0} |a_n|^2 \int_{Y'}^{\infty} e^{-4\pi(n+\kappa)y} y^{k-2} dy \leq \sum |a_n|^2 \int_0^{\infty}$$

$$= \frac{1}{(4\pi)^{k-1}} \sum \frac{|a_n|^2}{(n+\kappa)^{k-1}} \int_0^{\infty} e^{-t} t^{k-2} dt$$

$$= \frac{\Gamma(k-1)}{(4\pi)^{k-1}} \sum_{n+\kappa > 0} \frac{|a_n|^2}{(n+\kappa)^{k-1}}.$$

Suppose $a_n = O(n^\alpha)$, $n \to \infty$. Then

$$\sigma N(Y, z_0) \leq C(k) \sum_{n+\kappa > 0} (n+\kappa)^{2\alpha+1-k},$$

a sum that converges when $k-1-2\alpha > 1$, that is, when $\alpha < k/2-1$. Since $N(Y, z_0) \to \infty$ as $Y \to 0$, convergence of the sum would be contradictory. Thus $a_n = O(n^{k/2-1-\varepsilon})$ $\varepsilon > 0$, is impossible and we obtain $a_n = \Omega(n^{k/2-1-\varepsilon})$. That is, (ii) of the Theorem is proved.

Now suppose Γ has finite area. We assume $k > 0$ and we wish to establish the stronger lower bound (i). Recalling the definition

$$g_k(x) = \int_x^\infty e^{-t} t^{k-2} dt,$$

we shall find that we need to apply different estimates for g_k according as $k > 1$, $k < 1$ or $k = 1$.

If $k > 1$,

$$g_k(x) \leq e^{-x/2} \int_x^\infty e^{-t/2} t^{k-2} dt \leq e^{-x/2} \int_0^\infty e^{-t/2} t^{k-2} dt$$

$$= 2^{k-1} \Gamma(k-1) e^{-x/2}. \tag{4.1a}$$

For $k < 1$, as noted previously,

$$g_k(x) \leq \frac{x^{k-1} e^{-x}}{1-k}. \tag{4.1b}$$

For $k = 1$,

$$g_k(x) \leq e^{-x/2} \int_x^\infty e^{-t/2} t^{-1} dt = e^{-x/2} \{-e^{-x/2} \log x + \frac{1}{2} \int_x^\infty e^{-t/2} \log t dt\}$$

$$\leq e^{-x/2} \{e^{-x/2} |\log x| + \frac{1}{2} \int_0^\infty e^{-t/2} |\log t| dt\},$$

and we have

$$g_k(x) \leq e^{-x} |\log x| + C_0 e^{-x/2}. \tag{4.1c}$$

We apply (3.2) in the form

$$\sigma N(Y, z_0) \le (4\pi)^{-k+1} \sum_{n+\kappa > 0} \frac{|a_n|^2}{(n+\kappa)^{k-1}} g_k(4\pi(n+\kappa)Y')$$

and assume $a_n = O(n^{\frac{k-1}{2} - \varepsilon})$, for some $\varepsilon > 0$. This assumption will ultimately lead to a contradiction. At this point it yields, together with (4.1),

$$\text{(4.2)} \quad \sigma N(Y, z_0) \le \begin{cases} C_k \sum_{n+\kappa > 0} (n+\kappa)^{-2\varepsilon} e^{-2\pi(n+\kappa)Y'}, & k > 1 \\ C_k' Y'^{k-1} \sum_{n+\kappa > 0} (n+\kappa)^{k-1-2\varepsilon} e^{-4\pi(n+\kappa)Y'}, & k < 1 \\ C_k'' \sum_{n+\kappa > 0} (n+\kappa)^{-2\varepsilon} e^{-4\pi(n+\kappa)Y'} |\log 4\pi(n+\kappa)Y'| & \\ + C_k''' \sum_{n+\kappa > 0} (n+\kappa)^{-2\varepsilon} e^{-2\pi(n+\kappa)Y'}, & k = 1. \end{cases}$$

Here C_k, C_k', C_k'', C_k''' are positive constants depending only upon k. To estimate the right-hand side of (4.2) we apply the Lipschitz summation formula [6, p. 65]:

$$\text{(4.3)} \qquad \sum_{n=0}^{\infty} (n+\kappa)^{\lambda} e^{-2\pi(n+\kappa)t} = \frac{\Gamma(\lambda+1)}{(2\pi)^{\lambda+1}} \sum_{q=-\infty}^{\infty} (t+qi)^{-\lambda-1} e^{2\pi i q\kappa},$$

for $0 < \kappa < 1$, $\lambda > -1$ and Re $t > 0$.

Assume first that $k > 1$. Without loss of generality we may suppose $\varepsilon < \frac{1}{2}$, so that $\lambda = -2\varepsilon > -1$. Then, by (4.2) and (4.3),

$$\sigma N(Y, z_0) \le C_k \frac{(1-2\varepsilon)}{(2\pi)^{1-2\varepsilon}} \sum_{q=-\infty}^{\infty} (Y'+qi)^{-1+2\varepsilon} e^{2\pi i q\kappa},$$

in the case when $\kappa > 0$. If $\kappa = 0$, we rewrite (4.2):

$$\sigma N(Y, z_0) \le C_k \sum_{n=1}^{\infty} n^{-2\varepsilon} e^{-2\pi n Y'}$$

$$= C_k e^{\pi Y'} \sum_{n=1}^{\infty} (n+1/2)^{-2\varepsilon} e^{-2\pi(n+\frac{1}{2})Y'} \left(\frac{n+\frac{1}{2}}{n}\right)^{2\varepsilon}$$

$$\leq (\tfrac{3}{2})^{2\varepsilon} C_k e^{\pi Y'} \sum_{n=1}^{\infty} (n+\tfrac{1}{2})^{-2\varepsilon} e^{-2\pi(n+1/2)Y'} .$$

Thus we may apply (4.3) with $\kappa = 1/2$ to obtain

$$\sigma N(Y, z_0) \leq (\tfrac{3}{2})^{2\varepsilon} C_k \frac{\Gamma(1-2\varepsilon)}{(2\pi)^{1-2\varepsilon}} e^{\pi Y'} \sum_{q=-\infty}^{\infty} (Y'+qi)^{-1+2\varepsilon} e^{\pi iq} .$$

In either case we may apply partial summation to the sum on q to get, for $0 < Y' \leq 1$,

$$\sigma N(Y, z_0) \leq DY'^{-1+2\varepsilon} + \tilde{D} = D^* Y^{-1+2\varepsilon} + \tilde{D} , \tag{4.4}$$

where D, $\tilde{D}$ are positive constants depending upon k, ε and κ and $D^* = De^{2\rho(1-2\varepsilon)}$

But, since $\varepsilon > 0$, (4.4) contradicts Lemma (2.2) if we let $Y \to 0+$; thus $a_n = \Omega(n^{\frac{k-1}{2}-\varepsilon})$, for all $\varepsilon > 0$.

Suppose $0 < k < 1$. Without loss of generality assume that $\varepsilon < k/2$. By (4.2) and (4.3),

$$\sigma N(Y, z_0) \leq C_k' Y'^{k-1} \frac{\Gamma(k-2\varepsilon)}{(2\pi)^{k-2\varepsilon}} \sum_{q=-\infty}^{\infty} (2Y'+qi)^{-k+2\varepsilon} e^{2\pi iq\kappa},$$

provided $\kappa > 0$. For $\kappa = 0$ proceed as before to obtain

$$\sigma N(Y, z_0) \leq (\tfrac{3}{2})^{1-k+2\varepsilon} C_k' Y'^{k-1} e^{2\pi Y'} \frac{\Gamma(k-2\varepsilon)}{(2\pi)^{k-2\varepsilon}} \sum_{q=-\infty}^{\infty} (2Y'+qi)^{-k+2\varepsilon} e^{\pi iq}.$$

In either case, if $Y' \leq 1$, we have

$$\begin{aligned} \sigma N(Y, z_0) &\leq DY'^{k-1} Y'^{-k+2\varepsilon} + \tilde{D} Y'^{k-1} \\ &= D_1 Y^{-1+2\varepsilon} + \tilde{D}_1 Y^{k-1} , \end{aligned} \tag{4.5}$$

by partial summation. Since $\varepsilon > 0$ and $k > 0$, (4.6) contradicts Lemma (2.2) and the theorem is proved for the case $k < 1$.

The case $k = 1$ is precisely the same except for the sum involving the extra factor $|\log 4\pi(n+\kappa)Y'|$, and the logarithm may be replaced by a small power. Once this is done the proof carries through as before. This completes the proof of Theorem (4.1).

5. Fabry Gaps for Cusp Forms.

In this section we consider the possibility of Fabry gaps in the Fourier coefficients of a fuchsian form. By definition Fabry gaps occur when the Fourier coefficients are supported on a set of integers of density zero. We have no results for general fuchsian groups but give examples confined to subgroups of the modular group $\Gamma(1) = SL(2, Z)$. Two famous identities (Eulêr, Jacobi) come to mind immediately:

$$x^{-\frac{1}{24}}\eta(z) := \prod_{1}^{\infty}(1-x^m) = \sum_{-\infty}^{\infty}(-1)^{\lambda}x^{\frac{\lambda(3\lambda-1)}{2}}, \tag{5.1}$$

$$x^{-\frac{1}{8}}\eta^3(z) = \sum_{n=0}^{\infty}(-1)^n(2n+1)x^{\frac{n(n+1)}{2}}, \tag{5.2}$$

where $x = e^{2\pi iz}$. We recall that

$$\eta^{24} = \Delta \in \{\Gamma(1), 12\}_0, \text{ so } \eta \in \{\Gamma(1), \tfrac{1}{2}, v_0\}$$

with a certain multiplier system v_0.

Even integral powers of η have been considered by Newman [11] and van Lint [10], and Rankin [14] has treated more general forms. We follow Rankin's discussion (his Section 8).

For each k there are 6 MS on $\Gamma(1)$, denoted by $v(r, k) \equiv v^{(r)}$ and defined on the generators by

$$v^{(r)}(S) = \exp\{\pi i(k-r)/6\},\ v^{(r)}(T) = \exp\{-\pi i(k-r)/2\}, \tag{5.3}$$

$$S = (1, 1:\ 0, 1),\quad T = (0, -1:\ 1, 0).$$

$$r \in R = \{0, 4, 6, 8, 10, 14\}. \tag{5.4}$$

Define the Eisenstein series

$$E_r(z) = \sum (cz+d)^{-r}, \; r = 4, 6, \ldots ,$$

summed over a system of representative $\Pi = (a, b: \; c, d)$ in $\Gamma(1) = \Gamma_S \cdot \Pi$; $\Gamma_S = < -I, S >$. Let

$$M_k^{(r)} = \{\Gamma(1), k, v^{(r)}\}_0 .$$

With the abbreviation

$$F_{rs} = E_r \eta^{2s}, \; s \geq 1 ,$$

we have $F_{rs} \in M_{r+s}^{(r)}$.

Write $e(z) = e^{2\pi iz}$; then

$$F_{rs} = \sum_{n=0}^{\infty} a_{rs}(n) e((n+\tfrac{s}{12})z) = \sum_{h=0}^{\infty} b_{rs}(h) e(\tfrac{hz}{12}) , \tag{5.5}$$

with

$$b_{rs}(h) = a_{rs}(\tfrac{h-s}{12}) ,$$

and the usual convention $a_{rs}(x) = 0$ if $x \notin Z$.

For $f \in \{\Gamma, k, v\}$ we define the Hecke operator T_p by

$$f|T_p = p^{k-1} f(pz) + p^{-1} \sum_{j=0}^{p-1} f(\tfrac{z+12j}{p}) ,$$

p being a prime. (See [14, (3.4)].) Then [14, (8.12)]

$$F_{rs}|T_p \in M_{r+s}^{(r_p)} , \; p > 3, \tag{5.6}$$

where

$$r_p \equiv pr-(p-1)(r+s) \pmod{12}, \; r_p \in R. \tag{5.7}$$

On the other hand

(5.8) $$F_{rs}|T_p = \sum_{h=0}^{\infty} \{b_{rs}(ph) + p^{k-1} b_{rs}(\tfrac{h}{p})\}\, e(\tfrac{hz}{12}).$$

On p. 52 of [14] there is a table showing the forms that are mapped into 0 by T_p together with the corresponding values of p:

(5.9)

r	s	p ≡ ... (12)	r	s	p ≡ ... (12)	r	s	p ≡ ... (12)
0	1	5,7,11	4	1	7,11	6	7	5
0	2	5,11	4	2	5,11	8	1	7,11
0	3	7,11	4	3	7,11	8	3	7,11
0	4	5,11	4	5	7,11	8	5	7
0	5	7,11	6	1	5,11	10	2	5,11
0	7	5,11	6	2	5,11			
0	13	11	6	4	5,11			

For these values of (r, s, p) we can apply (5.8) and deduce

(5.10) $$b_{rs}(ph) = 0, \quad (h,p) = 1.$$

Here we insert a lemma, which will be applied to the proofs of Theorems 5.2 and 5.3.

<u>LEMMA 5.1</u>. Let $P = \{p_i : i = 1, 2, 3, \ldots\}$ be a set of primes and let $S(P) \equiv S = \{ph : p \in P, h \in Z^+, (h, p) = 1\}$. Let $S'(P) \equiv S'$ be the set of positive integers not in S. For $x \in Z^+$, let S(x) be the number of elements in S which do not exceed x and let S'(x) be the number of positive integers outside of S which do not exceed x. (Thus $S(x)+S'(x) = x$.) Then,

$$\overline{\text{dens}}\ S' := \limsup_{x \to \infty} S'(x)/x \le \prod_{p \in P} \left(1 - \frac{1}{p}\left(1 - \frac{1}{p}\right)\right).$$

<u>PROOF</u>. Fix a positive integer t. Let S_t be the set $\{h_i p_i : 1 \le i \le t,\ h_i \in Z^+,\ (h_i, p_i) = 1\}$. Let S_t' be the set of positive integers not in S_t. (Note that since $S_t \subset S$, it follows that $S' \subset S_t'$.) Let $S_t(x)$ be

the number of elements in S_t which do not exceed x and let $S_t'(x)$ be the number of positive integers outside of S_t which do not exceed x. (Of course, $S_t(x)+S_t'(x) = x$.)

Now, for $i \leq i \leq t$, the number of integers h_i such that $h_ip_i \leq x$ and $(h_i, p_i) = 1$ is $[x/p_i] - [x/p_i^2]$. It may happen that with $i \neq j$, $h_ip_i = h_jp_j \leq x$. Then $p_j|h_i$, $h_i = p_jh_i'$, say, and similarly $h_j = p_ih_j'$. So $h_i'p_ip_j = h_j'p_ip_j$ or $h_i' = h_j'$. Thus these multiples are multiples of p_ip_j. Moreover, $(h_i', p_i) = 1 = (h_i', p_j) = (h_i', p_j)$, so that $(h_i', p_ip_j) = 1$; then

$$S_t'(x) = x-\sum_{i=1}^{t} ([\frac{x}{p_i}]-[\frac{x}{p_i^2}]) + \sum_{i,j=1}^{t} ([\frac{x}{p_ip_j}]-[\frac{x}{p_i^2p_j^2}]) - \dots .$$

Now $[\alpha] - [\beta] = \alpha-\beta + \theta$, $|\theta| < 1$, for any real α, β. Thus,

$$S_t'(x) = x - \sum_{i=1}^{t} \frac{x}{p_i} (1- \frac{1}{p_i}) + \sum_{i,j=1}^{t} \frac{x}{p_ip_j} (1- \frac{1}{p_ip_j}) + \dots + \theta \cdot 2^t$$

$$= x \prod_{i=1}^{t} (1- \frac{1}{p_i} (1- \frac{1}{p_i})) + \theta\cdot 2^t, \ |\theta| < 1.$$

Hence, with fixed t,

$$(5.11) \qquad \text{dens } S_t' := \lim_{x\to\infty} S_t'(x)/x = \prod_{i=1}^{t} (1- \frac{1}{p_i}(1- \frac{1}{p_i})).$$

But since S' is contained in every S_t', the result follows.

For the application fix a pair (r, s) chosen from the table (5.9) and let $\{p_i\}$ be the primes in P, a fixed one of the arithmetic progressions that correspond to (r, s). Since $\sum_p p_i^{-1} = +\infty$ [1, p. 156, ex. 6], the product in (5.11) converges to 0 with $t \to \infty$, so that

$$\text{dens } S(P) = \text{dens } \{p_ih:p_i \in P, h \in Z^+, (h, p_i) = 1\} = 1.$$

But by (5.10), $b_{rs}(p_ih) = 0$ when $p_ih \in S(P)$. This gives

<u>THEOREM 5.2.</u> Let (r, s) be a pair in the list (5.9) and let $F_{rs} = E_r\eta^{2s}$. Then the Fourier series of F_{rs} has Fabry gaps.

The preceding discussion could presumably be repeated for any congruence subgroup of $\Gamma(1)$ for which a basis of modular forms is known. We shall not elaborate on this but turn instead to the subgroups $\Gamma_0(N)$, $N > 1$, defined by $c \equiv 0 \pmod N$ in $(a, b: c, d) \in \Gamma(1)$. We shall make use of the results of [2]. A (cusp) newform on $\Gamma_0(N)$ is a modular form F which is an eigenfunction of all the Hecke operators T_p, $p \nmid N$, defined previously, and of the "ramified" operators U_q, $q|N$, and which, moreover, arises for the first time on $\Gamma_0(N)$ and not on some $\Gamma_0(N')$ with $N'|N$. (Forms of "weight k" in [2] are forms of "exponent 2k" here.)

Let $F(z) = \sum_1^\infty a(n)x^n$, $x = e(z)$, where F is a newform on $\Gamma_0(N)$. Then if q is a prime such that $q^2|N$,

(5.12)
$$a(nq_i) = 0, \ n \geq 1$$

([2, Th. 3]). So if P is the set of primes q_i with $q_i^2|N$, a fortiori we have (5.12) with $(n, q_i) = 1$, and by Lemma 5.1 we can assert that

(5.13)
$$\underline{\text{dens}}\ S(P) = \liminf_{x \to \infty} S(x)/x \geq 1 - \prod_{q_i \in P} \left(1 - \frac{1}{q_i}\left(1 - \frac{1}{q_i}\right)\right).$$

By choosing P to be the primes in an arithmetic progression and

$$N = N_h = q_1^2 q_2^2 \cdots q_h^2,$$

we can make the right member of (5.13) exceed $1-\varepsilon$ for h sufficiently large. This proves the following Theorem 5.3 once we establish the existence of nonzero newforms (with fixed k) on $\Gamma_0(N)$ for a sequence $N = N_h \to \infty$. As we shall see, it is convenient for this purpose to let P be the set of primes $\equiv -1 \pmod{12}$.

THEOREM 5.3. Let $\varepsilon > 0$, $k \geq 1$. There is a positive integer N depending on ε but not on k and a nonzero newform F in $\{\Gamma_0(N), 2k\}_0$ such that the surviving exponents in the Fourier series of F have density $< \varepsilon$.

REMARK. In [16] Serre considers the density of the Fourier coefficients of modular forms that are eigenfunctions of the Hecke operators $T_p (p \nmid N)$ and

$U_q(q|N)$. He proves, in particular, that the density of $\{a(n) \neq 0\}$ is zero for forms F of exponent 1 but suggests that this density is positive for forms of integral exponent ≥ 2, which are not of "C-M (complex multiplication) type". In a recent letter Serre has pointed out that Theorem 5.3 can be improved to a choice of $\varepsilon = 0$ with N fixed.

To prove the existence of such newforms, we follow the reasoning of [2, p. 158] and obtain a recursive formula for $\nu(q^2m)$, where q is a prime, $(q, m) = 1$ and $\nu(t)$ is the dimension of the subspace of newforms in the space $\{\Gamma_0(t), 2k\}_0$. If $\delta(t)$ is the dimension of $\{\Gamma_0(t), 2k\}_0$, we have

$$\delta(t) = \sum_{t'|t} \nu(t')d(t/t'),$$

which in turn implies

(5.14) $$\nu(t) = \sum_{t'|t} \delta(t')\beta(t/t'),$$

with $\beta(n) = \sum_{d|n} \mu(d)\mu(n/d)$. It follows that β is a multiplicative function and $\beta(1) = 1$; furthermore, for p a prime we have $\beta(p) = -2$, $\beta(p^2) = 1$ and $\beta(p^\alpha) = 0$ when $\alpha > 2$.

Now let P be the set of all primes $q_i \equiv -1 \pmod{12}$, with $11 = q_1 < q_2 < \dots$. Suppose $q \in P$ and m is a product of squares of elements in P which are distinct from q, so that $(q, m) = 1$. With $t = q^2m$, we shall calculate $\delta(t)$. The dimensions δ are well known. (See, for example, [9, pp. 216, 293.) Write $\mu_0(t)$ for $[\Gamma(1): \Gamma_0(t)]$ and $\sigma_0(t)$ for the number of parabolic classes in $\Gamma_0(t)$. By the results in [5, pp. 810-11] there are no elliptic elements in $\Gamma_0(t)$ for $t = q_1^{\varepsilon_1} q_2^{\varepsilon_2} \dots q_r^{\varepsilon_r}$, $0 \leq \varepsilon_i \leq 2$, if each $q_i \equiv -1 \pmod{12}$. For such t, therefore,

(5.15) $$\delta(t) = (2k-1)(g-1)+(k-1)\sigma_0(t),$$

with $g = g(t)$ the genus of $H/\Gamma_0(t)$. Also ([5, p. 810]),

$$g(t)-1 = \frac{\mu_0(t)}{12} - \frac{1}{2}\sigma_0(t),$$

$$\mu_0(t) = t \prod_{p|t} (1+p^{-1}), \text{ p prime},$$

and

$$\sigma_0(t) = \sum_{d|t} \phi((d, t/d)) = \prod_{p^e \| t} (p^{[e/2]}+p^{[\frac{e-1}{2}]});$$

here $p^e \| t$ means $p^e | t$, $p^{e+1} \nmid t$. Hence

(5.16) $$\delta(t) = \mu_0(t) \frac{2k-1}{12} - \frac{1}{2}\sigma_0(t).$$

Next, define, for $\alpha = 0, 1, 2$,

(5.17) $$D_\alpha = \sum_{m'|m} \delta(m'q^\alpha)\beta(m/m') = \sum_{\gamma=0}^{\alpha} (\alpha-\gamma+1)\nu(mq^\gamma),$$

the equality following from (5.14) and the properties of β. We calculate that

(5.18) $$D_2-(q+1)D_1+(q+1)D_0 = \nu(mq^2)-(q-1)\nu(mq)-(q-2)\nu(m).$$

On the other hand, the left-hand side of (5.18) vanishes by (5.16), the definition of D_α and the multiplicative properties of μ_0 and σ_0. Hence

(5.19) $$\nu(mq^2) = (q-1)\nu(mq)+(q-2)\nu(m),$$

under the assumptions we have made on q and m.

If we now define $N_h = q_1^2 q_2^2 \dots q_h^2$, as before, it follows that $\nu(N_h) > 0$, as desired, if we can show that $\nu(q_1^2) > 0$. But (5.14) and (5.16) yield

$$\begin{aligned}\nu(q_1^2) &= \delta(1)+\delta(q_1^2)-2\delta(q_1) \geq \delta(q_1^2)-2\delta(q_1)\\ &= (q_1+1)q_1 \frac{2k-1}{12} - \frac{q_1+1}{2} - 2\{(q_1+1)\frac{2k-1}{12} -1\}\\ &= \frac{q_1+1}{12}\{(q_1-2)(2k-1)-6\} + 2 > 0,\end{aligned}$$

since $q_1 = 11$, $k \geq 1$. This completes the proof of Theorem 5.3.

6. Fabry Gaps for Powers of $\Theta(z)$. We now turn to the question of Fabry gaps in powers of the Θ-function, which are modular forms though not cusp forms. Let

$$\Theta(z) = \sum_{-\infty}^{\infty} e^{\pi i n^2 z} = 1+2 \sum_{n=1}^{\infty} e^{\pi i n^2 z}. \tag{6.1}$$

It is well known that $\Theta \in \{\Gamma_\theta, \frac{1}{2}, v_1\}$ where $\Gamma_\theta \subset \Gamma(1)$ is generated by $z \to z+2$, $z \to -1/z$ and Γ_θ is of index 3 in $\Gamma(1)$. Since Θ has no zeros in H, $\Theta^s(z)$ is well defined for s an arbitrary positive number. As is customary, we write

$$\Theta^s(z) = 1+ \sum_{m=1}^{\infty} r_s(m) e^{\pi i m z}; \tag{6.2}$$

if $s \in Z^+$, $r_s(m)$ is the number of ways in which m can be written as a sum of s squares.

By Lagrange's theorem $r_s(m) > 0$ for all m when s is an integer ≥ 4. Furthermore, it is a familiar fact that a positive proportion of the positive integers can be expressed as a sum of 3 squares. Thus the expansion (6.2) does not have Fabry gaps when s is an integer ≥ 3. In contrast, Landau's results [8] show that (6.2) does have Fabry gaps when $s = 2$. We shall show here that for arbitrary real $s > 4$ the expansion (6.2) does not have Fabry gaps.

The nonexistence of Fabry gaps for $s > 4$ follows directly from two results on $r_s(m)$, with $s > 4$, proved in [3]. (Cf. [15, pp. 238-243].) These are:

(6.3)

$$\text{(i) } A_s(x) := 1+ \sum_{1 \leq m \leq x} r_s(m) = \frac{\pi^{s/2}}{\Gamma(s/2+1)} x^{s/2} + O(x^{s/2-1}), \quad x \to \infty,$$

$$\text{(ii) } r_s(m) = O(m^{s/2-1}), \quad m \to \infty.$$

The summatory function $A_s(x)$ may be regarded as the number of integral lattice points in a "sphere of dimension s" and radius $\sqrt{x}$.

We begin by rewriting (6.2) in the form

$$\Theta^s(z) = 1+ \sum_{h=1}^{\infty} r_s(m_h)e^{\pi i m_h z} \tag{6.4}$$

with $r_s(m_h) \neq 0$ for all h. To say that (6.4) has Fabry gaps is to say that $m_h/h \to +\infty$ as $h \to \infty$. Let $N_s(x)$ be the number of $m_h \leq x$. Since $N_s(m_h) = h$, it is easy to see that $N_s(x)/x$ is bounded away from 0 as $x \to \infty$ if and only if m_h/h is bounded as $h \to \infty$.

We shall derive our result by showing that for $s > 4$, $N_s(x) \geq Cx$ for some $C > 0$.

By (6.3i),

$$\sum_{1 \leq m \leq x} |r_s(m)| \geq C_1 x^{s/2}, \text{ for } x > 0,$$

while (6.3ii) implies $|r_s(m)| \leq C_2 m^{s/2-1}$, where C_1 and C_2 are > 0. Thus

$$C_1 x^{s/2} \leq \sum_{1 \leq m \leq x} |r_s(m)| = \sum_{1 \leq m_h \leq x} |r_s(m_h)|$$

$$\leq C_2 \sum_{1 \leq m_h \leq x} m_h^{s/2-1} \leq C_2 N_s(x) x^{s/2-1}.$$

From this follows $N_s(x)/x \geq C_1/C_2 > 0$, and the proof is complete.

Although we expect that there are no Fabry gaps for Θ^s in the larger interval $s > 2$, the above proof fails when $s \leq 4$, since the estimates (6.3) must be replaced for $s \leq 4$ by

$$\text{(i)}\quad A_s(x) = \frac{\pi^{s/2}}{\Gamma(\frac{s}{2}+1)} x^{s/2} + O(x^{s/4+\varepsilon}),\ \varepsilon > 0,\ x \to \infty,$$

(6.5)

$$\text{(ii)}\quad r_s(m) = O(m^{s/4+\varepsilon}),\ \varepsilon > 0,\ m \to \infty.$$

(6.5i) is proved, in somewhat sharper form, in [3], while (6.5ii) follows by differencing (6.5i). While we are unable to rule out Fabry gaps when $s < 4$, we can apply (6.5) to prove the gap result that, in (6.4)

$$m_h \geq Ch^{4/s+\varepsilon},\ \varepsilon > 0, \tag{6.6}$$

is impossible, for $s \leq 4$. That this is weaker than the true result can be seen by comparing it with the known results for s = 1, 2, 3 and 4, which are $m_h = h^2$, $m_h \sim Ch(\log h)^{1/2}$, $m_h \sim Ch$ and $m_h = h$, respectively (see [8]).

To show that (6.6) is false, suppose that $s \leq 4$ and $m_h \geq Ch^\alpha$, with $\alpha > 0$ and $C > 0$. Then, by (6.5ii),

$$\sum_{1 \leq m_h \leq} |r_s(m_h)| \leq K \sum_{1 \leq m_h \leq x} m_h^{s/4+\varepsilon} \leq K \sum_{Ch^\alpha \leq x} {}_h \, x^{s/4+\varepsilon}$$

$$\leq K' x^{s/4+\varepsilon+1/\alpha}.$$

By (6.5i), however, this implies that $s/4 + 1/\alpha \geq s/2$, or $\alpha \leq 4/s$. Thus (6.6) is impossible.

This same method can be applied to cusp forms $F \in \{\Gamma, 2k, v\}_0$, with Γ a congruence subgroup of the modular group $\Gamma(1)$. For such forms Rankin [13] has derived the following estimates for the coefficients in the expansion (2.2):

$$\text{(6.7)} \quad \begin{aligned} &\text{(i)} \quad \sum_{1 \leq n \leq x} |a_n|^2 = \beta(\Gamma)x^{2k} + O(x^{2k-2/5}), \ x \to \infty, \text{ with } \beta(\Gamma) > 0; \\ &\text{(ii)} \quad a_n = O(n^{k-1/5}), \ n \to \infty . \end{aligned}$$

Using (6.7) with the same calculation as above shows that $m_h \geq Ch^{5/2+\varepsilon}$, $\varepsilon > 0$, is impossible, where m_h is defined by the expansion (2.2) of the cusp form F. If $\Gamma = \Gamma(1)$ itself and $4k \in Z$, then by [7] (6.7ii) may be replaced by $a_n = O(n^{k-1/4+\varepsilon})$, $\varepsilon > 0$, as $n \to \infty$, and the same calculation shows that $m_h \geq Ch^{2+\varepsilon}$, $\varepsilon > 0$, cannot occur in this case. The identities (5.1) and (5.2) - for which k = 1/4, 3/4, respectively - show that this result is virtually the best possible.

Finally, it is worth observing that if we combine (6.7i) with Deligne's deep result $\tau(n) = O(n^{11/2+\varepsilon})$, $\varepsilon > 0$, as $n \to \infty$, for the coefficients of $\Delta = \eta^{24} \in \{\Gamma(1), 12\}_0$, we find that $m_h \geq Ch^{1+\varepsilon}$, $\varepsilon > 0$, is impossible in the Fourier expansion of Δ.

Temple University, Philadelphia, PA.19122
314-N Sharon Way, Jamesburg,N.J, 08831

REFERENCES

1. T. Apostol, Introduction to analytic number theory, Springer-Verlag, New York, 1976.

2. A.O.L. Atkin and J. Lehner, Hecke operators on $\Gamma_0(m)$, Math. Ann. 185 (1970), 134-160.

3. P.T. Bateman and M. Knopp, On lattice points in spheres of fractional dimensions, in preparation.

4. K.G. Binmore, Analytic functions with Hadamard gaps, Bull. London Math. Soc. 1 (1969), 211-217.

5. E. Hecke, Mathematische Werke, Vandenhoeck and Ruprecht, Göttingen, 1959.

6. M. Knopp, Modular functions in analytic number theory, Markham Pub. Co., Chicago, 1970.

7. M. Knopp and J.R. Smart, On Kloosterman sums connected with modular forms of half-integral dimension, Illinois J. Math. 8 (1964), 480-487.

8. E. Landau, Über die Einteilung der positiven ganzen Zahlen in vier Klassen nach der Mindestzahl der zu ihrer additiven Zusammensetzung erforderlichen Quadrate, Archiv der Mathematik und Physik, 13 (1908), 305-312.

9. J. Lehner, Discontinuous groups and automorphic functions, American Math. Soc., Providence, 1964.

10. J.H. van Lint, Hecke operators and Euler products, Thesis, Utrecht, 1957.

11. M. Newman, Some theorems about $p_r(n)$, Canadian J. Math. 9 (1957), 68-70.

12. R.A. Rankin, An Ω-result for the coefficients of cusp forms, Math. Ann. 203 (1973), 239-250.

13. R.A. Rankin, Contributions to the theory of Ramanujan's function $\tau(n)$ and similar arithmetic functions II, Proc. Cambridge Phil. Soc. 35 (1939), 357-372.

14. R.A. Rankin, Hecke operators on congruence subgroups of the modular group, Math. Ann. 168 (1967), 40-58.

15. R.A. Rankin, Modular forms and functions, Cambridge Univ. Press, Cambridge, 1977.

16. J.-P. Serre, Divisibilité de certaines fonctions arithmétiques, L'Enseignement Mathematique 22 (1976), 227-260.

17. M. Tsuji, Potential theory in modern function theory, Maruzen Co. Ltd., Tokyo, 1959.

GAPS IN THE FOURIER SERIES OF AUTOMORPHIC FORMS II

Thomas A. Metzger

Dedicated to Emil Grosswald

0. Introduction: In this paper, a continuation of [K-L1], the question of gap series for automorphic functions and forms acting on the unit disk D in the complex plane will be discussed. In the main, we shall use the notation of [K-L1] with the following modification: g is said to be an automorphic form of weight k with respect to a Fuchsian group Γ if

$$(0.1) \qquad g(\gamma z)\gamma'(z)^k = g(z)$$

for all z in D and γ in Γ. The change in this convention is that "weight k" here is the same as "exponent 2k" in [K-L1]. Furthermore, multipler systems are excluded. It should be noted almost all of the results are valid even in the presence of multiplier systems but we shall not include this generalization here. With this convention the two papers can be read as a continuous work.

In section 1, the case of automorphic functions is considered; the main result asserts that if f is an automorphic function with respect to a Fuchsian group Γ which contains a parabolic element then Hadamard gaps cannot occur in the Taylor's series expansion for f. A similar result holds if f is allowed to have additive Eichler periods. In section 2 the case of automorphic forms of weight 1 is considered. If Γ has a parabolic element then again no Hadamard gaps can occur. However for groups of convergence type and f in the Bers space even weaker gap conditions than that of Hadamard gaps cannot occur. The emphasis here is on the fact that Γ need not contain parabolic elements.

The author wishes to thank Professors Marvin Knopp and Joseph Lehner for many interesting and informative discussions on the topics treated in this paper.

1. Taylor coefficients for Automorphic Forms. Let Γ be a Fuchsian group acting on the unit disk D. A holomorphic function f on the unit disk is said to be an automorphic function with respect to Γ if

(1.1) $$f(\gamma z) = f(z)$$

for all z in D and γ in Γ.

Since f is holomorphic in D it has a Taylor's series expansion about the origin that we write in the form

(1.2) $$f(z) = \sum_{k=0}^{\infty} c_k z^{n_k} \quad (c_k \neq 0 \text{ for all } k).$$

We first prove.

<u>THEOREM 1.1</u>: Suppose that f(z) satisfies (1.1) and (1.2) and assume that the Hadamard gap condition

(1.3) $$\frac{n_{k+1}}{n_k} \geq \theta > 1,$$

holds. If Γ has a parabolic element then $f(z) \equiv c_0$.

<u>PROOF</u>. The key idea of the proof is to apply Binmore's Theorem [Bi1] twice, first to f and then to f'. With R < 1 chosen sufficiently large, the set Δ_R defined by

$$\Delta_R = \bigcup_{\gamma \in \Gamma} \gamma\{z: |z| \leq R\}$$

contains a simply connected subset E_R of D whose boundary intersects the unit circle. The equation (1.1) implies that f is bounded on E_R and so Binmore's Theorem implies that the Taylor coefficients of f are bounded. As is well known (see [ACP-1]) (1.3) and the fact that $\{c_k\}$ are bounded implies that f is a Bloch function, i.e., for all z in D

$$|f'(z)| \leq M(1-|z|^2)^{-1} .$$

This last inequality is equivalent to the assertion that f' belongs to $B_1(\Gamma)$, the Bers space. Therefore, Theorem 3.2.3 of [L-1] implies that $|f'(z)| \to 0$ as $z \to \zeta$ inside a parabolic sector at ζ. A second application of Binmore's theorem, to f', shows that the Taylor coefficients of the automorphic form f' are bounded, since these coefficients also satisfy (1.3). This contradicts Corollary 2 of [M-1], since Γ is not cyclic hyperbolic; one concludes that $f'(z) \equiv 0$ and this completes the proof.

REMARK 1.1: P. Nicholls and L. Sons [N-S-1] have shown that if Γ is a finitely generated Fuchsian group of the first kind then even weaker gap conditions than (1.3) are sufficient to imply that $f(z) = c_0$.

It is easy to see that with the use of the transformation $z = e^{2\pi i\zeta}$, $\zeta \in H$, (H = the upper half-plane), the case of an automorphic function with respect to a Fuchsian group Γ acting on H can be treated. Thus one has

COROLLARY 1.2: Suppose that $F(\zeta)$ is an automorphic function with respect to a Fuchsian group Γ acting on the upper half-plane H. Assume that Γ is not cyclic parabolic and that Γ is normalized to contain the translation $Tz = z+1$. If $F(\zeta) = \sum_{k=0}^{\infty} b_k e^{2\pi i\zeta n_k}$ where the $\{n_k\}$ satisfy (1.3) then $F(\zeta) \equiv b_0$.

The above methods apply also to automorphic integrals (Eichler integrals), i.e., to functions f satisfying

$$f(\gamma z) = f(z) + c(\gamma) \tag{1.4}$$

for all z in D and γ in Γ. The periods $c(\gamma)$ obviously satisfy the consistency condition

$$c(\gamma_1\gamma_2) = c(\gamma_1) + c(\gamma_2) \tag{1.5}$$

for all γ_1, γ_2 in Γ. If f satisfies (1.4) then $f'(z)$ is a form of weight 1. Conversely, given any automorphic form g of weight 1, then the Eichler integral

$$f(z) = \int_0^z g(w)dw \tag{1.6}$$

will satisfy (1.4), as the obvious computation shows.

COROLLARY 1.3: Let $f(z)$ satisfy (1.4) with at least one $c(\gamma) = 0$, where γ is not an elliptic transformation or the identity. Assume that the Taylor series of f satisfies (1.3) and suppose that Γ contains a parabolic element. Then $f(z) \equiv c_0$.

PROOF: It suffices to show that the Taylor coefficients of f are bounded, since once that is accomplished the proof proceeds as in Theorem 1.1. Thus, let $R_1 = \frac{1}{2}(1-|\gamma o\ |)$ where γ is the hyperbolic or parabolic element of Γ for which $c(\gamma) = 0$, and define

$$\Delta_1 = \bigcup_{n=-\infty}^{\infty} \gamma^n \{z: |z| \leq R_1\}.$$

f is bounded on Δ_1 since $f(\gamma^n z) = f(z)$ for all integers n. Since Δ_1 contains $\{\gamma^n 0: -\infty < n < \infty\}$ it follows that the boundary of Δ_1 intersects the boundary of the unit disk. Moreover, the choice of R_1 insures that there exists a curve λ which lies in Δ_1, and goes out to the boundary. Thus, Binmore's Theorem can be applied to f and it follows that the Taylor coefficients of f are bounded and the proof is complete.

REMARK 1.2: A.M. Macbeath has pointed out that (1.5) implies $c(\gamma_1\gamma_2\gamma_1^{-1}\gamma_2^{-1}) = 0$ for all γ_1, γ_2 in Γ. Thus except for the simplest groups Γ there is always a commutator γ not elliptic or the identity for which $c(\gamma) = 0$ and so Corollary 1.3 can be applied.

2. Special results for automorphic forms. We first note that Binmore's Theorem and Corollary 2 of [M-1] imply

THEOREM 2.1: Suppose that g is a holomorphic automorphic form of weight $k \geq 1$ whose Taylor's series expansion satisfies (1.3). Then $g \equiv 0$ if at some ζ on the boundary of D there exists a curve λ ending at ζ with

$$\sup_{z \in \lambda} |g(z)| < \infty. \tag{2.1}$$

REMARK 2.1: We emphasize that the ζ in Theorem 2.1 does not even have to be a limit point of the group in order for the result to hold. If only (2.1) holds an automorphic form cannot have Hadamard gaps in its Taylor's series expansion. Thus if g is a cusp form or a regular form (assuming that Γ contains a parabolic element) the result follows.

Theorem 2.1 indicates that one should consider either gap conditions weaker than (1.3) and/or groups Γ which do not contain parabolic elements. At this time we do not have complete results but some special cases would seem to be of interest and they shall be proved below.

In the case Γ does not contain any parabolic elements, the disc algebra $\mathcal{A}$ comes into our proofs. We recall that f belongs to $\mathcal{A}$ if f is holomorphic on D and continuous on $\bar{D}$, the closure of D, and note

<u>LEMMA 2.2</u>: Suppose f belongs to $\mathcal{A}$ and satisfies either (1.1) or (1.4), then $f(z) \equiv c_0$.

<u>PROOF</u>: If a non-constant f in $\mathcal{A}$ satisfies (1.1), let z_0 and z_1 be points in D such that $f(z_0) \neq f(z_1)$. If ζ is fixed by some hyperbolic or parabolic element in Γ then the sequences $\{\gamma^n z_0\}$ and $\{\gamma^n z_1\}$ both converge to ζ. Thus, $f(\zeta) = \lim_{n \to \infty} f(\gamma^n z_0) = \lim_{n \to \infty} f(z_0) = f(z_0)$, and similarly $f(z_1) = f(\zeta)$. This contradicts the continuity of f in $\bar{D}$ unless $f \equiv c_0$.

If f satisfies (1.4) then either f also satisfies (1.1) and f is a constant or there exists a γ_0 in with $c(\gamma_0) \neq 0$. It follows that $|f(\gamma_0^n 0)| = |f(0) + nc(\gamma_0)| \to \infty$ and this contradicts the fact that f is bounded on D and the proof is complete.

<u>REMARK 2.3</u>: A review of the proof shows that (1.1) and the fact f is continuous at p from within D when p is a limit point of Γ is sufficient to imply that $f \equiv c_0$. Also, the hypothesis that f is in $H^\infty(D)$, the space of bounded holomorphic functions on D, and (1.4) with at least one $c(\gamma) \neq 0$ implies $f \equiv c_0$. It is not true that f in $H^\infty(D)$ and (1.1) imply that f is a constant as there exist Riemann Surfaces on which the space of bounded non-constant holomorphic functions is non-trivial.

The Bers spaces $A_k^p(\Gamma)$ $(1 \le p \le \infty,\ 0 < k < \infty)$ are defined to be the Banach spaces of holomorphic automorphic forms of weight k with the norm,

(2.2) $$||g||_p = \left(\int\int_{D/\Gamma} |g(z)|^p (1-|z|^2)^{pk-2} dxdy\right)^{1/p}.$$

We shall also need, in certain cases, the assumption that Γ is of convergence type, i.e.,

(2.3) $$\sum_{\gamma \in \Gamma} |\gamma'(0)| < \infty.$$

PROPOSITION 2.3: Suppose that Γ satisfies (2.3) and that the Taylor's series expansion of g in $A_1^1(\Gamma)$ satisfies (1.3), then $g \equiv 0$.

PROOF: It is well known (see [L1] Theorem 2.26) that (2.3) implies that

$$\sup_{z \in D} \sum_{\gamma \in \Gamma} (1-|\gamma z|^2) \leq M < \infty.$$

Thus if $g(z) = \sum_{k=0}^{\infty} c_k z^{n_k}$ is in $A_1^1(\Gamma)$ an obvious interchange of summation and integration yields

(2.4) $$\int_D\int |g(z)| dxdy < \infty .$$

Since (2.4) is equivalent to the statement that g belongs to $A_2^1(\mathrm{id})$, Corollary 2 of [DS1] implies that

(2.5) $$\sum_{k=0}^{\infty} \frac{|c_k|}{n_k} < \infty .$$

Thus, the f defined by (1.6) satisfies (1.4) and has an absolutely convergent Taylor's series expansion and therefore f belongs to $\mathcal{a}$. An application of Lemma 2.2 then yields the desired result.

REMARK 2.4: It is not clear that the space $A_1^1(\Gamma)$ is of real interest when (2.3) holds. This is due to the fact that $A_1^1(\Gamma) = \{0\}$ if Γ is any Fuchsian group of the second kind. (See [MR1] for the precise result).

PROPOSITION 2.4: Let g belong to $A_1^p(\Gamma)$, $1 < p \leq 2$, and suppose that Γ satisfies

(2.6) $$\sum_{\gamma \in \Gamma} (1-|\gamma o|^2) \log\left(\frac{1}{1-|\gamma o|^2}\right) < \infty.$$

If the Taylor series expansion for g satisfies (1.3) then $g \equiv 0$.

PROOF: It suffices to show that (2.4) holds and then repeat the last part of the proof of Proposition 2.3. The obvious switch of summation and integration with the fact that $(1-|z|^2)\,|\gamma'(z)| = (1-|\gamma z|^2))$ shows that the left hand side of (2.4) equals

$$\int\int_{D/\Gamma} |g(z)| \sum_{\gamma\in\Gamma} (1-|\gamma z|^2)\frac{dxdy}{(1-|z|^2)}$$

$$= \int\int_{D/\Gamma} |g(z)|(1-|z|^2) \sum_{\gamma\in\Gamma} (1-|\gamma z|^2)\frac{dxdy}{(1-|z|^2)^2}$$

$$< ||g||_p\, I_{p'}^{1/p'} \quad (\tfrac{1}{p}+\tfrac{1}{p'}=1),$$

where

$$I_{p'} = \int\int_{D/\Gamma} (\sum_{\gamma\in\Gamma} |\gamma'(z)|)^{p'} (1-|z|^2)^{p'-2}\, dxdy.$$

Since $p' \geq 2$ and (2.6) implies that (2.3) holds it follows that $I_{p'} \leq M\cdot I_2$. However

$$I_2 = \int\int_{D/\Gamma} (\sum_{\gamma\in\Gamma} |\gamma'(z)|)^2\, dxdy = \int\int_D \sum_{\gamma\in\Gamma} |\gamma'(z)|\, dxdy$$

$$= \sum_{\gamma\in\Gamma} (1-|\gamma^{-1}(o)|^2) \int_o^1\int_o^{2\pi} |1-z\overline{\gamma^{-1}(o)}|^{-2} r dr d\theta \quad .$$

An elementary computation using Parseval's formula and the Taylor's series expansion

$$(1-z\overline{\gamma^{-1}(o)})^{-1} = \sum_{n=o}^{\infty} \overline{\gamma^{-1}(o)}^{\,n} z^n$$

shows that I_2 equals the left member of (2.6) and the proof is complete.

REMARK 2.5: It would be of interest to know if (2.6) holds for all Fuchsian groups of convergence type. The best result in this direction is due to Ch. Pommerenke [Po1] who proved that (2.6) holds for all groups of Widom type.

The groups of Widom type were studied thoroughly by H. Widom in [W1] and [W2]. In these papers he gave a number of interesting and useful characterizations and properties of these groups. Perhaps, the most interesting characterization, at least for this study is the following: Let $\{\chi(\gamma): \gamma \in \Gamma\}$ be any character group for the Fuchsian group Γ, i.e. χ is a homomorphism of Γ into the circle group. Then Γ is of Widom type if and only if given any character group $\{\chi(\gamma)\}$ there exists a non-constant bounded holomorphic function $h(z)$ on D such that $h(\gamma z) = \chi(\gamma)h(z)$.

Next, gap conditions weaker than (1.3) will be considered. We shall first consider groups Γ whose exponent of convergence $\delta = \delta(\Gamma) < 1$, where

$$\delta(\Gamma) = \inf\{t: \sum_{\gamma \in \Gamma} |\gamma'(o)|^t < \infty\}.$$

It is known that many groups Γ have exponent of convergence strictly less than one, for example, all finitely generated Fuchsian groups of the second kind have this property and in [P2] S.J. Patterson exhibited an infinitely generated Fuchsian group of the first kind with $\delta(\Gamma) = 0$, (See [B1], [P1] and [P2] for these results and many other interesting facts about such groups). We have

<u>THEOREM 2.5</u>: Let g be in $A_1^2(\Gamma)$ and suppose that

$$1 > \alpha > \max(1/2, \delta).$$

If g has a Taylor's expansion of the form (1.2) with

$$\sum_{k=1}^{\infty} 1/n_k^{1-\alpha} < \infty \quad \text{then } g \equiv 0.$$

The proof uses two elementary computations given as Lemma 2.6 and 2.7 below. Both of these have also been noted by Ch. Pommerenke (private communication). In order to facilitate the statements we shall use the notation $a \doteq b$ to mean that there exists a positive constant m such that $m^{-1}a \leq b \leq ma$.

<u>LEMMA 2.6</u>: Let $1/2 < \alpha < 1$, if $\alpha > \delta(\Gamma)$ then

$$J_\alpha \equiv \int\int_{D/\Gamma} \Big(\sum_{\gamma \in \Gamma} (1-|\gamma z|^2)^\alpha\Big)^2 \frac{dxdy}{(1-|z|^2)^2}$$

$$\doteq \sum_{\gamma \in \Gamma} (1-|\gamma o|^2)^\alpha \log \frac{1}{(1-|\gamma o|^2)} < \infty .$$

PROOF: The usual switch of summation and integration yields

$$J_\alpha = \int\int_D (1-|z|^2)^{2\alpha-2} \sum_{\gamma\in\Gamma} |\gamma'(z)|^\alpha \, dxdy$$

$$= \sum_{\gamma\in\Gamma} (1-|\gamma o^{-1}|^2) \int_0^1 (1-r^2)^{2\alpha-2} \, rdr \int_0^{2\pi} |1-z\overline{\gamma o^{-1}}|^{2\alpha} \, d\theta.$$

Parseval's Theorem and the fact that

$$(1-z\gamma^{-1}(o))^{-\alpha} = \sum_{n=o}^{\infty} \frac{\Gamma(n+\alpha)}{\Gamma(\alpha)\Gamma(n+1)} \overline{\gamma^{-1}(o)}^n z^n$$

yields that

$$J_\alpha = \sum_{\gamma\in\Gamma} (1-|\gamma^{-1}o|^2)^2 \sum_{n=o}^{\infty} \frac{\Gamma^2(n+\alpha)}{\Gamma^2(\alpha)\Gamma^2(n+1)} |\gamma^{-1}(o)|^2 \int_0^1 (1-r^2)^{2\alpha-2} r^{2n+1} dr$$

$$= \sum_{\gamma\in\Gamma} (1-|\gamma^{-1}o|^2) \sum_{n=0}^{\infty} \frac{\Gamma^2(n+\alpha)}{\Gamma^2(\alpha)\Gamma^2(n+1)} B(2\alpha-1,\ n+1)|\gamma^{-1}(o)|^{2n}.$$

An elementary computation involving the gamma function, the beta function and the fact that $\Gamma(x)/\Gamma(x+a) \sim x^{-a}$ completes the proof of the Lemma.

LEMMA 2.7: Suppose that the J_α in Lemma 2.6 is finite and let $g \in A_1^2(\Gamma)$, with $g(z) = \sum_{k=o}^{\infty} a_k z^k$. Then $a_k = O(k^\alpha)$.

PROOF: We first note that $|g(z)|\,(1-|z|^2)$ is Γ-invariant, so

$$\int\int_D |g(z)|\,(1-|z|^2)^{\alpha-1} \, dxdy = \int\int_{D/\Gamma} |g(z)|\,(1-|z|^2) \sum_{\gamma\in\Gamma} (1-|\gamma z|^2)^\alpha \frac{dxdy}{(1-|z|^2)^2}$$

$$\leq ||g||_2 \, J_\alpha^{1/2} < \infty .$$

Since

$$|a_k| r^k \leq \int_0^{2\pi} |\, g(re^{i\theta})|\, d\theta ,$$

we have that

$$|a_k| \int_0^1 r^{k+1}(1-r)^{\alpha-1} \, dr \leq \int\int_D |g(z)|(1-|z|^2)^{\alpha-1} dxdy < \infty.$$

Since the integral on the left hand side equals $B(k+2,\alpha)$ which is $\Gamma(k+2)\Gamma(\alpha)/\Gamma(k+2+\alpha)$, the fact that $\Gamma(x)/\Gamma(x+a) \sim x^{-a}$ yields the desired result.

The proof of Theorem 2.5 proceeds by letting f be defined as in (1.6) so that $f(z) = \sum_{k=0}^{\infty} \frac{a_k}{n_k+1} z^{n_k+1}$. Because of the hypothesis on α, Lemma 2.6 and therefore Lemma 2.7 hold. We have

$$\sum_{k=0}^{\infty} \left|\frac{a_k}{n_k}\right| \leq \sum \frac{|a_k|}{n_k^{\alpha}} \quad \frac{1}{n_k^{1-\alpha}} \leq c\cdot\sum \ 1/n_k^{1-\alpha} \quad < \infty,$$

and so f belongs to $\mathcal{a}$. An application of Lemma 2.2 shows that f is a constant and the proof is complete.

Theorem 2.5 leaves open the possibility that gaps of the form k^{ρ} can occur for automorphic forms. Next we show that for certain groups Γ and certain forms g on Γ even these very mild gaps cannot occur.

Let Γ be a Fuchsian group of the second kind and assume that h(z) belongs to $\mathcal{a}$ the disc algebra. The absolute Poincaré series $\sum_{\gamma \in \Gamma} |\gamma'(z)|$ converges uniformly on compact subsets of $\mathcal{O}$, the ordinary set of Γ. Thus, it follows that for h in $\mathcal{a}$ the Poincaré series $\theta_1(h,\Gamma)(z)$ given by

$$\theta_1(h,\Gamma)(z) \equiv \sum_{\gamma \in \Gamma} h(\gamma z)\gamma'(z) \tag{2.7}$$

is continuous on compact subsets of $\mathcal{O}$ $\bar{D}$. Since Γ is of the second kind there is an interval J, which we shall assume is $[-\beta,\beta]$, on ∂D on which $\theta_1(h,\Gamma)(z)$ is continuous.

THEOREM 2.8: Let Γ be a Fuchsian group of the kind and suppose that h belongs to $\mathcal{a}$. If $\theta_1(h,\Gamma)(z)$ has a Taylor's series expansion of the form (1.2) which satisfies

(i) $n_k \geq Ak^{\rho}$ for some $\rho > 1$

(ii) $n_{k+1}-n_k \geq 4\pi/\beta$ where 2β is the length of some compact subarc of $\mathcal{O} \subset \partial D$, then

$$\theta_1(h,\Gamma)(z) \equiv 0.$$

The proof depends on a regularity result for Fourier series due to Bojanic and Tomic [BT1].

THEOREM B.T.: Let F be in $L^1(-\pi,\pi)$, $F(x) \sim \sum_{k=1}^{\infty} a_k e^{in_k x}$ on $J = [-\beta,\beta]$. Let $k(x) = \sum_{n_k \leq x} 1$ and let $W_J(\varepsilon, F) = \sup\{|F(x)-F(y)|: x,y \in J \text{ and } |x-y| \leq \varepsilon\}$.

Suppose that $n_{k+1}-n_k \geq 4\pi/\beta$ and that $\int_{-\beta}^{\beta} |dF(x)| < \infty$. If

$$\int_1^{\infty} W_J(1/t,F)\, k(t)^{1/2} \frac{dt}{t^{3/2}}$$

is finite then

$$\sum_{k=1}^{\infty} |a_k| < \infty.$$

The proof of Theorem 2.8 proceeds as follows: since h is in $\mathcal{a}$ it follows that h belongs to $H^2(D)$, the Hardy space. Thus, $\theta_1(h,\Gamma)$ belongs to $A_1^2(\Gamma)$ and by Lemma 3 of [M2] one has that the f(z) gotten in (1.6) by integrating $\theta_1(h,\Gamma)$ is in $H^2(D)$. If J is the compact subarc of $\mathcal{O} \subset \partial D$ given in (ii) it follows that

$$\int_{-\beta}^{\beta} |df(z)| \leq \int_J \Big(\sum_{\gamma \in \Gamma} |h(\gamma z)|\, |\gamma'(z)| \Big) |dz|$$

$$\leq 2\beta\, ||h||_{\infty} \Big(\sup_J \sum_{\gamma \in \Gamma} |\gamma'(z)|\Big) < \infty.$$

Moreover the continuity of f on J and (i) imply that

$$\int_1^{\infty} W_J(1/t,f) k(t)^{1/2} t^{-3/2} dt \leq K \int_1^{\infty} t^{\frac{1}{2\rho} - \frac{3}{2}}\, dt < \infty.$$

Thus, Theorem B.T. implies that f has an absolutely convergent Taylor's series expansion and $f \in \mathcal{a}$. Lemma 2.2 applied to this f completes the proof.

Most of the above results do not obviously generalize to the case k > 1. This is due to the fact that we have consistently used the integral (1.6), which automatically satisfies (1.1) or (1.4), in our proofs. If k is an integer strictly larger than one then one must integrate 2k-1 times in order to get a formula similar to (0.1). Some results can be stated in the general case and one such shall be stated below.

Given a sequence $\{n_k\}$, define

$$(2.8)\qquad P_\alpha(n_k) = \prod_{\substack{j\neq k\\ j=1}}^{\infty} \left|\frac{n_j+n_k}{n_j^{\ \alpha}-n_k}\right| .$$

We note (see [A1]) that (1.3) implies that $P_\alpha(n_k) = O(1)$.

PROPOSITION 2.9: Suppose that g is a holomorphic automorphic form with respect to Γ of weight $k > 2$. Assume that g has a Taylor's series expansion of the form (1.2) where $\sum_{k=1}^{\infty} 1/n_k^\alpha < \infty$ for some α in $[1/2,1)$. Assume that for each $\varepsilon > 0$, $P_\alpha(n_k) = O(n_k^\varepsilon)$ as $k \to \infty$. Then either $g(z) \equiv 0$ or else g is an annular function.

We recall that a holomorphic function h on the unit disk D is called annular if for every curve λ tending to ∂D

$$(2.9)\qquad \varlimsup_{\substack{z\in\lambda\\ |z|\to 1}} |h(z)| = \infty$$

Clearly, if g is any automorphic form then (2.9) is incompatible with (2.1) of Theorem 2.1. Thus we have

COROLLARY 2.10: Let g be an automorphic form of weight $k > 1$ with respect to a Fuchsian group Γ. If the Taylor's series expansion for g satisfies the conditions of Proposition 2.9 and if Γ has a point on the boundary at which (2.1) holds for g then $g(z) \equiv 0$.

The proof of Proposition 2.9 will be omitted since it is basically a restatement of Theorem 3 of [A-1]. This follows from the fact that if g is an automorphic form of weight k then

$$\overline{\lim_{r \to 1}} \frac{\log M(r,g)}{-\log(1-r)} \geq k$$

where

$$M(r,g) = \sup_{|z| = r} |g(z)|$$

(See [R1]).

REFERENCES

A1. J.M. Anderson, "Boundary properties of an analytic function with gap power series", Quart. J. Math. 21 (1970), pp. 247-256.

ACP1. J.M. Anderson, J. Clunie and Ch. Pommerenke "On Bloch Functions and Normal Functions", J. for reine and angewandt. Mat. 270 (1974), pp. 12-37.

B1. A. Beardon, "Inequalities for Certain Fuchsian Groups", Acta Math. 127 (1971), pp. 221-253.

B.1. K.G. Binmore, "Analytic Function with Hadamard Gaps", Bull. London Math. Soc. 1 (1969), pp. 211-217.

BT1. R. Bojanic and M. Tomic, "Absolute Convergence of Fourier Series with Small Gaps", Math. SB. (111) (1966), pp. 279-309 in <u>Mathematics International</u>, vol. 1 Gorden Breach Sci. Publ. N.Y. (1972), pp. 193-203.

DS1. P. Duren and A. Shields, "Properties of H^p, $0 < p < 1$, and its Containing Banach Space", Trans. Amer. Math. Soc. 141 (1969), pp. 255-262.

KL1. M.I. Knopp and J. Lehner. (Proceedings of this vol.).

L1. J. Lehner, "Automorphic Forms" in <u>Discrete Groups and Automorphic Functions</u> (W.J. Harvey ed.) Academic Press, N.Y. (1977), pp. 73-120.

M1. T.A. Metzger, "On the Growth of the Taylor Coefficients of Automorphic Forms", Proc. Amer. Math. Soc. 39 (1973), pp. 321-328.

M2. T.A. Metzger, "On Vanishing Eichler Periods and Carleson Sets", Mich. Math. J. 24 (1977), pp.197-202.

MR1. T.A. Metzger and K.V. Rajeswara Rao, "Fuchsian Groups of Convergence Type and Nontangential Growth of Automorphic Forms", Proc. Amer. Math. Soc. 48 (1975), pp. 135-139.

NS1 P. Nicholls and L. Sons, "Automorphic Functions with Gap Power Series", Ill. J. of Math. (to appear).

P1. S.J. Patterson, "The Exponent of Convergence of Poincaré Series", Monat. fur Math. 82 (1976), pp. 297-316.

P2. S.J. Patterson, "Some Examples of Fuchsian Groups, Proc. London Math. Soc. 39 (1979), pp. 276-298.

Po1. Ch. Pommerenke, "On the Green's Function for Fuchsian Groups", Ann. Acad. Sci. Fenn. 2 (1976), pp. 409-427.

R1. K.V. Rajeswara Rao, "Fuchsian Groups of Convergence Type and Poincaré Series of Dimension - 2", J. of Math and Mech. 19 (1968/69), pp. 629-644.

W1. H. Widom, "The maximum principle for multivalued analytic functions", Acta Math. 126 (1971), pp. 63-81.

W2. H. Widom, "H_p sections of vector bundles over Riemann surfaces", Ann. of Math. 94 (1971), pp. 304-324.

MODULAR FUNCTIONS REVISITED

Morris Newman*

Dedicated in friendship to Emil Grosswald

I. Introduction. We will be concerned with the classical modular group $\Gamma = PSL(2,Z)$ and some of its congruence subgroups. We will follow the usual practice of writing the elements of Γ as matrices, with the understanding that a matrix and its negative are to be identified. We set

$$S = \begin{bmatrix} 1 & 1 \\ 0 & 1 \end{bmatrix}, \quad T = \begin{bmatrix} 0 & 1 \\ -1 & 0 \end{bmatrix}, \quad W = \begin{bmatrix} 1 & 0 \\ 1 & 1 \end{bmatrix}, \quad U = ST = \begin{bmatrix} -1 & 1 \\ -1 & 0 \end{bmatrix}.$$

Then Γ is the free product $\{T\} * \{U\}$, where the cyclic group $\{T\}$ is of order 2 and the cyclic group $\{U\}$ of order 3. If n is a nonzero integer, $\Gamma_o(n)$ will denote the congruence subgroup of Γ consisting of all elements $\begin{smallmatrix} a & b \\ c & d \end{smallmatrix}$ of Γ such that $n|c$. K will denote the "Θ-subgroup" of Γ, generated by the elements T and S^2. Then K consists of all elements A of Γ such that $A \equiv I(\text{mod } 2)$ or $A \equiv T(\text{mod } 2)$, so that K is a congruence group as well. K is a conjugate of $\Gamma_o(2)$; in fact, $K = W\Gamma_o(2)W^{-1}$. The index of $\Gamma_o(n)$ in Γ is given by

$$(\Gamma : \Gamma_o(n)) = n \prod_{p|n} (1 + \frac{1}{p}).$$

Thus K is of index 3 in Γ.

The group we intend to study here is the subgroup of K consisting of all elements $\begin{bmatrix} a & b \\ c & d \end{bmatrix}$ of K such that $q|c$, where q is an odd prime. We denote this group by K(q). Then it is easy to see that in fact, $K(q) = W \ \Gamma_o(2q)W^{-1}$. Our ultimate purpose is to derive identities of number-theoretic interest for the coefficients of modular forms for this group. The results we obtain embody an infinite family of identities, from which the simplest and most interesting are chosen for display. For example, let the total number of

---

*This work was supported by NSF Grant MCS 76-8293

representations of n by the form

$$\sum_{i=1}^{r} x_i^2 + q\sum_{i=1}^{s} y_i^2, \quad x_i, y_i \in Z,$$

be denoted by R(n; r,s,q); and let the total number of representations of n by the form

$$\sum_{i=1}^{r} \frac{x_i^2 + x_i}{2} + q\sum_{i=1}^{s} \frac{y_i^2 + y_i}{2}, \quad x_i, y_i \in Z, \quad x_i, y_i \geq 0,$$

be denoted by T(n; r,s,q). Then the following results (some of which are classical) hold:

(I.1) $R(n;\ 1,1,3) = 2\prod_{p^e \| n} (1+(\frac{-3}{p}) + \ldots + (\frac{-3}{p})^e)$

$= 2\sum_{d|n} (\frac{-3}{d})$, n odd.

(I.2) $R(n;\ 2,2,3) = 4\{\sigma(n)-3\sigma(\frac{n}{3})\}$, n odd (Liouville).

(I.3) $R(n;\ 1,5,3) = 2\sigma_2(n)$, if each prime factor of n is of the form 3k+1.

(I.4) $R(n;\ 1,3,5) = 2\sigma(n)$, if each prime factor of n is of the form $5k\pm1$.

(I.5) $R(n;\ 1,1,5) = 2\prod_{p^e \| n} (1+(\frac{-5}{p}) + \ldots + (\frac{-5}{p})^e) = 2\sum_{d|n} (\frac{-5}{d})$,

if each prime factor of n is of the form 4k+1.

(I.6) $R(n;\ 1,1,7) = 2\prod_{p^e \| n} (1+(\frac{-7}{p}) + \ldots + (\frac{-7}{p})^e) = 2\sum_{d|n} (\frac{-7}{d})$, n odd.

(I.7) $T(\frac{n-1}{2};\ 1,1,3) = \sum_{d|n} (\frac{-3}{d})$, n odd.

(I.8) Put $T(n-1;\ 2,2,3) = c(n)$. Then $c(np) = c(n)c(p)-pc(\frac{n}{p})$,

p an odd prime.

(I.9) Put $T(n; 1,1,5) = c^*(n)$, $\delta = \frac{3}{4}(p-1)$, p a prime $\equiv 1 \pmod 4$).

Then $c^*(np+\delta) = c^*(n)c^*(\delta) - (\frac{-5}{p})c^*(\frac{n-\delta}{p})$.

(I.10) $T(n-1; 1,1,7) = \sum_{d|n} (\frac{-7}{d})$, n odd.

(I.11) Put $b(n) = R(n; 2,2,5)$. Then for all primes $p \neq 2,5$,

$$b(np) + \gamma b(n) + pb\,(\tfrac{n}{p}) = \beta\{r_4(n) + 5r_4(\tfrac{n}{5})\}$$

where

$$\gamma = \frac{p+1}{2} - \frac{3}{8}\,b(p), \quad \beta = \frac{p+1}{4} - \frac{1}{16}\,b(p)$$

and $r_4(n)$ is the number of representations of n as a sum of 4 squares.

(I.12) $R(p; 3,3,3) = 4(p^2-1)$, p prime, $p \equiv -1 \pmod 3$)

$$R(2p; 3,3,3) = \begin{cases} 6(p^2-1), & p \text{ prime}, p \equiv -1 \pmod 3 \\ 12(p^2+1), & p \text{ prime}, p \equiv +1 \pmod 3 \end{cases}$$

These results are connected with the theory of modular functions through the two theta series:

$$\Theta(\tau) = \sum_{n=-\infty}^{\infty} u^{n^2}, \quad u = \exp(\pi i\tau),$$

$$\phi(\tau) = \sum_{n=0}^{\infty} x^{(n^2+n)/2}, \quad x = u^2 = \exp(2\pi i\tau),$$

with Im $\tau > 0$.

(In what follows all summations will be from 0 to ∞ and all products from 1 to ∞, unless otherwise indicated.) As we shall observe in Section III $\Theta(\tau)$ is a modular form of dimension $-\frac{1}{2}$ on K and $\phi(\tau)$, related to $\Theta(\tau)$ by the identity (III.7), is a modular form of dimension $-\frac{1}{2}$ on $\Gamma_o(2) = W^{-1}KW$.

For r,s,q integers, q > 0, define

$$B(\tau) = B(\tau;\ r,s,q) = \Theta^r(\tau)\Theta^s(q\tau),$$

a modular form of dimension -(r+s)/2 for K(q), and

$$C(\tau) = C(\tau;\ r,s,q) = x^t\phi^r(\tau)\phi^s(qt),\ t = \frac{r+sq}{8},$$

a modular form of dimension -(r+s)/2 for the corresponding congruence subgroup of $\Gamma_o(2)$. If r,s > 0, then B(τ; r,s,q) and C(τ; r,s,q) are the generating functions of R(n; r,s,q) and T(n; r,s,q), respectively:

$$B(\tau;\ r,s,q) = \sum R(n;\ r,s,q)u^n,$$

$$C(\tau;\ r,s,q) = x^t \sum T(n;\ r,s,q)x^n.$$

Thus the identities (I.1)-(I.6) and (I.11)-(I.12) are statements about the modular form B(τ), while (I.7)-(I.10) express properties of the modular form C(τ). It is the modular character of B(τ) and C(τ) that proves decisive in the derivation of these identities.

Further results (involving negative values of r or s) follow:

(I.13) $R(n;\ 3,-1,3) = 6\prod_{p^e||n}\left(1+\left(\frac{-3}{p}\right) + \ldots + \left(\frac{-3}{p}\right)^e\right) = 6\sum_{d|n}\left(\frac{-3}{d}\right)$, n odd.

(I.14) $R(n;\ -1,3,3) = -2\prod_{p^e||n}\left(1+\left(\frac{-3}{p}\right) + \ldots + \left(\frac{-3}{p}\right)^e\right)$

$$= -2\sum_{d|n}\left(\frac{-3}{d}\right),\ n \text{ odd}.$$

(I.15) $T(n;\ 3,-1,3) = 3\sum_{d|n}\left(\frac{-3}{d}\right)$, n odd.

(I.16) Put c(n) = T(n; -1,3,3). Then $c(np) = c(n)c(p) - \left(\frac{-3}{p}\right)c\left(\frac{n}{p}\right)$,

n, p odd.

We also find the modular equations of level 3:

$$(\text{I}.17)\quad \Theta(\tau)\Theta(3\tau) - \Theta(\tau+1)\Theta(3\tau+3) = \Theta^3(3\tau+3)/\Theta(\tau+1) - \Theta^3(3\tau)/\Theta(\tau) = \frac{1}{3}\{\Theta^3(\tau)/\Theta(3\tau) - \Theta^3(\tau+1)/\Theta(3\tau+3)\}.$$

$$(\text{I}.18)\quad \phi^3(\tau)/\phi(3\tau) - \phi^3(\tau+\tfrac{1}{2})/\phi(3\tau+\tfrac{1}{2}) = 3\{\phi(\tau)\phi(3\tau) - \phi(\tau+\tfrac{1}{2})\phi(3\tau+\tfrac{1}{2})\}.$$

Some very general identities are also stated and proved in terms of a Hauptmodul for the group K(q) for q = 3,5. The precise statements are given in the last section.

The general procedure we employ is classical, and runs as follows: First we work out the topological structure of the fundamental region of K(q) in the upper half-plane, regarded as a group of linear fractional transformations. We then construct families of modular functions (i.e., modular forms of dimension 0) built up from basic modular forms, which are holomorphic in the upper half-plane and invariant with respect to the transformations of K(q). The next step is to determine the orders of the poles of these functions at the parabolic vertices in the proper uniformizing variables. Finally by choosing appropriate linear combinations of these functions, we obtain new ones which are pole-free in the upper half-plane, including the parabolic vertices, and so must be constant. (A modular function bounded in the upper half-plane is constant.) Comparison of coefficients then yields the desired identities.

A good deal of computation and routine technical work is required. This has been for the most part suppressed in the presentation, since the details tend to obscure the main flow of the paper, and since they are easily reproduced.

II. The groups K(q). It is readily seen that

$$K(q) = \Gamma_0(q)(T;\ 2),$$

where the notation means that K(q) consists of all elements of $\Gamma_0(q)$ which are congruent to a power of T (and so to I or T) modulo 2. Using this property, we find from the isometric circles that the fundamental region K(q)* of K(q) has the appearance given in Fig. 1 below:

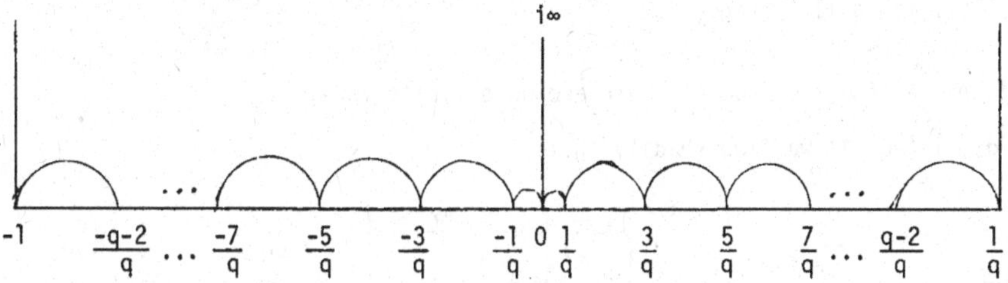

Figure 1 K(q)*

Thus the vertices of K(q)* (other than possible elliptic ones of period 2)† are all parabolic and consist of the points

$$0, \pm\frac{1}{q}, \pm\frac{3}{q}, \ldots, \pm 1, i\infty .$$

We now look for the elliptic vertices. Suppose that $\frac{2r+i}{q}$, $1 \leq r \leq \frac{q-1}{2}$, is an elliptic vertex and suppose that

$$A = \begin{bmatrix} a & b \\ qc & d \end{bmatrix}$$

has $\frac{2r+i}{q}$ as fixed point. Then

$$\text{(II.1)} \qquad \begin{aligned} A&: \frac{2r-1}{q} \to \frac{2r+1}{q} \\ A&: \frac{2r+1}{q} \to \frac{2r-1}{q} \end{aligned}$$

The pair (II.1) implies that

$$\text{(II.2)} \qquad \begin{aligned} a(2r-1)+bq &= c(4r^2-1)+d(2r+1) \\ a(2r+1)+bq &= c(4r^2-1)+d(2r-1) \end{aligned}$$

---

† There are no elliptic vertices of period 3, since a,d are of the same parity, and so $a+d = \pm 1$ is impossible.

from which we find that

(II.3) $a+d = 0$

and so

(II.4) $d^2+1 = -bcq.$

Hence if $q \equiv 3 \pmod 4$ there are no elliptic vertices. If $q \equiv 1 \pmod 4$ we find readily that

$$A = \pm \begin{bmatrix} 2r & -\frac{4r^2+1}{q} \\ q & -2r \end{bmatrix}$$

where r is the unique integer such that

(II.5) $4r^2+1 \equiv 0 \pmod q, \; 1 \leq r \leq \frac{q-1}{2}.$

Thus if $q \equiv 1 \pmod 4$ there are two elliptic vertices

(II.6) $\frac{2r+i}{q}, \; \frac{-2r+i}{q}$ where r satisfies (II.5)

Summaarizing we have shown that if $q \equiv 3 \pmod 4$ $K(q)^*$ has only parabolic vertices given by

$$0, \pm\frac{1}{q}, \pm\frac{3}{q}, \ldots, \pm\frac{q-2}{q}, \pm 1, i\infty$$

and has $2(\frac{q+3}{2})$ sides identified in pairs. If $q \equiv 1 \pmod 4$ $K(q)^*$ has in addition the two elliptic vertices (II.6) of period 2 and has $2(\frac{q+5}{2})$ sides identified in pairs.

We now look at the cycles of vertices. It is readily verified that

$$(0), (i\infty), (-1,1)$$

are parabolic cycles and it is also true that

(II.7) $(\pm \frac{1}{q}, \pm \frac{3}{q}, \ldots, \pm \frac{q-2}{q})$

constitutes a parabolic cycle. This is seen as follows:

Suppose that $\alpha = \pm 1, \pm 3, \ldots, \pm (q-2)$. Then $(\alpha,q) = 1$. Since q is odd we can determine α_o, q_o such that $2\alpha_o\alpha - q_o q = 1$. Then also q_o is odd. Put

$$M_\alpha = \begin{bmatrix} \alpha - qq_o & q_o \\ q(1-2\alpha_o) & 2\alpha_o \end{bmatrix}.$$

Since α is odd, it is readily verified that $M_\alpha \equiv T \pmod 2$, $M_\alpha \in K(q)$. But

$$M_\alpha \frac{1}{q} = \frac{(\alpha - qq_o)\frac{1}{q} + q_o}{q(1-2\alpha_o)\frac{1}{q} + 2\alpha_o} = \frac{\alpha}{q}.$$

Hence (II.7) is indeed a parabolic cycle. Also if $q \equiv 1 \pmod 4$ it is readily verified that $(\frac{2r+i}{q})$, $(\frac{-2r+i}{q})$ are each an elliptic cycle. Summarizing, we have proved that if $q \equiv 3 \pmod 4$ there are 4 cycles of vertices and if $q \equiv 1 \pmod 4$ there are 6 cycles of vertices. Thus we obtain

<u>THEOREM</u> II.1. <u>The genus of</u> K(q), q <u>an odd prime, is given by</u>

$$\nu(q) = \begin{cases} \frac{1}{2}\{\frac{q+5}{2} - 6 + 1\} = \frac{q-5}{4} & q \equiv 1 \pmod 4 \\ \frac{1}{2}\{\frac{q+3}{2} - 4 + 1\} = \frac{q-3}{4} & q \equiv 3 \pmod 4. \end{cases}$$

<u>A compact expression for</u> $\nu(q)$ <u>is given by</u>

$$\nu(q) = [\tfrac{q-2}{4}].$$

We also have the following elementary results:

Suppose that

$$p \neq q,\ p,\ q \text{ odd primes.}$$

Define

$$R = \begin{bmatrix} 1 & -2 \\ -2q_oq & p_op \end{bmatrix}, \quad p_op-4q_oq = 1 .$$

Then $R \equiv I \pmod 2$, $R \in K(q)$. Recall that $W = \begin{bmatrix} 1 & 0 \\ 1 & 1 \end{bmatrix}$.

It is known (see [2]) that $K(pq)$ is a subgroup of $K(q)$ of index $p+1$, and

$$R_k = W^{-2kq}, \ 0 \le k \le p-1, \ R_p = R$$

forms a complete set of right coset representatives for $K(q)$ modulo $K(pq)$. It is also known that $K(p^2q)$ is a subgroup of $K(pq)$ of index p, and

$$R_k^p = W^{-2kpq}, \ 0 \le k \le p-1$$

forms a complete set of right coset representatives for $K(pq)$ modulo $K(p^2q)$.

These results will be used to construct functions on $K(q)$. The general principle involved is that if $G \quad H$ are subgroups of Γ such that $(G : H)$ is finite, and if $R_1, R_2, \ldots, R_\mu$ are right coset representatives for G modulo H, then $g(R_1\tau)+g(R_2\tau)+ \ldots +g(R_\mu\tau)$ is a function on G, provided that $g(\tau)$ is a function on H (see [1]).

Thus if $g(\tau)$ is a function on $K(p^2q)$ then

$$h(\tau) = \sum_{k=0}^{p-1} g(W^{-2kpq}\tau) = \sum_{k=0}^{p-1} g(R_k^p\tau)$$

is a function on $K(pq)$, and

$$G(\tau) = \sum_{k=0}^{p} h(R_k\tau)$$

is a function on $K(q)$. We have that

$$G(\tau) = \sum_{k=0}^{p=1} h(R_k\tau)+h(R_p\tau)$$

$$= \sum_{k=0}^{p-1}\sum_{j=0}^{p-1} g(R_j^pR_k\tau) + \sum_{k=0}^{p-1} g(R_k^pR_p\tau)$$

$$= \sum_{0\le k,j\le p-1} g(W^{-2q(pj+k)}\tau) + \sum_{0\le k\le p-1} g(W^{-2qkp}R_p\tau),$$

$$G(\tau) = \sum_{k=0}^{p^2-1} g(W^{-2qk}\tau) + \sum_{k=0}^{p-1} g(W^{-2qkp}R\tau),$$

so that the intermediate function $h(\tau)$ does not appear in the final expression.

Define

$$G_1(\tau) = \sum_{k=0}^{p^2-1} g(W^{-2qk}\tau),$$

$$G_2(\tau) = \sum_{k=0}^{p-1} g(W^{-2qkp}R\tau),$$

$$F_1(\tau) = G_1(R^{-1}\tau),$$

$$F_2(\tau) = G_2(R^{-1}\tau).$$

Then, as a calculation shows,

$$G(\tau) = G_1(\tau)+G_2(\tau) = F_1(\tau)+F_2(\tau),$$

and $F_1(\tau)$, $F_2(\tau)$ are functions on $K(pq)$ while $G(\tau)$ is a function on $K(q)$.

III. <u>The basic modular forms.</u> We consider functions built up from the basic modular forms

$$\Theta(\tau) = \sum_{n=-\infty}^{\infty} e^{i\pi n^2\tau} = \frac{\eta^2(\frac{\tau+1}{2})}{\eta(\tau+1)},$$

$$\eta(\tau) = e^{\frac{i\pi\tau}{12}} \Pi (1-e^{2\pi in\tau}).$$

We let r, s, q be integers, $q > 0$. q will generally be chosen to be an odd prime. In Section I we have defined

$$B(\tau) = B(\tau;\ r,s,q) = \Theta^r(\tau)\Theta^s(q\tau).$$

Then, as we shall see, $B(\tau)$ is an entire modular form for $K(q)$ vanishing nowhere in the interior of the upper half-plane and of dimension $-\varepsilon$, where

$$\varepsilon = \frac{r+s}{2} .$$

Recall that

$$t = \frac{r+sq}{8}$$

If $f = f(r,s)$ is a function of the variables r, s then by f^* we shall understand $f(s,r)$.

Thus

$$\varepsilon^* = \varepsilon,\ t^* = \frac{s+rq}{8},\ B^*(\tau) = \Theta^s(\tau)\Theta^r(q\tau).$$

<u>Transformation Formulae for $\Theta(\tau)$.</u> The following formulae are classical.

Let $A = \begin{bmatrix} a & b \\ c & d \end{bmatrix}$, $c > 0$, $A \equiv T \pmod 2$, $A \in K$. Then

(III.1) $$\frac{\Theta(A\tau)}{\Theta(\tau)} = \{-i(c\tau+d)\}^{\frac{1}{2}}\, i^{\frac{1-c}{2}} \left(\frac{a}{c}\right) .$$

Let $A = \begin{bmatrix} a & b \\ c & d \end{bmatrix}$, $a > 0$, $c > 0$, $A \equiv I \pmod 2$, $A \in K$. Then

(III.2) $$\frac{\Theta(A\tau)}{\Theta(\tau)} = \{-i(c\tau+d)\}^{\frac{1}{2}}\, i^{1-\frac{a}{2}} \left(\frac{c}{a}\right) .$$

Let $A = \begin{matrix} a & b \\ c & d \end{matrix}$, $a < 0$, $c > 0$, $A \equiv I \pmod 2$, $A \in K$. Then

(III.3) $$\frac{\Theta(A\tau)}{\Theta(\tau)} = \{-i(c\tau+d)\}^{\frac{1}{2}}\, i^{\frac{a}{2}} \left(\frac{c}{-a}\right) .$$

Formulae (III.2) and (III.3) can be combined in one. Define

$$s(a) = \begin{cases} 1 & a > 0 \\ 0 & a < 0 \end{cases}$$

Then if $A = \begin{bmatrix} a & b \\ c & d \end{bmatrix}$, $c > 0$, $A \equiv I \pmod 2$, $A \in K$, we have

(III.4) $\frac{\Theta(A\tau)}{\Theta(\tau)} = \{-i(c\tau+d)\}^{\frac{1}{2}} i^{s(a)-\frac{|a|}{2}} (\frac{c}{|a|})$.

Further we have

(III.5) $\Theta(S^2\tau) = \Theta(\tau),\ S^2 = \begin{bmatrix} 1 & 2 \\ 0 & 1 \end{bmatrix}$;

(III.6) $\Theta(T\tau) = (-i\tau)^{\frac{1}{2}} \Theta(\tau),\ T = \begin{bmatrix} 0 & 1 \\ -1 & 0 \end{bmatrix}$;

(III.7) $\Theta(W\tau) = 2(\tau+1)^{\frac{1}{2}} x^{\frac{1}{2}} \phi(\tau),\ W = \begin{bmatrix} 1 & 0 \\ 1 & 1 \end{bmatrix}$,

$$\phi(\tau) = \Pi(1-x^{2n})(1+x^n) = \sum x^{\frac{n^2+n}{2}},$$

$$x = \exp 2\pi i\tau.$$

IV. Transformation Formulae for $B(\tau;\ r,s,q)$. Consider

$B(\tau) = B(\tau;\ r,s,q) = \Theta^r(\tau)\Theta^s(q\tau)$, r,s integers, with q odd, $q > 0$.

Let $A = \begin{bmatrix} a & b \\ qc & d \end{bmatrix} \in K(q)$, $c > 0$.

Put $A_0 = \begin{bmatrix} a & qb \\ c & d \end{bmatrix}$. Then $A \equiv A_0 \pmod 2$ since q is odd, and so $A_0 \in K$.

We have that

$$B(A\tau) = \Theta^r(A\tau)\Theta^s(A_0 q\tau).$$

Suppose first that $A \equiv T \pmod 2$. Then (III.1) implies after some calculation that

$$B(A\tau) = \{-i(qc\tau+d)\}^{\varepsilon-4ct^*} (\frac{a}{q})^r (\frac{a}{c})^{2\varepsilon} B(\tau).$$

Thus we have that if

$$A = \begin{bmatrix} a & b \\ qc & d \end{bmatrix} \in K(q),\ c > 0,\ A \equiv T \pmod 2, \text{ then}$$

(IV.1) $\frac{B(A\tau)}{B(\tau)} = \{-i(qc\tau+d)\}^{\varepsilon}\, i^{\varepsilon-4ct^*} (\frac{a}{q})^r (\frac{a}{c})^{2\varepsilon}$.

Now suppose that $A \equiv I \pmod 2$ and $a > 0$. Then (III.2) implies that

(IV.2) $\frac{B(A\tau)}{B(\tau)} = \{-i(qc\tau+d)\}^{\varepsilon}\, i^{\varepsilon(2-a)}(\frac{q}{a})^{r}(\frac{c}{a})^{2\varepsilon}$

If $a < 0$, then (III.3) implies that

(IV.3) $\frac{B(A\tau)}{B(\tau)} = \{-i(qct+d)\}^{\varepsilon} i^{a\varepsilon}(\frac{q}{-a})^{r}(\frac{c}{-a})^{2\varepsilon}$

These can be combined onto one formula, which we state as a theroem:

THEOREM IV.1. Let $A = \begin{bmatrix} a & b \\ qc & d \end{bmatrix} \in K(\bar{q})$, $c > 0$, $A \equiv I \pmod 2$. Then

(IV.4) $\frac{B(A\tau)}{B(\tau)} = \{-i(qc\tau+d)\}^{\varepsilon}\, i^{(2s(a)-|a|)\varepsilon}(\frac{q}{|a|})^{r}(\frac{c}{|a|})^{2\varepsilon}$

Further we have that

(IV.5) $B(S^2\tau) = B(\tau)$,

(IV.6) $T_qB(\tau) = (-i\tau)^{\varepsilon}\, q^{\frac{r}{2}}\, B^*(\tau)$, where

$T_q : \tau \to -1/q\tau$.

(N.B. We have adopted the convention here and in what follows that $T_qF(\tau) = F(T_q\tau) = F(-\frac{1}{q\tau})$.)

We have as well (after some elementary calculation) that

(IV.7) $B(W^q\tau) = (q\tau+1)^{\varepsilon}\, 2^{2\varepsilon}\, C(\tau)$,

with $C(\tau) = C(\tau; r,s,q) = x^t\phi^r(\tau)\phi(q\tau)$.

V. Functions on $K(n)$, n odd. Let $\{r_\delta\}$ be a sequence of integers indexed by the positive divisors δ of n with $r_1 = 0$. Define

$$\phi_\delta = \frac{\Theta(\delta\tau)}{\Theta(\tau)},$$

$$g(\tau) = \prod_{\delta|n} \phi_\delta^{r_\delta} = \prod_{\delta|n} \frac{\Theta(\delta\tau)}{\Theta(\tau)}^{r_\delta}.$$

Let $A = \begin{bmatrix} a & b \\ nc & d \end{bmatrix} \in K(n)$, $c > 0$, $A \equiv T \pmod 2$, $T = \begin{bmatrix} 0 & 1 \\ -1 & 0 \end{bmatrix}$. Since $K(n) = \Gamma_o(n)(T;2)$ it is only necessary to verify that $g(A\tau) = g(\tau)$ in order to show that $g(\tau)$ is a function on $K(n)$. We have

$$\delta A\tau = \delta\begin{bmatrix} a & b \\ nc & d \end{bmatrix}\tau = \begin{bmatrix} a & \delta b \\ \delta' c & d \end{bmatrix}\delta\tau = A_\delta\,\delta\tau,\ \delta\delta' = n.$$

Then also $A_\delta \equiv T \pmod 2$, and

$$\phi_\delta(A\tau) = \frac{\Theta(A_\delta\delta\tau)}{\Theta(A\tau)}.$$

Using (III.1) we find that

$$\phi_\delta(A\tau) = i^{\frac{1}{2}(n-\delta')c}\left(\frac{a}{\delta}\right)\phi_\delta(\tau)$$

Hence

$$g(A\tau) = \lambda(A)g(\tau), \text{ where}$$

$$\lambda(A) = i^N \prod_{\delta|n} \left(\frac{a}{\delta}\right)^{r_\delta}, \text{ and}$$

$$N = \frac{1}{2}c\sum_{\delta|n}(n-\delta')r_\delta.$$

Thus $g(\tau)$ is a function on $K(n)$ if and only if

(V.1) $\sum_{\delta|n}(n-\delta')r_\delta \equiv 0 \pmod 8$,

(V.2) $\prod_{\delta|n}\delta^{r_\delta}$ is a rational square.

Since n is odd, condition (V.1) is identical with

(V.3) $\sum_{\delta|n}(\delta-1)r_\delta \equiv 0 \pmod 8$

Now suppose that p,q are odd primes. Then the preceding result implies

<u>THEOREM</u> V.1.

$\frac{B(p\tau)}{B(\tau)}$ <u>is an entire modular function on</u>

$K(pq)$, <u>provided that</u> $t(p-1) \equiv \varepsilon \equiv 0 \pmod 1$

<u>Furthermore</u>, $\frac{B(p^2\tau)}{B(\tau)}$ <u>is an entire modular function on</u>

$K(p^2q)$ <u>for all integral</u> r,s.

VI. The function $\frac{B(p\tau)}{B(\tau)}$. We now make the choice

$$g(\tau) = \frac{B(p\tau)}{B(\tau)}, \ p \text{ prime}, \ p \neq q, \ p \text{ odd},$$

where r+s is even. Let

$$\varepsilon = \frac{r+s}{2} \equiv 0 \pmod 1,$$

$$\delta = t(p-1) = \frac{r+sq}{8}(p-1) \equiv 0 \pmod 1.$$

Then $g(\tau)$ is an entire modular function on $K(pq)$. Setting

$$R_k = W^{-2kq}, \ 0 \leq k \leq p-1$$

$$R_p = R = \begin{bmatrix} -1 & -2 \\ -2q_0q & p_0p \end{bmatrix}, \ p_0p-4q_0q = 1$$

we have that

$$G(\tau) = \sum_{k=0}^{p} g(R_k\tau)$$

is an entire modular function on $K(q)$.

Thus

$$G(\tau) = \sum_{k=0}^{p-1} \frac{B(pR_k\tau)}{B(R_k\tau)} + \frac{B(pR\tau)}{B(R\tau)}.$$

Suppose that $1 \le k \le p-1$. We have

$$pR_k\tau = \begin{bmatrix} -p & 0 \\ 2kq & -1 \end{bmatrix} \tau = \begin{bmatrix} -p & 2k_0 \\ 2kq & -d \end{bmatrix} \begin{bmatrix} 1 & 2k_0 \\ 0 & p \end{bmatrix} \tau$$

where $dp-4kk_0q = 1$. Thus k and k_0 simultaneously run over a reduced set of residues modulo p.

Put $$M_k = \begin{bmatrix} -p & 2k_0 \\ 2kq & -d \end{bmatrix}.$$ Then

$$g(R_k\tau) = \frac{B(M_k \frac{\tau+2k_0}{p})}{B(R_k\tau)}, \quad 1 \le k \le p-1$$

Since $M_k \equiv I \pmod 2$, $R_k \equiv I \pmod 2$ and ε is an integer, formula (17) implies that

$$g(R_k\tau) = p^{-\varepsilon}(-1)^{\frac{\varepsilon(p-1)}{2}} \left(\frac{q}{p}\right)^r \frac{B(\frac{\tau+2k_0}{p})}{B(\tau)}$$

Also

$$pR\tau = \begin{bmatrix} p & -2p \\ -2q_0q & p_0p \end{bmatrix} \tau = \begin{bmatrix} -p & 2 \\ 2q_0q & -p_0 \end{bmatrix} \frac{\tau}{p}$$

$$= M_p \frac{\tau}{p}, \quad M_p \equiv I \pmod 2, \quad M_p \in K(q).$$

Thus

$$g(R\tau) = \frac{B(M_p \frac{\tau}{p})}{B(R\tau)} = (-1)^{\frac{\varepsilon(p-1)}{2}} \left(\frac{q}{p}\right)^r \frac{B(\frac{\tau}{p})}{B(\tau)}.$$

We thus obtain

$$\text{(VI.1)} \quad G(\tau) = \frac{B(p\tau)}{B(\tau)} + p^{-\varepsilon}\left(\frac{(-1)^{\varepsilon}q^r}{p}\right) \sum_{k=0}^{p-1} \frac{B(\frac{\tau+2k}{p})}{B(\tau)}$$

Formula (VI.1) implies that

$G(\tau)$ <u>is pole-free at</u> $\tau = i\infty$.

We now find out the effect of the transformation T_q on G. Suppose $0 \leq k \leq p-1$. Then

$$R_k T = TS^{2qk}$$

$$\sum_{k=0}^{p-1} g(R_k T\tau) = \sum_{k=0}^{p-1} g(TS^{2qk}\tau) = \sum_{k=0}^{p-1} \frac{B(T\frac{\tau+2qk}{p})}{B(T(\tau+2qk))}.$$

Thus

$$T_q \sum_{k=0}^{p-1} g(R_k\tau) = \sum_{k=0}^{p-1} \frac{T_q B(\frac{\tau+2k}{p})}{T_q B(\tau+2k)} = \sum_{k=0}^{p-1} \frac{\{-i\frac{\tau+2k}{p}\}^{\varepsilon} q^{\frac{r}{2}} B^*(\frac{\tau+2k}{p})}{\{-i(\tau+2k)\}^{\varepsilon} q^{\frac{r}{2}} B^*(\tau+2k)} = p^{-\varepsilon} \sum_{k=0}^{p-1} \frac{B^*(\frac{\tau+2k}{p})}{B^*(\tau)} .$$

Further

$$pRT\tau = \begin{bmatrix} -p & 2p \\ 2q_0 q & -p_0 p \end{bmatrix} T\tau = M_p Tp\tau ,$$

$$T_q g(R\tau) = \frac{B(M_p Tpq\tau)}{B(R\ T\ q\tau)} = (-1)^{\frac{\varepsilon(p-1)}{2}} (\frac{q}{p})^r \frac{B^*(p\tau)}{B^*(\tau)} .$$

Thus

$$T_q G(\tau) = p^{-\varepsilon} \sum_{k=0}^{p-1} \frac{B^*(\frac{\tau+2k}{p})}{B^*(\tau)} + (\frac{(-1)^{\varepsilon} q^r}{p}) \frac{B^*(p\tau)}{B^*(\tau)} ,$$

$$\text{(VI.2)} \quad T_q G(\tau) = \frac{(-1)^{\varepsilon} q^r}{p} G^*(\tau)$$

Formula (VI.2) implies that

$$G(\tau) \ \underline{\text{is pole-free at}} \ \tau = 0$$

Put $\alpha = p^{-\varepsilon}(\frac{(-1)^{\varepsilon} q^r}{p})$. Then

$$G(\tau) = \frac{B(p\tau)}{B(\tau)} + \alpha \sum_{k=0}^{p-1} \frac{B(\frac{\tau+2k}{p})}{B(\tau)} .$$

We are now interested in $G(W^q\tau)$.

We have

$$\tau_k = \frac{W^q\tau+2k}{p} = \begin{bmatrix} 2kq+1 & 2k \\ pq & p \end{bmatrix} \tau$$

Suppose first that $(2kq+1, p) = 1$. Determine k_o so that

$$(1+2kq)(1-8k_oq) \equiv 1 \pmod p, \; k_o \geq 0,$$

and set

$$(1+2kq)(1-8k_oq) = 1 + 2\beta pq.$$

Then

$$\tau_k = \begin{bmatrix} 1+2kq & 2\beta \\ pq & 1-8k_oq \end{bmatrix} \begin{bmatrix} 1 & 8k_o \\ 0 & p \end{bmatrix} \tau$$

$$= \begin{bmatrix} 1+2q(k-\beta) & 2\beta \\ q(p-1+8k_oq) & 1-8k_oq \end{bmatrix} W^q \frac{\tau+8k_o}{p}$$

$$= M_k W^q \frac{\tau+8k_o}{p},$$

where

$$M_k \equiv I \pmod 2, \; M_k \in K(q),$$

$$q(p-1+8k_oq) > 0, \quad 1+2q(k-\beta) > 0.$$

Then

$$\frac{B(\tau_k)}{B(W^q\tau)} = B(M_k W^q \frac{\tau+8k_o}{p}) / B(W^q\tau)$$

and formulas (IV.2) and (IV.7) imply after some calculation that

$$\frac{B(\tau_k)}{B(W^q\tau)} = \frac{C(\frac{\tau+8k_o}{p})}{C(\tau)}.$$

Now suppose that $2kq+1 \equiv 0 \pmod p$, and set $2kq+1 = \beta p$. Then

$$\tau_k = \begin{bmatrix} \beta p & 2k \\ pq & p \end{bmatrix} \tau = \begin{bmatrix} \beta & 2k \\ q & p \end{bmatrix} p\tau$$

$$= \begin{bmatrix} \beta & 2k \\ q & p \end{bmatrix} \begin{bmatrix} 1 & 0 \\ -q & 1 \end{bmatrix} W^q p\tau$$

$$= \begin{bmatrix} 2kq-\beta & -2k \\ q(p-1) & -p \end{bmatrix} W^q p\tau = M_k W^q p\tau$$

Here $M_k \in K(q)$, $M_k \equiv I \pmod 2$, $q(p-1) > 0$, $2kq-\beta > 0$.

Formula (IV.2) applies and we find that

$$\frac{B(\tau_k)}{B(W^q\tau)} = p^{\varepsilon} \left(\frac{(-1)^{\varepsilon} q^r}{p}\right) \frac{C(p\tau)}{C(\tau)} .$$

We are now left with $\frac{B(pW^q\tau)}{B(\tau)}$. We have

$$pW^q\tau = \begin{bmatrix} p & 0 \\ q & 1 \end{bmatrix} \tau = \begin{bmatrix} p & -8c \\ q & -d \end{bmatrix} \begin{bmatrix} 1 & 8c \\ 0 & p \end{bmatrix} \tau, \quad -pd+8qc = 1,\ c > 0,\ d > 0.$$

Thus

$$pW^q\tau = \begin{bmatrix} p & -8c \\ q & -d \end{bmatrix} \begin{bmatrix} 1 & 0 \\ -q & 1 \end{bmatrix} W^q \frac{\tau+8c}{p} ,$$

$$\begin{bmatrix} p+8cq & -8c \\ q(1+d) & -d \end{bmatrix} W^q \frac{\tau+8c}{p} = MW^q \frac{\tau+8c}{p} ,$$

where $M \in K(q)$, $M \equiv I \pmod 2$, $p+8cq > 0$, $q(1+d) > 0$.

Formula (IV.2) applies and we find that

$$\frac{B(pW^q\tau)}{B(W^q\tau)} = p^{-\varepsilon}(-1)^{\frac{\varepsilon(p-1)}{2}} \left(\frac{q}{p}\right)^r \frac{C(\frac{\tau+8c}{p})}{C(\tau)} .$$

Summarizing, we have shown that

$$G(W^q\tau) = \alpha \sum_{\substack{0\le k\le p-1 \\ k\ne c}} \frac{C(\frac{\tau+8k}{p})}{C(\tau)} + \alpha \cdot p^{\varepsilon}(\frac{(-1)^{\varepsilon}q^r}{p}) \frac{C(p\tau)}{C(\tau)} + p^{-\varepsilon}(\frac{(-1)^{\varepsilon}q^r}{p}) \frac{C(\frac{\tau+8c}{p})}{C(\tau)} .$$

Hence

$$\text{(VI.3)}\quad G(W^q\tau) = \frac{C(p\tau)}{C(\tau)} + \alpha \sum_{k=0}^{p-1} \frac{C(\frac{\tau+8k}{p})}{C(\tau)} .$$

We now interpolate some general remarks. As representative parabolic points of the fundamental region of $K(q)$ we may choose the points (one from each cycle) $i\infty$, 0, $\frac{1}{q}$, -1. Now T_q permutes $i\infty$ and 0 and $\frac{1}{q}$ and -1; while W^q takes $i\infty$ into $\frac{1}{q}$. Thus we have enough information to determine the behaviour of $G(\tau)$ completely throughout the fundamental region of $K(q)$. We make the observation that in general $G(\tau)$ is an entire modular function on $K(q)$ if any only if $T_qG(\tau)$ is such a function. This is proved by observing that if

$$A = \begin{bmatrix} a & b \\ qc & d \end{bmatrix} \in K(q)$$

then

$$A_1 = \begin{bmatrix} 0 & -1 \\ q & 0 \end{bmatrix} A \begin{bmatrix} 0 & \frac{1}{q} \\ -1 & 0 \end{bmatrix} = \begin{bmatrix} d & -c \\ -bq & a \end{bmatrix} ,$$

so that

$$A_1 \equiv A \pmod 2,\ A_1 \in K(q).$$

We note that

$$\text{(VI.4)}\quad G(T_qW^q\tau) = (\frac{(-1)^{\varepsilon}q^r}{p})\, G^*(W^q\tau) = (\frac{(-1)^{\varepsilon}q^r}{p})\{\frac{C^*(p\tau)}{C^*(\tau)} + \alpha \sum_{k=0}^{p-1} \frac{C^*(\frac{\tau+8k}{p})}{C^*(\tau)}\}$$

and that T_qW^q takes $i\infty$ into -1.

We can now state our first principal result.

THEOREM VI.1 <u>$G(\tau)$ is pole-free at $i\infty$ and 0. If $t \ge 0$ then $G(\tau)$ has a pole of order not exceeding $[\frac{\delta}{p}]$ at $\tau = \frac{1}{q}$, and if $t < 0$ then $G(\tau)$ has a pole of order</u>

$-\delta$ at $\tau = \frac{1}{q}$. If $t^* \geq 0$ then $G(\tau)$ has a pole of order not exceeding $[\frac{\delta^*}{p}]$ at $\tau = -1$, and if $t^* < 0$ then $G(\tau)$ has a pole of order $-\delta^*$ at $\tau = -1$.

Here

$$t = \frac{r+sq}{8},\ \delta = t(p-1),\ t^* = \frac{s+rq}{8},\ \delta^* = t^*(p-1).$$

Now set

$$B(\tau) = \sum b(n)e^{i\pi n\tau},\ C(\tau) = x^t \sum c(n)x^n = x^t\gamma(\tau).$$

$$(b(n) = R(n;\ r,s,q),\ c(n) = T(n;\ r,s,q).)$$

Then

$$\text{(VI.5)}\quad B(\tau)G(\tau) = \sum \{b(\tfrac{n}{p}) + \alpha p b(np)\}e^{i\pi n\tau},$$

and

$$\text{(VI.6)}\quad \gamma(\tau)G(W^q\tau) = \sum c(\tfrac{n-\delta}{p})x^n + \alpha p \sum c(np+\delta)x^n,$$

where

$$\alpha = p^{-\varepsilon}(\frac{(-1)^{\varepsilon}q^r}{p}),$$

as is easily deduced from (VI.1) and (VI.3).

Thus if $G(\tau)$ is constant, a calculation shows that

$$\text{(VI.7)}\quad \frac{b(np)-b(n)}{b(n)-b(\frac{n}{p})} = p^{\varepsilon-1}(\frac{(-1)^{\varepsilon}q^r}{p}) = p^{\varepsilon-1}\gamma,$$

and

$$\text{(VI.8)}\quad c(np+\delta) = \beta c(n) - p^{\varepsilon-1}\gamma\, c(\tfrac{n-\delta}{p}),$$

where

$$\beta = \begin{cases} c(\delta) & \delta > 0 \\ 1+\gamma p^{\varepsilon-1} & \delta = 0 \end{cases}.$$

It is simple to determine from Theorem VI.9 when $G(\tau)$ is free of poles at all of the parabolic vertices ($i\infty$, 0, $\frac{1}{q}$, -1) and is thus constant for all p. This certainly happens if

(VI.9) $0 \leq \tau \leq 1,\ 0 \leq \tau^* \leq 1.$

All instances of (VI.9) are given in the following table.

(VI.10)

q	3				5	7
r	1	2	3	-1	1	1
s	1	2	-1	3	1	1

$G(\tau)$ will also be constant in certain other cases, that is, if the coefficients of the pole terms vanish; this happens, for example, when r = 1, s = 5, q = 3, $p \equiv 1 \pmod 3$ and in many other instances. The cases summarized in (VI.10) and these "accidental" cases lead directly to the results stated in the first section. The identities for R(n; r,s,q) ((I.1)-(I.6) and (I.11)-(I.12)) follow from (VI.7) and those for T(n; r,s,q) ((I.7)-(I.10) from (VI.8).

We now make the choice r = -s. Then

$$B(\tau) = \left\{\frac{\Theta(q\tau)}{\Theta(t)}\right\}^s .$$

Thus if s is even and $s(q-1) \equiv 0 \pmod 8$, $B(\tau)$ is a function on K(q). Because of the formula

$$B(W^q\tau) = x^{\frac{s(q-1)}{8}} \frac{\phi(q\tau)}{\phi(q\tau)}^s ,$$

we find the following table:

(VI.11)

	s > 0	s < 0
$i\infty$	pole and zero free	pole and zero free
0	pole and zero free	pole and zero free
-1	pole of order $\frac{s(q-1)}{8}$	zero of order $\frac{s(q-1)}{8}$
$\frac{1}{q}$	zero of order $\frac{s(q-1)}{8}$	pole of order $\frac{s(q-1)}{8}$

Thus for $q = 3$, $s = 4$; $q = 5$, $s = 2$ $B(\tau)$ is a Hauptmodul for $K(q)$. We set

$$M = M_q(\tau) = \left\{\frac{\Theta(q\tau)}{\Theta(\tau)}\right\}^{\frac{8}{q-1}} , \quad q = 3,5.$$

Let $\nu(P)$ denote the order of the pole of $G(\tau)$ at $\tau = P$, $P = 0, i\infty, \frac{1}{q}, -1$. Then Theorem VI.1 implies that

$$\nu(0) = 0$$

$$\nu(i\infty) = 0$$

$$\nu(\tfrac{1}{q}) \leq \begin{cases} [\frac{\delta}{p}] & t \geq 0 \\ -\delta & t < 0 \end{cases}$$

$$\nu(-1) \leq \begin{cases} [\frac{\delta^*}{p}] & t \geq 0 \\ -\delta & t < 0 . \end{cases}$$

Since M is a Hauptmodul for $K(q)$, $q = 3,5$ we conclude

<u>THEOREM</u> VI.2. <u>Let</u> $q = 3,5$. <u>Then</u> <u>there</u> <u>exists</u> <u>polynomials</u> f,g <u>such</u> <u>that</u>

$$\deg f \leq \nu(\tfrac{1}{q}), \ \deg\ g \leq \nu(-1)$$

<u>and</u>

$$G(\tau) = f(\frac{1}{M(\tau)}) + g(M(\tau)).$$

This is so since there is no interaction between the poles.

VII. <u>The</u> <u>function</u> $\frac{B(p^2\tau)}{B(\tau)}$.

In order to treat the case 2ε odd, we choose

$$g(\tau) = \frac{B(p^2\tau)}{B(\tau)} ,$$

which is an entire modular function on $K(p^2q)$. We set

$$G_1(\tau) = \sum_{k=0}^{p^2-1} g(W^{-2qk}\tau)$$

$$G_2(\tau) = \sum_{k=0}^{p-1} g(W^{-2qkp}R\tau)$$

$$F_1(\tau) = G_1(R^{-1}\tau)$$

$$F_2(\tau) = G_2(R^{-1}\tau),$$

$$G(\tau) = G_1(\tau) + G_2(\tau).$$

Since $R \in K(q)$ and $G(\tau)$ is a function on $K(q)$ we also have that

$$G(\tau) = F_1(\tau) + F_2(\tau).$$

Thus $F_1(\tau)$, $F_2(\tau)$ are entire modular functions on $K(pq)$ and $G(\tau)$ is an entire modular function on $K(q)$.

The same techniques used previously apply here and we find

$$F_1(\tau) = p^{-2\varepsilon} \sum_{k=0}^{p^2-1} \frac{B(\frac{\tau+2k}{p^2})}{B(\tau)},$$

$$F_2(\tau) = \frac{B(p^2\tau)}{B(\tau)} + \rho p^{-\varepsilon} \sum_{k=1}^{p-1} (\frac{k}{p})^{2\varepsilon} \frac{B(\tau+\frac{2k}{p})}{B(\tau)},$$

$$\rho = (\frac{-2}{p})^{2\varepsilon} (\frac{q}{p})^s \, i^{\varepsilon(1-p)},$$

$$T_q G(\tau) = G^*(\tau)$$

and corresponding formulae for $G(W^q\tau)$ which we do not bother quoting.

Define $\Delta = t(p^2-1)$. We find our second principal result:

THEOREM VII.1. $G(\tau)$ _is pole-free at_ 0 _and_ $i\infty$. _At_ $\tau = \frac{1}{q}$, $G(\tau)$ _has a pole of order not exceeding_ $[\frac{\Delta}{p^2}]$ _if_ $t \geq 0$, _and a pole of order_ $-\Delta$ _if_ $t < 0$. _At_ $\tau = -1$, $G(\tau)$ _has a pole of order not exceeding_ $[\frac{\Delta^*}{p^2}]$ _if_ $t^* \geq 0$ _and a pole of order_ $-\Delta^*$ _if_ $t^* < 0$.

Hence $G(\tau)$ is constant certainly when $0 \leq t \leq 1$, $0 \leq t^* \leq 1$ as well as for certain other values, including those below:

(VII.1)

r	1	2	1	2
s	2	1	2	1
q	3	3	5	5

We have the expansion (setting $B(\tau) = \sum c(n)e^{i\pi n\tau}$)

$$B(\tau)G(\tau) = p^{2-2\varepsilon}\sum c(np^2)e^{i\pi n\tau} + \sum c(\tfrac{n}{p^2})e^{i\pi n\tau} + \rho\alpha_p p^{\frac{1}{2}-\varepsilon} \sum (\tfrac{n}{p})c(n)e^{i\pi n\tau}$$

where $\alpha_p = i^{(\frac{p-1}{2})^2}$ and $\rho = (\frac{-2}{p})^{2\varepsilon}(\frac{q}{p})^s i^{\varepsilon(1-p)}$.

It follows that for the cases listed in (VII.1), the coefficients c(n) must satisfy the recurrence

$$\frac{c(np^2) - c(n)}{c(n) - c(\frac{n}{p^2})} = p - (\frac{-nq^s}{p}).$$

We have from Theorem VII.1 that

$$\nu(0) = 0$$

$$\nu(i\infty) = 0$$

$$\nu(\tfrac{1}{q}) \leq \begin{cases} \left[\frac{\Delta}{p^2}\right], & t \geq 0 \\ -\Delta, & t < 0 \end{cases}$$

$$\nu(-1) \leq \begin{cases} \left[\frac{\Delta^*}{p^2}\right], & t \geq 0 \\ -\Delta^*, & t < 0 . \end{cases}$$

As our final result, we have

THEOREM VIII.2. Let $q = 3,5$. Then polynomials f,g exist such that

$\deg f \leq \nu(\frac{1}{q})$, $\deg g \leq \nu(-1)$ and $G(\tau) = f(\frac{1}{M(\tau)}) + g(M(\tau))$, where

$$M(\tau) = \left\{\frac{\Theta(q\tau)}{\Theta(\tau)}\right\}^{\frac{8}{q-1}}.$$

We end our discussion here. It is clear that there are many other interesting number-theoretic results which may be determined as consequences of the formulas given in this paper.

REFERENCES

1. Morris Newman, Structure theorems for modular subgroups, Duke Math J. 22, 25-32 (1955).

2. Morris Newman, Subgroups of the modular group and sums of squares, Amer. J. Math. 82, 761-778 (1960).

Department of Mathematics
University of California
Santa Barbara, California 93106
U.S.A.

BOUNDING THE NORM OF THE POINCARÉ θ-OPERATOR

by

L.A. Parson and Mark Sheingorn*

Dedicated to our good friend Emil Grosswald,

a mathematician's mathematician.

§1. Definition of the theta operator

Let H be the upper half plane; let Γ be a Fuchsian group acting on H and containing the translation $Sz = z+1$. Then $\Gamma_\infty = \langle S \rangle$ is the stabilizer of ∞. A function f defined in H is called a q-form (automorphic form of weight q) if for all $z \in H$, $A \in \Gamma$

$$f(Az)A'^q(z) = f(z).$$

Here q is an integer which is two or larger. (Since we shall consider non-meromorphic q-forms, no smoothness requirement is placed on f.)

$A_q(\Gamma)$, the Bers space of integrable analytic q-forms, is a Banach space with norm $||f||_{A_q(\Gamma)} = \int\int_R |f| y^{q-2} dxdy$, where R is a Ford fundamental region for Γ in H. $B_q(\Gamma)$, the Bers space of bounded analytic q-forms, has norm $||f||_{B_q(\Gamma)} = \sup\limits_R y^q |f(z)|$. The basic facts about these spaces are established in Kra [6], Chapter III. If $\Gamma = \langle S \rangle$, the spaces are denoted by A_q and B_q. Further, the Poincaré θ-operator, θ_q, which is defined by

$$\theta_q(f) = \sum_{A \in \Gamma/\Gamma_\infty} f(Az)A'^q(z),$$

is a surjective continuous linear operator from A_q to $A_q(\Gamma)$. This paper concerns the norm of θ_q. That this norm does not exceed one may be seen as follows.

PROPOSITION 1: $||\theta_q|| \le 1$.

PROOF: For any $f \in A_q$

* Second author partially supported by NSF grant MCS 79-03519.

$$||\theta_q(f)||_{A_q(\Gamma)} = \int\int_R |\theta_q(f)(z)| y^{q-2} dxdy \leq$$

$$\sum_{A \in \Gamma/\Gamma_\infty} \int\int_R |f(Az)| \mathrm{Im}(Az)^{q-2} |A'(z)|^2 dxdy =$$

$$\sum_A \int\int_{AR} |f(w)| v^{q-2} dudv = ||f||_{A_q}. \text{ Thus}$$

$$||\theta_q|| \leq 1.$$

It is interesting to note that inequality occurs only once in the above proof. Further, if $||\theta_q|| = 1$, the supremum in the norm definition is never assumed as, for each analytic f, the above inequality is strict.

§2. The connection with Teichmüller theory

The question of when the norm of the theta operator is strictly less than one was brought to our attention about two years ago by Earle and Kra. Their interest stemmed from a connection when $q = 2$ with the Teichmüller metric, which we now describe. We are grateful to F. Gardiner for providing us, non-experts, with an exposition on which the following is entirely based. (Any mistakes are, of course, our own.) Our notation is that of Earle in [4].

Let μ and σ be in $L^\infty(\mathcal{U})$ where $\mathcal{U}$ is the unit disk, with $||\mu||_\infty$, $||\sigma||_\infty < 1$. After extension by reflection and normalization, these give rise to quasiconformal homeomorphisms of $\mathbb{C}$ which are denoted by W^μ and W^σ. W^μ, for example, is the unique quasiconformal map of $\mathbb{C}$ onto itself such that $W_{\bar{z}} = \mu W_z$, $W_{\bar{z}} \equiv 0$ in $\mathbb{C}-\mathcal{U}$ and $W(z)-z \to 0$ as $|z| \to \infty$. $\mu \sim \sigma$ means $W^\mu|_{\partial\mathcal{U}} = W^\sigma|_{\partial\mathcal{U}}$. Let $[\mu]$ be the equivalence class of μ. The Teichmüller space for $\mathcal{U}$ may be thought of as the set of equivalence classes $[\mu]$. The Teichmüller metric $\rho_\mathcal{U}$ for this space is given by

(1) $$\rho_\mathcal{U}(0,[\mu]) = \frac{1}{2} \inf_{\sigma \sim \mu} \left(\log \frac{1 + ||\sigma||_\infty}{1 - ||\sigma||_\infty}\right).$$

There is an "infinitesimal formula" for $\rho_\mathcal{U}(0,[t\mu])$ given by Royden [13]:

$$(2)\qquad \rho_U(0,[t\mu]) = |t| \sup_{\substack{f \in A_2 \\ ||f||_{A_2} = 1}} \left| \int\int_U \mu f\, dxdy\right| + o(t);\ \text{as } t \to 0.$$

(The transition from U to H proceeds with no difficulty. The main difference is the absence of necessity for an analogue of Γ_∞.)

There are similar formulae for the Teichmüller metric ρ_Γ on the Teichmüller space $T(\Gamma)$ of the surface $S = U/\Gamma$ where Γ is a Fuchsian group. A Beltrami differential μ for S lifts to an $L^\infty(U)$ function satisfying

$$(3)\qquad \mu(Az)\,\overline{A'(z)} = \mu(z)A'(z) \text{ for all } A \in \Gamma,\ z \in U.$$

Equivalence of μ and σ is as before. The corresponding formulae are:

$$(1')\qquad \rho_\Gamma(0,[\mu]) = \frac{1}{2} \inf_{\substack{\sigma \sim \mu \\ \sigma \text{ satisfies (3)}}} \left(\log \frac{1 + ||\sigma||_\infty}{1 - ||\sigma||_\infty}\right)$$

and

$$(2')\qquad \rho_\Gamma(0,[t\mu]) = |t| \sup_{\substack{\phi \in A_2(\Gamma) \\ ||\phi||_{A_2(\Gamma)} = 1}} \left| \int\int_R \mu\phi\, dxdy\right| + o(t).$$

Examination of (1) and (1') shows immediately that $\rho_U(0,[\mu]) \le \rho_\Gamma(0,[\mu])$. In [15] Strebel provides several examples for which $\rho_U < \rho_\Gamma$. This same inequality holds if it is known that $||\theta_2|| < 1$.

<u>PROPOSITION 2</u>: Suppose $\mu \in L^\infty(U)$, $||\mu||_\infty < 1$, and satisfies (3). If $||\theta_2|| \le C < 1$, then $\rho_U(0,[\mu]) < \rho_\Gamma(0,[\mu])$.

<u>PROOF</u>:

$$\frac{\rho_U(0,[t\mu])}{|t|} = \sup_{\substack{f \in A_2 \\ ||f||_{A_2} = 1}} \left| \int\int_U \mu f\, dxdy\right| + o(1)$$

$$= \sup_{||f||_{A_2} = 1} \left| \int\int_R \mu\, \theta_2(f)\, dxdy\right| + o(1)$$

$$\leq \sup_{\substack{\phi \in A_2(\Gamma) \\ ||\phi||_{A_2(\Gamma)} \leq C}} \left| \int\int_R \mu \, \phi \, (f) \, dxdy \right| + o(1)$$

$$= C \sup_{||\phi||_{A_2(\Gamma)} = 1} \left| \int\int_R \mu \, \phi \, dxdy \right| + o(1).$$

In other words,

$$(4) \qquad \frac{\rho_\mu(0,[t\mu])}{|t|} \leq \frac{C \, \rho_\Gamma(0,[t\mu])}{|t|} + o(1).$$

Inequality (4) says that the derivative at zero (with respect to t) of $\rho_U(0,[t\mu])$ is less than or equal to that of $C\rho_\Gamma(0,[t\mu])$. Thus $\rho_U(0,[t\mu]) < C\rho_\Gamma(0,[t\mu])$ if t is sufficiently small. Hence, if one measures a Teichmüller geodesic $\mu_t = \frac{t|\phi|}{\phi}$ in $T(\Gamma)$ (where ϕ is in $A_2(\Gamma)$) with respect to the infinitesimal metric (2') from $t = 0$ to $t = t_o$, one obtains $\rho_\Gamma(0,[t_o|\phi|/\phi]) = \frac{1}{2} \log \frac{1+t_o}{1-t_o}$. On the other hand, if one measures the same path with respect to the infinitesimal metric (2), one obtains a smaller value by inequality (4). Since $\rho_U(0,[\mu])$ is the infimum of the lengths of paths joining 0 to $[\mu]$, $\rho_U < \rho_\Gamma$.

§3. Connections with number theory

The problem of bounding the norm of the theta operator has one foot in modern complex analysis. Our approach to this problem shows that it has another foot planted in a celebrated and venerable pathway of number theory, namely, the study of the Fourier coefficients of automorphic forms and the Dirichlet series associated with them.

We have arrived at this conclusion in two separate ways. The first starts with $f \in A_q$ written as a Fourier series: $f(z) = \sum_{n=1}^{\infty} \alpha_n e^{2\pi inz}$. $\theta_q(f)$ also has a Fourier expansion:

$$(5) \qquad \theta_q(f) = \sum_{n=1}^{\infty} b_n e^{2\pi inz} .$$

If Γ is finitely generated of the first kind, it is well known that $b_n = O(n^q)$; and since R is (for practical purposes) bounded away from the real axis, the value of $||\theta_q(f)||_{A_q(\Gamma)}$ should be determined by the first few terms of (5). We have then only to bound the Fourier coefficients of $\theta_q(f)$ in terms of $||f||_{A_q}$. This procedure is applied successfully to two distinct bases of A_q. (See §5). However, we have not yet managed to apply it successfully to linear combinations of these basis elements. The difficulty lies in determining the exact effect of the value of $||f||_{A_q}$ on the α_n. Even though we have improved an earlier result of Duren, Romberg, and Shields [2] (see Lemma 3), our information is still not sufficient to deal with the most general function in A_q. For example, when $q = 6$ and Γ is the modular group, we must show convergence of (and bound in terms of $||f||_{A_q}$) $\sum_{n=1}^{\infty} \frac{\alpha_n \tau(n)}{n^{11}}$ where $\tau(n)$ is Ramanujan's function. In this case, $\alpha_n = o(n^5)$ and the Dirichlet series $\sum\tau(n)/n^s$ converges for $\sigma > 6 - 1/6$. (See [8]). Thus this method succeeds for a much wider class of f than basis elements - say $f = \sum\alpha_n e^{2\pi inz}$ with $\frac{\alpha_n}{n^{5+\frac{1}{6}-\varepsilon}}$ monotone. But, of course, the set of such f are not dense.

To illustrate our second approach let us assume that $A_q(\Gamma)$ is one-dimensional. (Lemma 1 is used to generalize to the finite dimensional case.) Let f be the q-form that generates $A_q(\Gamma)$, normalized so that its first Fourier coefficient is one. Then for any $g \in A_q$ we have $\theta_q(g) = c_g \cdot f$ where $c_g \in \mathbb{C}$. It is easy to see that $g \to c_g$ is a bounded linear functional on A_q, which we denote by Λ. Also $||\theta_q|| = ||\Lambda||\ ||f||_{A_q(\Gamma)}$. As noted earlier, once the first few Fourier coefficients of f are known, $||f||_{A_q(\Gamma)}$ can be accurately estimated. (Lemma 2 does this for the modular group $\Gamma(1)$ when $q = 6$.) It remains to compute $||\Lambda||$.

By the Bers Isomorphism Theorem (Kra [6], p. 89)
$c_g = \Lambda(g) = \int\int_{H/\Gamma_\infty} g\cdot\bar{h}y^{2q-2}dxdy$ where $h \in B_q$ is uniquely determined. In fact in this case it is easy to show that $h = K\cdot f$ where K is computable once the first few coefficients of f are known. (Lemma 4 carries this out for $q = 6$,

$\Gamma = \Gamma(1)$). We then have $||\Lambda|| \leq |K|\ ||f||_{B_q}$. Unfortunately, $|K|\ ||f||_{B_q} \geq 1$; in fact, when $q = 6$, $\Gamma = \Gamma(1)$, $|K|\ ||\Delta||_{B_6} > 1.1$. (See Lemma 5).

The difficulty here is that the Bers Isomorphism Theorem is not isometric. We can, however, give an isometric version. Specifically,
$||\Lambda|| = \min_{h \in A_q^\perp} ||Kf + h||_{B_q}$ where $A_q^\perp = \{h | h$ is measurable, $h(z+1) = h(z)$, $||h||_{B_q} < \infty$, and $\iint_{H/\Gamma_\infty} g\, \bar{h}\, y^{2q-2} dxdy = 0\ \forall g \in A_q\}$. (Note that we are using the B_q-norm to denote the sup norm even though the functions involved are only assumed to be measurable.) The introduction of an annihilating space also occurs in Duren [1] and Reich [12] although in neither case is the function to be altered as complicated as f, a q-form. To calculate $||\Lambda||$ we must find h so as to minimize the B_q-norm of Kf+h. Theorem 1, as stated for $q = 6$, $\Gamma = \Gamma(1)$, solves this problem under the additional assumption that h is either a q-form for Γ or analytic. In this case, Λ is represented by $h_1 = y^{-q} \dfrac{f}{|f|\ ||f||_{A_q(\Gamma)}}$.

As an immediate consequence, $||\theta_q|| = ||\Lambda||\ ||f||_{A_q(\Gamma)} \leq 1$. Also, since h_1 is a q-form for Γ, if $||\theta_q|| = 1$, then the distance from Kf to $A_q^\perp$ is the same as the distance from Kf to $A_q^\perp(\Gamma)$. This seems highly unlikely and parallels an analogous argument against the case for equality of (1) and (1').

Therefore, to make any additional progress toward showing that the theta operator norm is less than one, we must deal with functions in $A_q^\perp$ that are neither q-forms for Γ nor analytic. Our paper ends with a discussion of the sort of function which will further reduce the norm. For example, if $f(z) = \sum_{n=1}^{\infty} b_n e^{2\pi i n z}$, set $h(z) = -Kf(z) + y^{-6} C \sum_{n=1}^{\infty} \frac{b_n}{n^q} e^{2\pi i n x}$ where C is a specific computable constant depending on f, q and Γ. If h has finite B_q-norm, then $h \in A_q^\perp$; and we know how to show that $||\theta_q|| < 1$. The difficulty here comes in handling the series $\sum_{n=1}^{\infty} \frac{b_n}{n^q} e^{2\pi i n x}$. However, such series and other related series have been considered by many theorists, including Walfisz [17],

Wilton [18], Guinand [5], and Rankin [11]. We close with some remarks on the evidence for the success of this procedure.

§4. Detailed description of our approach

If Γ is a finitely generated Fuchsian group of the first kind containing the translation by one, $A_q(\Gamma)$ is then the finite dimensional space of cusp forms of weight q. As we have already seen, θ_q maps A_q onto $A_q(\Gamma)$. For any f in $A_q(\Gamma)$ denote by T_f the subspace of A_q spanned by $\theta_q^{-1}(f)$. Then for g in T_f, $\theta_q(g) = c_g f$, $c_g \in \mathbb{C}$. Define the linear functional Λ_f on T_f by $\Lambda_f(g) = c_g$ and extend Λ_f to all of A_q by the Hahn-Banach Theorem. We begin with the following simple observation.

LEMMA 1: $||\theta_q|| \le 1-\varepsilon$ if, and only if, $||\Lambda_f|| \le \dfrac{1-\varepsilon}{||f||_{A_q(\Gamma)}}$ for all $f \in A_q(\Gamma)$.

PROOF: Assume $||\theta_q|| \le 1-\varepsilon$. For any fixed $f \in A_q(\Gamma)$,

$$||\Lambda_f|| = \sup_{\substack{||g||_{A_q} = 1, \\ g \in \theta_q^{-1}(f)}} |c_g| . \quad \text{Since } |c_g| = \frac{||\theta_q(g)||_{A_q(\Gamma)}}{||f||_{A_q(\Gamma)}} ,$$

$$||\Lambda_f|| = \frac{\left\{ \sup\limits_{\substack{||g||_{A_q} = 1, \\ g \in \theta_q^{-1}(f)}} ||\theta_q(g)|| \right\}}{||f||_{A_q(\Gamma)}} \le \frac{1-\varepsilon}{||f||_{A_q(\Gamma)}} .$$ Conversely, choose $g \in A_q$ and let $\theta_q(g) = f$. Then

$$||\Lambda_f|| = \sup_{\substack{||g||_{A_q} = 1 \\ g \in \theta_q^{-1}(f)}} |c_g| \le \frac{1-\varepsilon}{||f||_{A_q(\Gamma)}}$$

from which it follows that

$$||\theta_q(g)||_{A_q(\Gamma)} \le 1-\varepsilon.$$

This lemma reduces the bounding of the norm of θ_q to the bounding of the norms of a set of linear functionals. Obviously the situation is simplest when $\dim A_q(\Gamma) = 1$, as then $\overline{\theta_q^{-1}(f)} = A_q$, for all $f \in A_q(\Gamma)$; and we need consider only one functional.

For this reason we now turn to the one dimensional case and, for the purpose of illustration, study the modular group and weight $q = 6$. Also, for notational convenience, we write $||\cdot||$ instead of $||\cdot||_{A_6(\Gamma(1))}$. $A_6(\Gamma(1))$ is spanned by the discriminant function $\Delta(z) = e^{2\pi iz} \prod_{m=1}^{\infty} (1-e^{2\pi imz})^{24}$ whose Fourier expansion at ∞ is $\Delta(z) = \sum_{n=1}^{\infty} \tau(n)e^{2\pi inz}$, where $\tau(n)$ is the celebrated Ramanujan τ-function. The properties of this function have been studied extensively by number theorists for more than fifty years. (See Rankin [8] for a complete summary of results up to 1970.) For $g \in A_6$, $\theta_6(g) = c_g\ \Delta(z)$ and $||\theta_6(g)|| = |c_g|\ ||\Delta||$. $||\Delta||$ is easily estimated.

LEMMA 2: $.0006100996 \leq ||\Delta|| \leq .0007134263$.

PROOF: We begin with the upper bound.

$$||\Delta|| = \int\int_R |\Delta(z)|y^4dxdy = \int\int_R \Big|\sum_{n=1}^{\infty} \tau(n)e^{2\pi inz}\Big|y^4dxdy$$

$$\leq \sum_{n=1}^{\infty} |\tau(n)| \int\int_R y^4e^{-2\pi ny}dxdy. \tag{6}$$

Set $I_n = \int\int_R y^4e^{-2\pi ny}dxdy$. For $n \leq 4$ we evaluate I_n numerically. For $n \geq 5$,

$$I_n \leq 11.14263551\ \frac{e^{-\pi\ n\sqrt{3}}}{n^2}\ .$$

This estimate is obtained by replacing the circular arc on the boundary of R with a straight line segment and integrating by parts. For $n \leq 4$, the values of $\tau(n)$ are $\tau(1) = 1$, $\tau(2) = -24$, $\tau(3) = 252$, and $\tau(4) = -1472$. For $n \geq 5$, we use $|\tau(n)| \leq n^{13/2}$. Substituting this information into (6) gives

$$||\Delta|| \leq .0007134024 + .0000000239 = .0007134263.$$

Note that the first summand in the preceding sum represents the contribution from the terms with $n \leq 4$.

To produce a lower bound on $||\Delta||$ we use a special case of a cusp from coefficient estimate. Specifically, since $1 = \tau(1) = \int_0^1 \Delta(z)e^{-2\pi iz}dx$,

$$.0006100996 = \int_1^\infty y^4 e^{-2\pi y}dy = \int_1^\infty\int_0^1 \Delta(z)\, y^4 e^{-2\pi ix}\, dxdy$$

$$\leq \int\int_R y^4|\Delta(z)|dxdy = ||\Delta||.$$

It remains to estimate $|c_g|$ in terms of $||g||_{A_6}$. We describe in detail two different approaches to this problem. They are as yet unsuccessful in producing an upper bound of less than one for $||\theta_6||$ but do provide some interesting results. In fact, the first approach gives an improved lower bound on $||\theta_6||$. We start from the simple observation that since $\tau(1) = 1$, c_g is the first coefficient in the Fourier expansion of $\theta_6(g)$. We also make extensive use of the Petersson scalar product which is defined by

$$(f,h;R) = \int_R\int f(z)\, \overline{h(z)}\, y^{10}\, dxdy$$

for f and h in $A_6(\Gamma(1))$.

§5. Method one

To illustrate this approach we consider two families of functions in A_6 each of which span A_6.

(i) $g_n = e^{2\pi inz}$, $n = 1,2,3,\ldots$

Then $\theta_6(g_n) = G_n$ is the so-called n-th Poincaré series. By the scalar product formula ([9], p. 149)

$$(G_n,\Delta;R) = \frac{\tau(n)\ 10!}{(4\pi n)^{11}}.$$

On the other hand, $(G_n,\Delta;R) = c_n(\Delta,\Delta;R)$ (Here we abbreviate c_{g_n} by c_n.); and

$c_n = \frac{10!\ \tau(n)}{(\Delta,\Delta;R)(4\pi n)^{11}}$. Since $||g_n||_{A_6} = 4!/(2\pi n)^5$,

$|c_n| = \frac{|\tau(n)|10!}{(\Delta,\Delta;R)(2\pi n)^6 2^{11} 4!} ||g_n||_{A_6}$. Fortunately, D.H. Lehmer has calculated

$(\Delta,\Delta;R)$ rather exactly in [7] as being .0000010354. By Hecke's classical cusp form coefficient estimate $|\tau(n)| \leq e^{2\pi} n^6 ||\Delta||_{B_6}$. An upper bound for $||\Delta||_{B_6}$ is provided in Lemma 5. Combining this information with Lemma 2, we have

$$\frac{||\theta_6(e^{2\pi inz})||}{||e^{2\pi inz}||_{A_6}} \leq .64253 \ldots \frac{||\Delta||_{B_6}||\Delta||}{(\Delta,\Delta;R)}$$

$$\leq .89748\ldots$$

Also

$$\frac{||\theta_6(e^{2\pi iz})||}{||e^{2\pi iz}||_{A_6}} \geq .70703\ldots$$

This is an improvement on the existing lower bound for $||\theta_6||$ which is 5/11. (See Kra [6], p. 91).

In attempting to extend this approach to all of A_6, certain difficulties arise. We need to consider $g(z) = \sum_{n=1}^{\infty} \alpha_n e^{2\pi inz} \in A_6$. It is not hard to show that $|\alpha_n| \leq ||g||_{A_6} \frac{(2\pi n)^5}{4!}$. This is best possible, as the above examples show. Once again, $(\theta_6 g,\Delta;R) = c_g(\Delta,\Delta;R)$. <u>If one could interchange the sum and the integral</u> in the scalar product, it would be true that

$$(\theta_6 g,\Delta;R) = (\sum_{n=1}^{\infty} \alpha_n G_n,\Delta;R) = \sum_{n=1}^{\infty} \alpha_n (G_n,\Delta;R) \tag{7}$$

and

$$c_g = \frac{10!}{(4\pi)^{11}(\Delta,\Delta;R)} \sum_{n=1}^{\infty} \alpha_n \frac{\tau(n)}{n^{11}}. \tag{8}$$

Now, as has been pointed out in [14], the space A_6 and the space $B_{1/6}$ of the Duren, Romberg, Shields analysis of H^p, $p < 1$, are the same ([2]). If $g \in B_{1/6}$,

α_n as above, [2, p. 41] shows that $\alpha_n = o(n^5)$. However, more is true.

LEMMA 3: If $g(z) = \sum_{n=1}^{\infty} \alpha_n e^{2\pi inz} \in A_6$, then $\sum_{n=1}^{\infty} \frac{|\alpha_n|}{n^6} < \infty$.

PROOF: $\infty > ||g||_{A_6} = \int_0^\infty \int_0^1 |\sum_{n=1}^{\infty} \alpha_n e^{-2\pi ny} e^{2\pi inx}| y^4 \, dxdy =$

$\int_0^\infty y^4 (\int_0^1 |\sum_{n=1}^{\infty} \alpha_n e^{-2\pi ny} e^{2\pi inx}| dx) dy \geq \frac{1}{\pi} \int_0^\infty y^4 \sum_{n=1}^{\infty} \frac{|\alpha_n| e^{-2\pi ny}}{n+1} \, dy,$

by Hardy's Inequality [1, p. 48]. Thus $\infty > \frac{1}{\pi} \sum_{n=1}^{\infty} \frac{|\alpha_n|}{n+1} \int_0^\infty y^4 e^{-2\pi ny} dy =$

$\frac{4!}{(2\pi)^5 \pi} \sum_{n=1}^{\infty} \frac{|\alpha_n|}{(n+1)n^5}$ which easily implies the result.

Ubfortunately, given what is known about the exact order of $\tau(n)$, Lemma 3 does not help to establish the convergence of (8). More specifically, the Dirichlet series $\sum \frac{\tau(n)}{n^s}$ converges for $\sigma > 6 - \frac{1}{6}$ and converges absolutely for $\sigma > 6 + \frac{1}{2}$. (See Rankin, [8].) Thus the series in (8) may be written as

$\sum_{n=1}^{\infty} \frac{\alpha_n}{n^{5+\frac{1}{6}-\varepsilon}} \cdot \frac{\tau(n)}{n^{6-\frac{1}{6}+\varepsilon}}$; this is a conditionally convergent series multiplied term-by-term by a null sequence. Because of the generality of the α_n - all we have is that $g \in A_6$ - we have not been able to progress. Indeed Rankin ([10], p. 208) requires $\sum_{n<x} |\alpha_n|^2 = O(x^\theta)$, $\theta < 10$, before establishing the validity of (7).

(ii) Let $K(z,\zeta)$ be the kernel function defined for all z, ζ in the strip $S = \{w = u+iv \in H |\ 0 < u < 1,\ v > 0\}$. (See Kra [6], p. 90) Set $g_\zeta(z) = K(z,\zeta)$. Then $g_\zeta \in A_6$ and the set $\{g_\zeta | \zeta \in S\}$ spans A_6. (See Earle [3].) Set $\zeta = \xi + i\eta$. Using the reproducing property and the A_6-norm of the kernel function ([6], p. 89) together with the representation formula for c_g which appears later in Lemma 4, we have

$$\frac{||\theta_6(g_\zeta)||}{||g_\zeta||_{A_6}} = \frac{5}{11}\pi^6|c_{g_\zeta}|\ ||\Delta|| = \frac{5}{11}\pi^6\frac{|\Delta(\zeta)|\ ||\Delta||}{(\Delta,\Delta;R)}$$

$$\leq \frac{5}{11}\ \frac{||\Delta||_{B_6}||\Delta||}{(\Delta,\Delta;R)}$$

$$\leq .63490...$$

by Lemmas 2 and 5. Again, when dealing with lienar combinations of the g_ζ, the same sorts of problems arise as in (i).

In summary, we have an improved lower bound for $||\theta_6||$ and have shown that $||\theta_6||$ is less than one on the two most obvious bases for A_6. However, the problem of extending this result to the whole space A_6 seems difficult.

§6. Method two

The basis of this method is the observation that the mapping $g \to c_g$ is a linear functional on A_6 and thus the Bers Isomorphism Theorem (Kra [6], p. 89) applies. If we refer to this linear operator as Λ (in keeping with Lemma 1), then trivially $||\theta_6|| = ||\Lambda||\ ||\Delta||$. We must calculate $||\Lambda||$.

LEMMA 4: $\Lambda(g) = \frac{1}{(\Delta,\Delta;R)}\int_0^\infty\int_0^1 g\cdot\bar{\Delta}\, y^{10}dxdy.$

PROOF: By the Bers Isomorphism Theorem there is a unique function $h \in B_6$ with

$$\Lambda(g) = \int_0^\infty\int_0^1 g\bar{h}\, y^{10}dxdy.$$

Write $h(z) = \sum_{n=1}^{\infty} h_n e^{2\pi inz}$. Now $\Lambda(e^{2\pi imz}) = h_m\int_0^\infty y^{10}e^{-4\pi my}dy = \frac{h_m 10!}{(4\pi m)^{11}}$.

However, we also know that $\Lambda(e^{2\pi imz}) = \frac{\theta_6(e^{2\pi imz})}{\Delta(z)} = \frac{\tau(m)\ 10!}{(4\pi m)^{11}(\Delta,\Delta;R)}$.

Therefore, $h_m = \frac{\tau(m)}{(\Delta,\Delta;R)}$ and $h(z) = \frac{\Delta(z)}{(\Delta,\Delta;R)}$.

Since the Bers isomorphism is not isometric, all that can be concluded from it is that $||\Lambda|| \le \frac{||\Delta||_{B_6}}{(\Delta,\Delta;R)}$. As the next lemma shows, this is not sufficient to prove $||\theta_6|| < 1$.

LEMMA 5. $e^{-2\pi} = .0018674\ldots \le ||\Delta||_{B_6} \le .00202717\ldots$

REMARK. Using the lower bounds in Lemmas 2 and 5, we have that

$\frac{||\Delta||\ ||\Delta||_{B_6}}{(\Delta,\Delta;R)} > 1.1.$

PROOF: We begin with the upper bound. Since Δ is a modular form,

$$||\Delta||_{B_6} = \sup_R y^6|\Delta(z)| \le \sup_R \sum_{n=1}^{\infty} |\tau(n)|y^6 e^{-2\pi ny}.$$

As in Lemma 2, we use the exact value of $\tau(n)$ for the first six terms and the estimate $|\tau(n)| \le n^{13/2}$ for the remaining terms. This gives the bound quoted above where the contribution from the tail of the series is approximately 2.5×10^{-9}.

To get the lower bound we use Hecke's cusp form coefficient estimate $|b_n| \le e^{2\pi} n^6 ||f||_{B_6}$ with $n = 1$ and $f = \Delta$.

By itself the Bers Isomorphism Theorem fails to produce a subtle enough bound for $||\Lambda||$. However, if we define

$$A_6^{\perp} = \{h | h \text{ is measurable, } h(z+1) = h(z),\ \sup_S y^6|h(z)| < \infty,$$
$$\int_0^{\infty}\int_0^1 g\,\bar{h}\,y^{10}dxdy = 0\ \ \forall\, g \in A_6\},$$

then

(9) $$||\Lambda|| = \inf_{h \in A_6^{\perp}} ||K\bar{\Delta}+h||_{B_6},\ K = 1/(\Delta,\Delta;R).$$

Further, the infimum in (9) is actually a minimum. This approach is described in Duren [1, p.110]. It is exploited by Reich in [12]; but in that study Reich deals with functions far simpler than Δ.

It might be thought that we could take advantage of (9) by modifying the leading terms in the Fourier series of $K\bar{\Delta}$ so as to reduce its B_6-norm. The following lemma shows that this is not the case.

LEMMA 6: Let $\Delta_t(z) = \sum_{n=t}^{\infty} \tau(n)e^{2\pi inz}$. Then $||\Delta||_{B_6} \leq ||\Delta_t||_{B_6}$.

PROOF: $y^6|\Delta(z)| = y^6|\sum_{n=1}^{t-1} \tau(n)e^{2\pi inz} + \Delta_t(z)| = |y^6 \sum_{n=1}^{t-1} \tau(n)e^{2\pi inz} + y^6 \Delta_t(z)|$.

Since $y^6|\Delta(z)|$ is invariant under the modular group and vanishes at ∞, it assumes a maximum in R at z_0. Choose a sequence z_j approaching the real axis, $Re(z_j) \in [0,1)$, z_j equivalent to z_0 under $\Gamma(1)$. If $z_j = x_j+iy_j$, then

$$||\Delta||_{B_6} = y_j^6|\Delta(z_j)| = |y_j^6 \sum_{n=1}^{t-1} \tau(n)e^{2\pi inz_j} + y_j^6 \Delta_t(z_j)|.$$

In the preceding expression, the first term approaches zero as $y_j \to 0$. This gives $||\Delta||_{B_6} \leq ||\Delta_t||_{B_6}$.

Before proceeding we need some additional definitions.

DEFINITIONS: (i) $L_6 = \{f | f$ is measurable in H, $f(z+1) = f(z)$, and $\sup_S y^6|f(z)| < \infty\}$

(ii) $L_6(\Gamma) = \{f | f \in L_6$ and $f(Az)A'^6(z) = f(z)\ \forall A \in \Gamma,\ z \in H\}$.

(Note that L_6 and $L_6(\Gamma)$ are the analogues of B_6 and $B_6(\Gamma)$ with analyticity replaced by measurability).

(iii) $(f,g;S) = \int_0^\infty \int_0^1 f\,\bar{g}\,y^{10} dxdy$.

(This is the Petersson scalar product for $\Gamma(1)_\infty$.)

LEMMA 7: If $f \in A_6$ and $g \in L_6(\Gamma(1))$, then $(f,g;S) = (\theta_6(f),g;R)$.

PROOF: This is a standard computation and will be omitted.

THEOREM 1: $\Lambda(g) = \frac{1}{||\Delta||} \int_0^1 \int_0^\infty g\cdot(y^{-6} \frac{\bar{\Delta}(z)}{|\Delta(z)|})\, y^{10}\, dxdy$.

In other words, the function $h_1(z) = \frac{y^{-6}\Delta(z)}{|\Delta(z)|\ ||\Delta||}$) represents Λ. Also, $h_1 \in L_6(\Gamma(1))$ with $||h_1||_{B_6} = \frac{1}{||\Delta||}$. In addition, any other function representing Λ which is either holomorphic or a 6-form for $\Gamma(1)$ cannot have smaller B_6-norm.

REMARK: This theorem immediately implies that $||\theta_6|| \leq 1$. Further, if $||\theta_6|| = 1$ we would have the distance from $K\bar{\Delta}$ to $A_6^{\perp}$ equaling the distance from $K\bar{\Delta}$ to $A_6^{\perp} \cap L_6(\Gamma(1))$, a much smaller subspace of L_6. (This distance would be $||K\bar{\Delta}-(K\bar{\Delta}-h_1)||_{B_6}$ and we cannot do better, if $||\theta_6|| = 1$, by (9).)

PROOF: That $\frac{y^{-6}\ (z)}{|\Delta(z)|\ ||\Delta||}$ is a 6-form for $\Gamma(1)$ follows from the fact that $y^6|\Delta(z)|$ is invariant under the modular group and Δ is a modular 6-form. The norm calculation is trivial.

To prove the representation formula, it suffices to show

$$h(z) = -K\Delta(z) + y^{-6}\frac{\Delta(z)}{|\Delta(z)|\ ||\Delta||} \in A_6^{\perp}.$$

Since h is in $L_6(\Gamma(1))$, it suffices to show that $(g,h;S) = 0$ for all $g \in A_6$. Now, by Lemmas 4 and 7

$$(g,h;S) = -K(g,\Delta;S) + (g, \frac{y^{-6}\Delta}{||\Delta||\ |\Delta|}\ ; S)$$

$$= -K(\Delta,\Delta;R)c_g + (\theta_6(g), \frac{y^{-6}\Delta}{||\Delta||\ |\Delta|}\ ; R)$$

$$= -c_g + \frac{c_g}{||\Delta||}\int_R\int\ y^4|\Delta|\ dxdy = 0.$$

Let us now consider other possible representations of $\Lambda(g)$. If $\Lambda(g) = (g,h;S)$ where $h \in L_6(\Gamma(1))$, $\Lambda(g) = (\theta_6(g),h;R)$ by Lemma 7; and

(10) $$|\Lambda(g)| \leq ||\theta_6(g)||\ ||h||_{B_6}.$$

On the other hand, by definition

(11) $$|\Lambda(g)| = \frac{||\theta_6(g)||}{||\Delta||} .$$

Combining (10) with (11) gives $||h||_{B_6} \geq \frac{1}{||\Delta||}$.

Next suppose that $\Lambda(g) = (g,h;S)$ with $h \in B_6$. By the preceding argument, it suffices to show that $(g,h;S) = (\theta_6(g),h;R)$. Set $\theta_6(g) = G$. Define $\hat{G}$ on H so that $\hat{G} = G$ on R and is zero elsewhere. Obviously, $\hat{G} \in L_6$. Let β_6 be the projection of L_6 onto A_6. Since $\theta_6(\beta_6\hat{G}) = G$ (Kra [6], p. 100), $\Lambda(g) = \Lambda(\beta_6\hat{G})$ or $(g,h;S) = (\beta_6\hat{G},h;S)$. By the self-adjointness of β_6, $(\beta_6\hat{G},h;S) = (\hat{G},\beta_6 h;S) = (\hat{G},h;S)$ since β_6 is a projection. It then follows that $(g,h;S) = (\hat{G},h;S) = (\theta_6(g),h;R)$ by definition of $\hat{G}$.

§7. Reducing the B_6-norm

Is there a function h in L_6 that represents Λ and has B_6-norm less than $1/||\Delta||$? We believe that there is. By Theorem 1 such a function can be neither a form for $\Gamma(1)$ nor analytic. We proceed as follows in our search. We set $h_2(z) = K\Delta(z) + h(z)$ where $h \in A_6^{\perp}$ and write $h_2(z) = \sum_{n=-\infty}^{\infty} a_n(y)e^{2\pi inx}$. J. Sturm [16] treats such functions; but his are C^∞. If we split $h_2(z)$ into the two sums, $\sum_{-\infty}^{0}$ and $\sum_{1}^{\infty}$, it is clear that the first sum is in $A_6^{\perp}$. However, since this sum appears likely to increase the B_6-norm of h_2, we assume that $h_2(z) = \sum_{n=1}^{\infty} a_n(y)e^{2\pi inx}$. We also lose nothing by assuming $a_n(y)$ is real since $\tau(n)$ is real.

Next, since h is assumed to be in $A_6^{\perp}$, $(e^{2\pi inz}, h;S) = 0$ for all $n \geq 1$. It then follows that

(12) $$(e^{2\pi inz},K\Delta;S) = (e^{2\pi inz},h_2;S) \text{ for all } n \geq 1.$$

Since $(e^{2\pi inz},K\Delta;S) = c_n$ where $\theta_6(e^{2\pi inz}) = c_n\Delta$, (12) becomes

(13) $$c_n = \frac{10!\ \tau(n)}{(4\pi n)^{11}(\Delta,\Delta;R)} = \int_0^\infty a_n(y)e^{-2\pi ny}y^{10}dy.$$

After much deliberation we feel that one reasonable choice for $a_n(y)$ is $a_n(y) = a_n y^{-6}$. When this choice for $a_n(y)$ is substituted into (13), we have

$$a_n = \frac{c_n}{||e^{2\pi inz}||_{A_6}} = \frac{10!\ \tau(n)}{4!\pi^6 2^{17}(\Delta,\Delta;R)n^6} \equiv C\,\frac{\tau(n)}{n^6}$$

and $h_2(z) = Cy^{-6} \sum_{n=1}^{\infty} \frac{\tau(n)}{n^6} e^{2\pi inx}$.

Now h_2 does represent Λ since (12) implies $(K\bar{\Delta}-h_2) \in A_6^{\perp}$. The remaining questions are

(i) is h_2 in L_6? I.e., is $||h_2||_{B_6} < \infty$?

and

(ii) how big is $||h_2||_{B_6}$?

At this point we make the following

<u>ASSUMPTION</u>. $T(x) = \sum_{n=1}^{\infty} \frac{\tau(n)}{n^6} e^{2\pi inx}$ converges uniformly for $x \in [0,1)$.

If the assumption is true, $h_2 \in L_6$. To complete the computation of $||\Lambda||$ we argue as follows. Since the contribution from the tail of $T(x)$ is small, the B_6-norm of h_2 is essentially determined by the leading terms in the series. Note that the situation for h_2 is in marked contrast to that of Δ. (See Lemma 6.) The last step is to adjust the choice of $a_n(y)$ for these leading terms in order to further reduce their contribution. We know how to do this. It amounts to choosing $a_n(y) \equiv 0$ if $y \leq 3/2\ n$; $a_n(y) = \alpha_n$ if $y > 3/2\ n$ where α_n is chosen to satisfy (13).

§8. <u>Appendix</u> (Added April 1, 1981)

We have recently determined that Assumption of §7 is false. Indeed, if $T(x) = O(1)$, then obviously $T(x) \in L^2(0,1)$ and, by Parseval's Identity,

$$||T||_2^2 = \sum_{n=1}^{\infty} \frac{\tau^2(n)}{n^{12}} < \infty.$$

But this last series diverges, as is easily seen using partial summation and Rankin's asymptotic formula for $\sum_{n \leq x} \tau^2(n) = \alpha x^{12} + O(x^{12-2/5})$ given in [8], equation (9).

On the other hand, we have discovered some modifications of $\Delta(z)$ which may be helpful. Let

$$a_n^{(k)}(y) = \begin{cases} 0 & y \geq \frac{1}{n} \\ \tau(n) y^k b^k, & \frac{1}{n+b} \leq y < \frac{1}{n+b-1};\ b = 1,2,3,\ldots \end{cases}$$

$$c_n^{(k)}(y) = \begin{cases} 0 & y \geq \frac{1}{n} \\ \dfrac{\tau(n) y^k (b+1)\ldots(b+k)}{\prod_{a \leq k} (1+ay)}, & \frac{1}{n+b} \leq y < \frac{1}{n+b-1};\ b = 1,2,3,\ldots . \end{cases}$$

Write

$$f^{(k)}(x+iy) = \sum_{n=1}^{\infty} a_n^{(k)}(y) e(nx)$$

where $e(nx) = e^{2\pi i n x}$.

$$\phi^{(k)}(x+iy) = \sum_{n=1}^{\infty} c_n^{(k)}(y) e(nx)$$

We can show that:

(i) For each $k \geq 1$, $f^{(k)}$ and $\phi^{(k)}$ represent θ_6 (apart from multiplicative constants depending on k and not on n,x and y). That is, (13) is satisfied.

(ii) If one evaluates $y^6|f^{(k)}(x+iy)|$ (resp. $y^6|\phi^{(k)}(x+iy)|$) at $y = \frac{1}{m}$, one gets the m-th partial sum of the k-th Riesz (resp. Césaro) mean of $\sum_{n=1}^{\infty} \tau(n) e(nx)$, divided by m^6.

(iii) In each case, $f^{(k)}$ and $\phi^{(k)}$ are in L_6.

Whether any of these means reduce the L_6 norm of Δ sufficiently, or reduce it at all, we do not yet know. But they certainly provide fairly short "locally (in m) finite" representations of θ_6 and perhaps they themselves are more tractable candidates for modification than $\Delta(z)$.

Taking another tack, we may consider $\frac{\Delta}{|\Delta|}$, the function playing a prominent role in §6. The modular form $\Delta^{1/24}$ is denoted η. Furthermore, it is well known that $\eta^{-1}(z) = e(\frac{-z}{24}) \sum_{m=0}^{\infty} p(m)e(mz)$, where $p(m)$ is the partition function. (See Knopp, Modular Functions in Analytic Number Theory, Markham, Chicago, 1970, page 34.) Now $\frac{\Delta}{|\Delta|} = \Delta\cdot\eta^{-12}\cdot\bar{\eta}^{-12}$, and from this we can calculate the Fourier series of $\frac{\Delta}{|\Delta|}(x+iy)$, y fixed. (The coefficients will be certain expressions involving $p(\cdot)$ and $\tau(\cdot)$. They are real.) Perhaps a partial sum (or a partial sum of a Césaro mean) would have the property that $||\frac{\Delta}{|\Delta|} - \sum_{-m_\varepsilon}^{m_\varepsilon}||_6 < \varepsilon$, for all y. (Note that this is clearly true if y is restricted to a compact subset of H^+.) Proving something of this sort would seem to rest on the modulus of continuity of $\frac{\Delta}{|\Delta|}$, for each $y > 0$, viewed as a function of x. If one could do it, this would lead, exactly as in the end of §7, to a means of computing $||\theta_6||$. Lastly, let us remark that the presence of negative Fourier coefficients in the expansion of $\frac{\Delta}{|\Delta|}$ suggests that discarding negative coefficients, as mentioned in §7, may be inappropriate.

BIBLIOGRAPHY

1. P. Duren, Theory of H^p-Spaces, Academic Press, New York, 1970.
2. P. Duren, B. Romberg, A. Shields, Linear functionals on H^p spaces with $0 < p < 1$, J. Riene Angew Math. 238 (1969), 32-60.
3. C. Earle, The integrable automorphic forms as a dual space, preprint.
4. C. Earle, Teichmüller Theory, 143-162 in W. Harvey, Ed. Discrete Groups and Automorphic Functions, Academic Press, New York, 1977.
5. A. Guinand, Integral modular forms and summation formulae, Proc. Cam. Phil. Soc., 43 (1947), 127-129.
6. I. Kra, Automorphic Forms and Kleinian Groups, Benjamin, Reading, MA., 1972.

7. D.H. Lehmer, Ramanujan's function $\tau(n)$, Duke J. 10 (1943), 483-492.

8. R. Rankin, Ramanujan's function $\tau(n)$, Symposia on Theoretical Physics and Mathematics, Vol. 10, (Inst. Math. Sci., Madras), 1969, 37-45, Plenum, New York, 1970.

9. ______, Modular Forms and Functions, Cambridge Univ. Press, Cambridge, 1977.

10. ______, The scalar product of modular forms, Proc. London Math. Soc. (3) 2 (1952), 198-217.

11. ______, Contributions to the theory of Ramanujan's function $\tau(n)$ and similar arithmetical functions (III), Proc. Camb. Phil. Soc. 36 (1940), 150-151.

12. E. Reich, An extremum problem for analytic functions with area norm, Ann. Acad. Sci. Fenn. (A.I.) 2 (1976), 429-445.

13. H. Royden, Automorphisms and isometries of Teichmüller space, Ann. of Math. Studies 66 (1971), 369-384.

14. M. Sheingorn, Characterizations of $A^q(U)^*$, J. of Research Nat. Bur. Standards Sect. B 77B (1973), no. 3-4, 85-92.

15. K. Strebel, On lifts of extremal quasiconformal mappings, J. D'Analyse Math. 31 (1977), 191-203.

16. J. Sturm, Projections of C^∞ automorphic forms, BAMS (New Ser.) 2 (1980), 435-439.

17. A. Walfisz, Uber die Koeffizientensummen einiger Modulformen, Math. Ann. 108 (1933), 75-90.

18. J. Wilton, A note on Ramanujan's arithmetical function $\tau(n)$, Proc. Camb. Phil. Soc., 25 (1929), 121-129.

Ohio State University
Columbus, Ohio 43210

Baruch College -CUNY
New York, N.Y. 10010

ANALYSIS ON POSITIVE MATRICES AS IT MIGHT HAVE OCCURRED TO FOURIER

Audrey Terras

ABSTRACT

Analysis on matrix groups and their homogeneous spaces is in a period analogous to that of Fourier, thanks to work of Harish-Chandra, Helgason, Langlands, Maass, Selberg, and many others. Here we try to give a simple discussion of Fourier analysis on the space P_n of positive n×n matrices, as well as on the Minkowski fundamental domain for P_n modulo the discrete group $GL(n,\mathbb{Z})$ of integer matrices of determinant ± 1. The main idea is to use the group invariance to see that the Plancherel or spectral measure in the Mellin inversion formula comes from the asymptotics and functional equations of the special functions which appear in the Mellin transform on P_n or $P_n/GL(n,\mathbb{Z})$, as analogues of the power function y^s in the ordinary Mellin transform. For P_n, these functions are matrix argument generalizations of K-Bessel and spherical functions. For $P_n/GL(N,Z)$. these special functions are generalizations of Epstein zeta functions known as Eisenstein series.

Dedicated to Emil Grosswald

§0. Introduction

It is appropriate to dedicate this paper to Emil Grosswald, not only because of his kind encouragement through the years, but also because several of his papers contain applications of Mellin transforms as well as K-Bessel functions and Epstein zeta functions (cf. Grosswald [15], [16], [17], plus Bateman and Grosswald [3]). Generalizations of Mellin transforms to matrix space as well as generalizations of K-Bessel functions and Epstein zeta functions are the stars of the present work. We should perhaps apologize to Fourier for emphasizing number-theoretic rather than physical applications. Still number theory and physics are not so far apart.

EXAMPLE 1. The Mellin transform in the study of the asymptotics of number-theoretical functions.

Many examples are found in Grosswald's paper [17]. See also Anderson and Stark [1] . The typical example of such a Mellin transform argument is found in proofs of the prime number theorem (cf. Grosswald's book [18]). Our goal here is to study matrix Mellin transforms. Hopefully this will lead to asymptotics for the functions of matrix number theory.

EXAMPLE 2. The Mellin transform and its use by Riemann, Hecke and Weil to study the correspondence between Dirichlet series with functional equations and modular forms.

The ordinary Mellin transform can be viewed as a Fourier transform on the space $\mathbb{R}^+ = P_1$ of positive real numbers. Riemann noticed that it allows one to study the Riemann zeta function using Jacobi's theta function:

$$\left(\zeta(s) = \sum_{n \geq 1} n^{-s}, \ \mathrm{Re}\ s > 1\right) \quad \longleftrightarrow \quad \left(\theta(z) = \sum_{n \in Z} \exp(i\pi n^2 z), \ \mathrm{Im}\ z > 0\right).$$

The explicit Mellin transform is:

$$2\pi^{-s}\Gamma(s)\zeta(2s) = \int_0^\infty y^{s-1}(\theta(iy)-1)\,dy.$$

One can use this formula and properties of theta (such as the transformation formula) to derive properties of zeta (such as the analytic continuation and functional equation). See [58] for a development of this theory. The Mellin inversion formula allows one to study theta conversely in terms of known results about zeta. It is Mellin inversion as well as explicit Mellin transforms that we seek to generalize to the space of positive matrices P_n.

Hecke generalized Riemann's work in the 1930's in [22] and Weil continued this process in the 1960's in [68] and [70]. But these extensions stayed in the realm of the ordinary Mellin transform on P_1 or a vector transform on P_1^r. The vector transform arises out of the theory of Hilbert modular forms over totally real algebraic number fields.

Many number-theoretic problems are matrix problems. This motivates one to begin a systematic study of matrix Mellin transforms. For example, Hecke shows in [21, pp. 198-207] that the Dedekind zeta function of an algebraic number field is a finite sum over the ideal class group of integrals over unit cubes involving Epstein zeta functions of certain $n \times n$ positive real matrices built up out of the units of the number field and an integral basis for an ideal in the ideal class. The Epstein zeta function will appear in §2 as the simplest analogue of y^s in the Mellin transform on $P_n/GL(n,\mathbb{Z})$.

On another level, Langlands' philosophy indicates that non-abelian Galois groups of extensions of number fields should have Artin L-functions coming from generalizations of modular forms to other matrix groups (cf. Tate [57, pp. 318-322] and Gerardin and Labesse in [6, II, p. 119]).

The study of matrix Mellin transforms and connections with generalizations of Hecke theory to Siegel modular forms seems to begin with Koecher's work [34] in the 1950's. However, not enough complex variables were present in Koecher's Mellin transform on $P_n/GL(n,\mathbb{Z})$, for an inverse transform to be possible. The proper generalization of Mellin transform (as in §§2) has enabled Kaori Imai to replace Riemann's example by one involving zeta functions of several variables and Siegel modular forms (cf. Imai [29]).

Fourier or Exxon might be less interested in zeta functions than in the solution of partial differential equations such as the wave equation on a symmetric space. Helgason has described how to do this using the techniques that we study here and an analogue of the Radon transform (cf. his readable lectures [26]).

There are however many cases in which number theorists and physicists have been led to the same problem in matrix analysis. One of the most pertinent examples is the search for computer solutions to the problem of the location of the non-trivial zeros of the Selberg zeta function of [47]. The latter has its non-trivial zeros corresponding to eigenvalues for the noneuclidean Laplacian on the upper half plane modulo a discrete subgroup Γ of $SL(2,\mathbb{R})$. The real parts of the zeros are equal to 1/2. And earlier computer calculations for $\Gamma = SL(2,\mathbb{Z})$ had included zeros of $\zeta(s)$ and an L-function. D. Hejhal has shown these eigenvalues to be spurious in [24]. The corresponding eigenfunctions of the Laplacian have logarithmic singularities. There is a curious moral here that physicists should also find interesting. The continuous spectrum can often mess up the discrete one in a numerical calculation of eigenvalues. The Selberg trace formula [47] and harmonic analysis on $P_2/SL(2,\mathbb{Z})$ is necessary for the analysis of Selberg's zeta function in this case. Good references are the books of Kubota [35], Lang [36], Hejhal [23], and the paper of Marie-France Vignéras [74]. Interesting number-theoretic applications of the Selberg trace formula can be found in Sarnak [46]. §§2 considers related matters.

Both number theorists and physicists have also been studying the statistics of the eigenvalues of random Hermitian matrices (cf. Mehta [42] and Montgomery [44, p. 184]). And there is an interesting story here. Chapter 4 of Mehta's book [42] concerns a conjectured value for a certain multi-dimensional analogue of the beta function. Selberg had already evaluated this integral in [48].

Yet another example of the common interests of number theorists and physicists is the Epstein zeta function of §§2, which gives information about algebraic number fields as we mentioned earlier (also see [59] or Stark [54]). Physicists see the Epstein zeta function as giving the potential function of a crystal lattice (cf. Born and Huang [8, p. 389]). It is interesting to note that when physicists needed the analytic continuation of the Epstein zeta function they used the same method that Riemann used (cf. Born and Huang loc. cit.).

Other applications of matrix analysis occur in multivariate statistics (see Farrell [13] or James [33]). Here again it has sometimes happened that number theorists and statisticians have proved the same result for very different reasons; e.g., the evaluation of the matrix gamma function (cf. Herz [28]). We have found for example that a formula of the statistician Wishart is useful in developing certain integral tests for series of positive matrices -- integral tests needed to study the generalizations of Epstein's zeta function which star in §§2. See [62] for more details, and Wishart [72].

In what follows we discuss two Mellin inversion formulas for the space P_n. Section 1 gives a simplified discussion of Mellin inversion for compactly supported functions on P_n. The original work is that of Harish-Chandra and Helgason (cf. [26]). Section 2 concerns Mellin inversion on the Minkowski fundamental domain for $P_n/GL(n,\mathbb{Z})$. The original work here is by Roelcke for n = 2 in [45], Selberg for general n in [47], Langlands for arbitrary reductive groups in [37], and Venkov for n = 3 in [73]. Our main object is to show that the determination of the Plancherel or spectral measure in the Mellin inversion formula proceeds in a similar and simple way in all cases by looking at the asymptotics and functional equations of the special functions generalizing y^s in the matrix Mellin Mellin transform. Our secondary object is the study of the special functions involved here. For P_n these special functions are matrix argument analogues of K-Bessel and spherical functions. For $P_n/GL(n,\mathbb{Z})$ these functions are generalizations of Epstein zeta functions known as Eisenstein series. The special case n = 2 was treated in [64] and the reader should look at that paper before reading this one. A more leisurely discussion of some of these things will appear in [61] hopefully. Section 2 also contains Fourier expansions of Eisenstein series for GL(3) first obtained by Kaori Imai in [30].

§§. MELLIN INVERSION ON P_n.

§1.1 GEOMETRY OF P_n.

We shall be interested in the general linear group $GL(n,\mathbb{R}) = G$ of $n \times n$ non-singular real matrices, as well as in the associated symmetric space P_n of positive $n \times n$ real symmetric matrices. We seek the spectral resolution of compactly supported functions $f: P_n \to \mathbb{C}$ in eigenfunctions of the $GL(n,\mathbb{R})$ - invariant differential operators on P_n. More information on these invariant differential operators will be found in this section. We are seeking to generalize Mellin inversion, since when $n = 1$, the differential operators on $\mathbb{R}^+ = P_1$ which commute with multiplication are polynomials in $y\,d/dy$ and y^s is an eigenfunction for $s \in \mathbb{C}$, $y \in P_1$. Before saying any more about Mellin inversion on P_n, we need to consider a few facts about the geometry of P_n. More details can be found in Helgason [25], Maass [40], and my notes [61].

The action of g in $GL(n,\mathbb{R})$ on Y in P_n is given by:

(1.1.1) $$Y \to {}^t gYg = Y[g] \text{ , if } {}^t g = \text{transpose of } g.$$

Then let $K=O(n)$ denote the orthogonal group of $n \times n$ orthogonal matrices. It is easily seen that $K\backslash G$ can be identified with P_n by sending the coset Kg to $Y = {}^t gg$ in P_n.

Now P_n is an open cone in $\mathbb{R}^{n(n+1)/2}$ and it is a Riemannian manifold, because it has an arc length ds defined by:

(1.1.2) $$ds^2 = Tr((Y^{-1}dY)^2) \quad , \quad dY = (dy_{ij}), \text{ if } Y = (y_{ij}) \in P_n.$$

Moreover, the arc length is invariant under $GL(n,\mathbb{R})$. The geodesics through I_n = the identity matrix in P_n can be shown to have the form

$$\exp(tX) = \sum_{n \geq 0} (tX)^n/n! = \text{the matrix exponential, for } t \in \mathbb{R}, X = {}^t X \in \mathbb{R}^{n \times n}$$

(cf. Maass [40, §3]). There is a geodesic-reversing isometry σ_Y at each point Y in P_n; e.g., $\sigma_{I_n}(X) = X^{-1}$, for X in P_n. This makes P_n a symmetric space.

The G-invariant volume $d\mu_n$ on P_n is defined by:

$$(1.1.3)\qquad d\mu_n(Y) = |Y|^{-(n+1)/2} \prod_{1 \le i \le j \le n} dy_{ij}\ ,\ dy_{ij} = \text{usual Lebesgue measure.}$$

Here $|Y|$ denotes the determinant of Y in P_n. It is rather easy to pass back and forth between P_n and the determinant one surface SP_n defined by:

$$(1.1.4)\qquad SP_n = \{Y \in P_n \mid |Y| = 1\}$$

Defining $SL(n,\mathbb{R})$ = the special linear group of determinant one elements of $GL(n,\mathbb{R})$, and $SO(n)$ as the special orthogonal group of determinant one elements of $O(n)$, it is clear that

$$SP_n \cong SO(n)\backslash SL(n,\mathbb{R})\ .$$

The Lie group $SL(n,\mathbb{R})$ is simple and thus nicer at times than the Lie group $GL(n,\mathbb{R})$, which is only reductive. But the difference between the two is just the abelian group $\mathbb{R}^+$ under multiplication. For example, the relation between the $SL(n,\mathbb{R})$ - invariant measure dW on SP_n and the $GL(n,\mathbb{R})$ - invariant measure $d\mu_n$ on P_n is given by setting:

$$(1.1.5)\qquad Y = t^{1/n}\, W \text{ for } Y \in P_n,\ t \in \mathbb{R}^+,\ W \in SP_n,\ d\mu_n(Y) = dW\, t^{-1}dt.$$

The G-invariant differential operators or generalized Laplacians on P_n are polynomials in the algebraically independent operators:

$$(1.1.6)\qquad Tr((Y\partial/\partial Y)^i),\ \text{where } \partial/\partial Y = (\tfrac{1}{2}(1 + \delta_{ij})\partial/\partial y_{ij}),\ \text{for } i = 1,2,\ldots,n.$$

We need two different types of coordinates to do analysis on P_n. The first coordinate system will be called partial Iwasawa coordinates and it is defined by

$$(1.1.7)\qquad Y = \begin{pmatrix} V & 0 \\ 0 & W \end{pmatrix} \begin{bmatrix} I_p & 0 \\ X & I_q \end{bmatrix}\ ,\ \text{for } V \in P_p,\ W \in P_q,\ X \in \mathbb{R}^{q\times p}, p+q=n.$$

One need only multiply out the matrices to check that this can be done. In partial Iwasawa coordinates the invariant arc length becomes

(1.1.8) $$ds_Y^2 = ds_V^2 + ds_W^2 + 2\ \mathrm{Tr}(V^{-1}\ {}^t dX\ W\ dX).$$

And the invariant volume becomes:

(1.1.9) $$d\mu_n(Y) = |V|^{-q/2}\ |W|^{p/2}\ d\mu_p(V)\ d\mu_q(W)\ dX,\ dX\text{=Lebesgue measure on } \mathbb{R}^{p\times q}.$$

One can also compute what happens to the invariant differential operators under this change of coordinates (cf. Maass [40,§6]). We have chosen the name "partial Iwasawa coordinates", because induction on (1.1.7) allows one to obtain the full Iwasawa decomposition:

(1.1.10) $Y = a[n]$, where a is positive diagonal, n is upper triangular,1 on diagonal.

This coordinate system corresponds to the Iwasawa decomposition $G = KAN$ of the general linear group (cf. Helgason [26, pp. 35-36]). Siegel has called this change of coordinates the Jacobi transformation in [50, p. 29] and Weil has called it Babylonian reduction in [69, p. 7].

A second useful coordinate system might be called polar coordinates and defined by:

(1.1.11) $$Y = a[k],\ \text{for a positive diagonal, } k \text{ in } K = O(n).$$

The existence of this decomposition follows from the spectral theorem for Y in P_n. My favorite proof of this can be found in Courant and Hilbert [11, Ch.1]. This proof is nice because it generalizes to Sturm-Liouville boundary value problems and gives the mini-max principle for computing the eigenvalues (cf. Courant and Hilbert [11, Ch. 6]). This leads to the numerical methods known as Rayleigh-Ritz and finite element methods. The main object of the present work is a related problem which is complicated by the presence of a continuous spectrum--namely the generalization of the spectral theorem to the $GL(n,\mathbb{R})$ - invariant differential operators on P_n or $P_n/GL(n,\mathbb{Z})$. Physicists often call the group - level level version of polar coordinates -- $G = KAK$ -- Euler angles (cf. Wigner [71]). And numerical analysts have called it the singular value decomposition (cf. Strang [55, p. 135]).

In polar coordinates the arc length becomes:

$$ds_Y^2 = \prod_{j=1}^{n} a_j^{-2} \; da_j^2 + \text{terms involving } dk. \tag{1.1.12}$$

And the invariant volume becomes:

$$d\mu_n(Y) = \prod_{i=1}^{n} a_i^{-(n+1)/2} \prod_{1 \le i \le j \le n} |a_i - a_j| \, da \, dk. \tag{1.1.13}$$

Here dk is a Haar measure on the compact matrix group O(n) and da denotes the ordinary Euclidean measure on $(\mathbb{R}^+)^n$. To use these formulas one must note that polar coordinates give a $2^n n!$ to one covering of P_n. For the entries of a are the eigenvalues of Y = a[k]. Thus they are unique up to the action of the Weyl group W of permutations of the a_j (j = 1,..., n) times the orthogonal diagonal matrices, which form a group M consisting of the matrices which are diagonal with entries +1 or -1 on the diagonal.

Polar coordinates have seen much use in multivariate statistics; for example, in the work of James [33]. The coordinate system is also central to the work of Harish-Chandra and Helgason in [26], as we shall see.

§1.2. SPECIAL FUNCTIONS ON P_n.

This section contains the matrix analogues of many of the functions to be found in the tool chests of number theorists and physicists -- tool chests such as the book of Lebedev [39] or the Bateman Manuscript Project [12]. We shall discuss gamma, incomplete gamma, K-Bessel and spherical functions of matrix argument.

The most basic special function for us is the power function $p_s(Y)$ defined for $s = (s_1,\ldots,s_n)$ in $\mathbb{C}^n$ and Y in P_n by:

$$p_s(Y) = \prod_{j=1}^{n} |Y_j|^{s_j}, \tag{1.2.1}$$

where Y_j in P_j denotes the j × j upper left hand corner of Y. It is also possible to view p_s as a homomorphism on the group of upper triangular matrices

$$T = \begin{pmatrix} t_1 & & * \\ & \ddots & \\ 0 & & t_n \end{pmatrix} \quad , \text{ with } t_j \text{ positive .}$$

For one can use the Iwasawa decomposition (1.1.10) to write $Y = {}^tTT$ and

(1.2.2.) $$p_s(Y) = \prod_{j=1}^{n} t_j^{r_j} \quad , \quad \text{with } r_j = 2(s_j + \ldots + s_n).$$

This gives an easy proof that p_s is an <u>eigenfunction</u> for all the generalized Laplacians on P_n (cf. Maass [40, pp. 68-70]).

We need a certain symmetry property of the power functions, which also finds applications in the study of K-Bessel functions. The <u>symmetry property</u> is:

(1.2.3) $$p_s(Y^{-1}[U]) = p_{s*}(Y), \quad \text{if } U = \begin{pmatrix} 0 & & 1 \\ & \cdot^{\cdot^{\cdot}} & \\ 1 & & 0 \end{pmatrix}, \; s* = (s_{n-1},\ldots,s_1,-(s_1+\ldots+s_n)).$$

Note that for y in $P_1 = \mathbb{R}^+$, the power function is $p_s(y) = y^s$, s in $\mathbb{C}$. So we really do have a generalization of the power function that appears in the Mellin transform. The power functions were introduced by Selberg in [47, pp. 57 - 58]. Tamagawa [56, p. 369] calls them right spherical functons. In the language of Harish-Chandra and Helgason [26, p. 52], the power function appear in the guise exp $\lambda(H(gk))$. This really involves exp of log. For if x in G has Iwasawa decomposition x = kan with k in K, a positive diagonal, n upper triangular and one on the diagonal, then H(x) = log a. Thus H maps the positive diagonal matrices into arbitrary diagonal matrices; i.e. into R^n. Now λ is a linear functional on R^n, which can be identified with an n-tuple s of complex numbers. Thus we have a power function.

The <u>correct normalization of variables</u> is:

(1.2.4) $$r_j = 2r_j^* + j - (n+1)/2$$

because then the eigenvalues for the generalized Laplacians acting on the $p_s(Y) = p_s({}^tTT) = f_{r*}(T)$ can be computed to be all the symmetric polynomials in the r_j^* (cf. Maass [40, pp. 71ff]). This is very useful.

Next we want to use the power function to build up the basic special functions of matrix space. The gamma function $\Gamma_n(s)$ for s in $\mathbb{C}^n$ is defined by the following matrix Mellin transform:

$$(1.2.5)\qquad \Gamma_n(s) = \int_{Y \in P_n} p_s(Y) \exp(-\mathrm{Tr}(Y))\, d\mu_n(Y),$$

for Re s_j sufficiently large, j = 1,..., n. The function actually factors as a product of ordinary gamma functions $\Gamma_1 = \Gamma$:

$$(1.2.6)\qquad \Gamma_n(s) = \pi^{n(n-1)/4} \prod_{j=1}^{n} \Gamma(s_j + \ldots + s_n - (j-1)/2).$$

The factorization is easily proved using the Iwasawa decomposition (1.1.10). Special cases of the gamma function for P_n appear in work of the statistician Wishart [72], as well as in Siegel's paper on quadratic forms [49, Vol. I, pp. 326-405]. One sees $\Gamma_n(s)$ in the functional equations of Eisenstein series. And here one uses the fact that

$$(1.2.7)\qquad p_s(A)\, \Gamma_n(s) = \int_{P_n} p_s(Y) \exp\{-\mathrm{Tr}(YA^{-1})\}\, d\mu_n(Y), \text{ for A in } P_n.$$

There are also useful matrix incomplete gamma functions which appear in the analytic continuation of Dedekind zeta functions as well as that of Eisenstein series for GL(n,$\mathbb{Z}$) in §§2 (cf. my papers [59] and [62]). Such incomplete gamma expansions of Dirichlet series with functional equations have been considered by Lavrik [38]. Theorem 7 of [62] shows that in special cases these incomplete gamma functions are ordinary Mellin transforms of the function $\Gamma(w/n)^n/(w-ns)$, as a function of w. Here the power s is a single complex variable. Thus incomplete gamma functions are rather closely related to the Meijer G-functions obtained by Grosswald [16] (cf. the Bateman Manuscript Project [12]). For Meijer G-functions are inverse Mellin transforms of powers of the ordinary gamma function (with no term such as 1/(w-ns)). Kaori Imai has obtained Meijer G-functions in answering a question of Jacquet concerning Mellin transforms of modular forms for GL(3) in [31].

The asymptotics of incomplete gamma functions are important for the Brauer -

Siegel theorem on the growth of the product of the regulator and class number with the discriminant of an algebraic number field (cf. [59, pp. 8-9]). Another asymptotic expansion of incomplete gamma functions plays a role in the paper of Goldfeld and Viola [14], as well as in the analytic continuation of Eisenstein series in §2. Very elegant algorithms for the computation of these matrix incomplete gamma functions were developod by R. Terras in [66], where several other applications in number theory and physics are mentioned. Unfortunately there is no time to go into more detail on incomplete gamma functions and Meijer G-functions here.

Matrix beta functions are also of interest and we will evaluate a special case in §1.3 in the computation of the Plancherel or spectral measure for Mellin inversion on $\mathcal{P}_n$. Selberg evaluated such an integral in his Norwegian paper [48] mentioned in the introduction.

Next we consider matrix K-Bessel functions. These are eigenfunctions f(Y) for the generalized Laplacians on $\mathcal{P}_n$ such that for all Y in $\mathcal{P}_n$

$$(1.2.8) \qquad f(Y[U]) = \exp[2i\mathrm{Tr}({}^tNX)]f(Y), \quad \text{for any} \quad U = \begin{pmatrix} I_m & X \\ 0 & I_{n-m} \end{pmatrix} .$$

in the abelian group

$$(1.2.9) \qquad N(m,n-m) = \left\{ \begin{pmatrix} I_m & X \\ 0 & I_{n-m} \end{pmatrix} \middle| \quad X \in \mathbb{R}^{m\times(n-m)} \right\} .$$

Thus we are looking for a matrix entry corresponding to the induced representation from the character $\exp[2i\mathrm{Tr}({}^tNX)]$ on the subgroup N(m,n-m) of $GL(n,\mathbb{R})$. Note that the group N(m,n-m) is isomorphic to the Euclidean space $\mathbb{R}^{m\times(n-m)}$, the latter as an additive group and the former multiplicative. We need the matrix K-Bessel functions in order to study the Eisenstein series (generalized Epstein zeta functions) in §§2.

The literature on matrix K-Bessel functions is rather sparse. However a special case as well as other related Bessel and hypergeometric functions was studied by Herz in [28]. The motivation was classical lattice point problems, such as the study of the asymptotics of

$$\{T \text{ in } Z^{k\times n} \mid {}^tTT < A, \text{ i.e., } A-{}^tTT \in P_m\}, \text{ as } A \text{ "goes to infinity".}$$

There were also motivations from statistics.

Most of the theory of K-Bessel functions that we shall need is developed by Tom Bengtson in his thesis [4], so we shall only sketch these matters. We need two integral formulas for K-Bessel functions which will be related by Bengtson's formula (1.2.17).

The first definition of a matrix K-Bessel function is:

$$(1.2.10)\qquad k_{m,n-m}(s|Y,N) = \int_{X\in\mathbb{R}^{m\times(n-m)}} p_{-s}\left(Y^{-1}\begin{bmatrix} I_m & 0 \\ {}^tX & I_{n-m}\end{bmatrix}\right)\exp[2i\,\mathrm{Tr}({}^tNX)]dX,$$

for $s \in \mathbb{C}^m$ with s_j in a suitable half plane, and Y in P_n, $N \in \mathbb{R}^{m\times(n-m)}$, $0 < m < n$. Here $p_s(Y)$ denotes a power function with the last variables $s_{m+1},\ldots, s_n$ all equal to zero. The second K-Bessel function is definied to be:

$$(1.2.11)\qquad K_m(S|A,B) = \int_{X\in P_m} p_s(X)\exp\{-\mathrm{Tr}(AX + BX^{-1})\}\,d\mu_m(X),$$

for A,B in P_m and any s in $\mathbb{C}^m$ or for B singular and Re s_j sufficiently large.

EXAMPLE. When n=2, m=1, we have the ordinary K-Bessel function K_s (cf. Lebedev [39, Ch. 5], since for a,b positive, $s \in \mathbb{C}$, $r,q \in \mathbb{R} - 0$:

$$(1.2.12)\qquad K_1(s|a,b) = \int_0^\infty x^{s-1}\exp\{-(ax + b/x)\}dx = 2\,K_s(2\sqrt{ab})(a/b)^{-s/2},$$

$$(1.2.13)\qquad \Gamma(s)k_{1,1}\left(s\,\middle|\begin{pmatrix}1/y & 0\\ 0 & y\end{pmatrix}, \pi n\right) = \begin{cases} 2\pi^s\,|n|^{s-1/2}y^{1/2}K_{s-1/2}(2\pi|n|y), & n \neq 0 \\ y^{1-s}\Gamma(1/a)\Gamma(s-1/2), & n = 0.\end{cases}$$

The first matrix K-Bessel function definition (1.2.10) is useful because it is obviously an eigenfunction of the generalized Laplacians with the invariance property (1.2.8). The second definition (1.2.11) is useful for the study of convergence. Note that

(1.2.14) $$K_m(s|A,0) = \Gamma_m(s)\, p_s(A^{-1}) ,$$

when Re s_j sufficiently large. Thus when A or B is singular the integral (1.2.11) will only converge if the variables s_j lie in some half plane.

Suppose that T is an upper triangular matrix with positive diagonal entries. Then it is easy to see that

(1.2.15) $$K_m(s|A[{}^tT],\ B[T^{-1}])\, p_s({}^tTT) = K_m(s|A,B) .$$

Suppose that a is the smallest element of the set of all eigenvalues of the positive matrices A and B. Then it is easy to see that

(1.2.16) $K_m(s|A,B) = O(a^{-m/2}\, e^{-2ma})$, as a goes to infinity, s fixed real.

Here one uses the Iwasawa decomposition (1.1.10) and the fact that the ordinary K-Bessel function satisfies $K_s(x) \leq (\pi/2x)^{1/2}\, e^{-x}$, for s real. From (1.2.16) it is easy to deduce the convergence of $K_m(s|A,B)$ for all s if $A,B \in P_m$. This also allows one to see the convergence of series of such K-Bessel functions.

Bengtson's formula relating the two K-Bessel functions of matrix argument is:

(1.2.17) $$\Gamma_m(-s^*)k_{m,n-m}\left(s\middle| \begin{pmatrix} G & 0 \\ 0 & H \end{pmatrix},N\right) = \pi^{m(n-m)/2}|H|^{m/2}\, K_m(\tilde{s}|H[{}^tN],\ G),$$

$$\tilde{s} = -s + (0,\ldots,\ 0,\ \ (n-m)/2)$$

where $s^* = (s_{m-1},\ldots,\ s_1\ ,\ -(s_1\ +\ldots+\ s_m))$. For a proof see Bengtson [4].

Bengtson also proves the useful formula:

(1.2.18) $$K_{m,1}\left(s\middle| \begin{pmatrix} G & 0 \\ 0 & h \end{pmatrix},N\right) = p_{-s}(G^{-1})|G|^{-1/2}h^{m/2}k_{m,1}(s|I_{m+1},h^{1/2}\ TN),\ \text{if } G={}^tTT,$$

T upper triangular with positive diagonal.

Question: Are these K-Bessel functions products of ordinary K-Bessel functions, as was the case for the gamma function? The answer appears to be "NO" -- except in certain very special cases. Another question is: Are there any relations between $k_{m,n-m}$ and $k_{n-m,m}$? If there were perhaps one could reduce $k_{m,1}$ to an ordinary K-Bessel function using (1.2.17). But this appears unlikely.

If one of the matrix arguments of K_m is singular but not zero, sometimes we can evaluate the function. For example, let $s = (s_1,s_2) \in \mathbb{C}^2$, $a > 0$, and

$$B = \begin{pmatrix} b_1 & * \\ * & * \end{pmatrix} \in P_2$$

Then we have

$$(1.2.19) \quad K_2\left(s \,\middle|\, \begin{pmatrix} a & 0 \\ 0 & 0 \end{pmatrix}, B\right) = 2\sqrt{\pi}\; b_1^{(s_1-s_2-1)/2} \; |B|^{\frac{s_2}{2}} \; \Gamma(-s_2-3/2) a^{-(s_1+s_2+1)/2}$$

$$\times\, K_{s_1+s_2+1}\left(2\sqrt{ab_1}\right) .$$

A similar example occurs in [63, Lemma 4].

Harmonic analysis on P_n in partial Iwasawa coordinates (1.1.7) involves these matrix K-Bessel functions. For SP_2, this reduces to the Kontorovich-Lebedev inversion formula (cf. [64], Lebedev [39, p. 131], and Sneddon [51, Ch. 6]):

$$(1.2.20) \quad f(y) = 2\pi^{-2} \int_0^\infty t \sinh \pi t \; y^{-1/2} K_{it}(y) \int_0^\infty f(u)\, u^{-1/2} K_{it}(u)\, du\, dt.$$

Lebedev uses this formula to find the electrostatic field for a point charge near the edge of a thin conducting sheet held at zero potential in [39, pp. 153-155].

To find the spectral measure $t \sinh \pi t$, with the exact constant $2\pi^{-2}$ and the line of integration for K_s; i.e. Re $s = 0$, one needs to know the asymptotics and functional equation of $K_s(y)$. The asymptotic formula is:

$$(1.2.21) \quad K_s(y) \sim 2^{s-1}\, \Gamma(s)\, y^{-s} \text{ , as } y \to 0, \text{ if Re } s > 0 \text{ (s fixed).}$$

This is clear from (1.2.12). The functional equation is:

$$(1.2.22) \quad K_s(y) = K_{-s}(y) \quad .$$

This is also clear from (1.2.12). Note that the K_s in (1.2.20) is integrated over the line fixed by the functional equation (1.2.22). And the spectral measure is chosen to cancel out $\Gamma(it)\,\Gamma(-it) = (t \sinh \pi t)^{-1}$, so that ordinary Mellin inversion gives (1.2.20) asymptotically for y and u near 0.

The asymptotics/functional equations principle seems to work in great generality, as we saw in [64]. Unfortunately it does not appear to have been formulated explicitly by people working in PDEs or harmonic analysis. Thus books such as that of Sneddon give different proofs for each transform formula (cf. [5], Chs. 6,7]).

It is clear how to generalize (1.2.21) to the matrix case, using (1.2.14). However, it is not clear how to generalize (1.2.22). Thus it appears that the generalization of (1.2.20) is still an open question. Similarly the explicit Mellin inversion formula for P_n in partial Iwasawa coordinates remains to be developed. Instead we shall follow Harish-Chandra and Helgason and use polar coordinates in §1.3.

The last special function to be considered here is the spherical function $f\colon P_n \to \mathbb{C}$ which is defined to be an eigenfunction of all the generalized Laplacians which is invariant under $Y \to Y[k]$ for all k in $O(n) = K$. Usually spherical functions are normalized to take the value 1 at the identity matrix I_n. Clearly spherical functions are analogues of spherical harmonics (cf. [61]). Harish-Chandra's formula for spherical functions says:

$$(1.2.23) \qquad h_s(Y) = \int_{k \in O(n)} p_s(Y[k])\, dk, \text{ for } s \text{ in } \mathbb{C}^n,\ Y \text{ in } P_n,\ dk\text{=Haar measure.}$$

Moreover the spherical function h_s satisfies the functional equations:

$$(1.2.24) \qquad h_s = h_{s'} \quad \text{if and only if } r^{*\prime} \text{ is a permutation of } r^* ,$$

using the normalized power variables (1.2.4).

EXAMPLE. When n = 2 and we look at SP_2 (the determinant one surface), the spherical functions are the Legendre functions $P_{it-1/2}(\cosh r)$, also known as Mehler's conical functions. Fourier analysis on SP_2 in polar coordinates amounts to the Mehler-Fock inversion formula for the Mehler transform involving the Legendre functions. This inversion formula can be discussed from the asymptotics/functional equations point of view. The details are in [64]. See also Sneddon [51, Ch. 6] and Lebedev [39] for other discussions and applications such as the determination of the electrostatic field due to a thin charged conductor in the shape of two intersecting spheres. In the next section we will discuss Mellin inversion on P_n in polar coordinates for arbitrary n by a method which specializes to that of [64] when n = 2.

Note that, in fact, separation of variables in polar coordinates leads actually to associated Legendre functions which look more like formula (1.2.10) than (1.2.23) in that an exponential appears in the integral as well as a power function (in the case that n = 2). However Fourier analysis in the angle variable is trivial and so one is reduced to the spherical function integral transform.

QUESTION. Are the spherical functions genuine new functions and not products of Legendre functions? Surely the answer to this question must be "YES". Why?

§1.3. HARMONIC ANALYSIS ON P_n IN POLAR COORDINATES.

Suppose that f: $P_n \to \mathbb{C}$ is nice; e.g., compactly supported. Define the Helgason-Mellin transform of f by:

$$\text{(1.3.1)} \qquad Hf(s,k) = \int_{Y \in P_n} f(Y)\, p_s(Y[k])\, d\mu_n(Y), \text{ for } s \in \mathbb{C}^n,\ k \in O(n).$$

Then Harish-Chandra's inversion formula says:

$$\text{(1.3.2)} \quad f(Y) = (n!\,(2\pi i)^n)^{-1} \int_{\mathrm{Re}\, s = -\frac{1}{2}} \int_{\bar{k} \in K/M} Hf(s,k)\, \overline{p_s(Y[k])}\, dk\, |c_n(s)|^{-2}\, ds.$$

Here $c_n(s)$ denotes the Harish-Chandra c-function defined in (1.3.9). The quotient space K/M is often called the boundary of P_n (cf. Helgason [26]), where M is defined in the discussion following formula (1.1.13).

The main goal of the present section is to discuss the Harish-Chandra inversion formula from the point of view used in explaining the Kontorovich-Lebedev inversion formula (1.2.20) -- the asymptotics/functional equations principle. Note that in fact formula (1.3.2) does express compactly supported functions on P_n as superpositions of eigenfunctions of the generalized Laplacians (the power functions). Thus we have obtained harmonic analysis on P_n in polar coordinates. This inversion formula of Harish-Chandra can also be viewed as a generalization of Mellin inversion clearly.

First note that ordinary Mellin inversion allows us to pull out the determinant of Y and see that the spectral measure $|c_n(s)|^{-2}$ must be independent of the last variable, s_n.

The next step of the discussion comes from Helgason [26, pp. 60-61]. Let S denote the <u>inverse transform</u> on functions F: $\mathbb{C}^n x(K/M) \to \mathbb{C}$ defined by:

$$(1.3.3) \qquad SF(Y) = c \int_{\mathrm{Re}\, s = -\frac{1}{2}} \int_{\bar{k} \in K/M} F(s,k)\, p_s(Y[k])\, |c_n(s)|^{-2}\, d\bar{k}\, ds\ .$$

Simple computations show that SH is self-adjoint and commutes with the action of $GL(n, \mathbb{R})$ on P_n. Thus it suffices to prove the inversion formula SH = Identity on a Dirac sequence of functions (i.e., a sequence of functions approaching the Dirac delta distribution). Furthermore, it suffices to prove SHf(Y) = f(Y) at any fixed Y in P_n.

Thus it suffices to prove the inversion formula (1.3.2) for O(n) - invariant functions in a Dirac sequence: F(Y), for Y diagonal with quotients of adjacent diagonal entries approaching zero. Then we require the <u>asymptotic formula</u> of Harish-Chandra and Bhanu-Murti (cf. [5] and [20]):

$$(1.3.4) \quad h_s(Y) \sim c_n(s)\, p_s(Y), \text{ if } Y = \begin{pmatrix} y_1 & & 0 \\ & \ddots & \\ 0 & & y_n \end{pmatrix}, \text{ as } y_i/y_{i+1} \to 0, |Y|=1, \mathrm{Re}\, s_j \text{ large, } s \text{ fixed}\ .$$

We shall discuss the computation of the Harish-Chandra c-function $c_n(s)$ in the remainder of this section. But let us just assume (1.3.4) for the moment. What would this imply about (1.3.2) for f in a Dirac sequence peaking as y_i/y_{i+1} all approach zero ($i = 1,\ldots, n-1$)? We would obtain:

$$f(Y) = (n!(2\pi i)^n)^{-1} \int_{\mathrm{Re}\, s = -\frac{1}{2}} \int_{\bar{k} \in K/M} \int_{0 < a_i < a_{i+1}} f(a) \int_{\bar{k}_1 \in K/M}$$

$$\times\ p_s(a[k_1])d\bar{k}_1\, D(a) p_s(Y[k])d\bar{k}\, |c(s)|^{-2}\, ds\ .$$

Here a is a diagonal matrix with entries a_i, $i = 1,\ldots, n$, D(a) denotes the Jacobian of polar coordinates (cf. (1.1.13)), and $\bar{k} = kM$ = a coset in K/M. This gives, using the formula for spherical functions (1.2.23):

$$f(Y) = (n'(2\pi i)^n)^{-1} \int_{\operatorname{Re} s = -\frac{1}{2}} \int_{0 < a_i < a_{i+1}} f(a)\, h_s(a)\, D(a)\, \overline{h_s(Y)}\, |c(s)|^{-2}\, ds.$$

If f runs through a Dirac sequence peaking as y_i/y_{i+1} approaches zero (Y diagonal), then we can approximately plug (1.3.4) into the last formula for $h_s(Y)$ and $h_s(a)$. Thus the integral in the last formula approaches:

$$(n'(2\pi i)^n)^{-1} \int_{\operatorname{Re} s = -\frac{1}{2}} \int_{0 < a_i < a_{i+1}} f(a)\, c_n(s) p_s(a)\, \overline{c_n(s) p_s(Y)}\, |c_n(s)|^{-2} ds.$$

Thus the spectral measure $|c_n(s)|^{-2}$ is chosen to cancel out the $c_n(s)\, \overline{c_n(s)}$ coming from the asymptotics of the spherical function. The constants in front of the integral sign are explained by the constants needed for ordinary Mellin inversion on the a_i, as well as the number (n!) of functional equations of the spherical function h_s given by (1.2.24).

Thus we are reduced to the computation of the Harish-Chandra c-function. This is most easily done by changing variables from K/M to $\overline{N}$ in formula (1.2.23), where $\overline{N}$ is the group of all lower triangular $n \times n$ matrices with 1 on the diagonal. To produce this change of variables, recall the Iwasawa decomposition of $G = SL(n, \mathbb{R}) = KAN$, $K = SO(n)$, A = diagonal $n \times n$ determinant one positive matrices with 1 on the diagonal. Set P = MAN = the minimal parabolic subgroup of all upper triangular matrices in G. Then K/M = G/P. It is easy to see that $\overline{N}P$ is an open subset of G with lower dimensional complement. For $\overline{N}P$ consists of all elements of G with nonsingular upper left $k \times k$ corners, $k = 1, \ldots, n$. Then, upon computing the exact change of variables, one obtains the new formula for spherical functions:

$$(1.3.5) \qquad h_s(I[a]) = c \int_{\overline{N}} p_s(I[ak(\bar{n})])\, \underline{p_{2\rho}}(I[\bar{n}])\, d\bar{n} \quad .$$

Here a denotes the diagonal matrix with entries $a_i > 0$, and $|a| = 1$, $\bar{n}$ is a lower triangular matrix with 1 on the diagonal and entries x_{ij} below the diagonal, $d\bar{n} = \prod dx_{ij}$, for $1 \leq j < i \leq n$. The vector $\rho = (1/2, \ldots, 1/2)$ comes from the Jacobian, which is derived from the Jacobian of the Iwasawa decomposition and the Jacobian of conjugation of an element of N by an element of A (as a function on N).

To find the asymptotic formula for h_s, we need a <u>power function identity</u>:

$$(1.3.6)\qquad p_s(I[a],\bar{n}(kM)) = p_{2s}(a)p_s(I[\bar{n}^a])p_{-s}(I[\bar{n}]), \quad \text{with } \bar{n}^a = a\bar{n}a^{-1} .$$

For Re s_j sufficiently large, it is legal to let a_i/a_{i+1} approach zero inside the integral in (1.3.5). Using (1.3.6), one sees that $I[\bar{n}^a]$ approaches the identity matrix, as a_i/a_{i+1} goes to zero ($i = 1,\dots, n-1$). Thus (1.3.4) holds with the <u>Harish-Chandra c-function given by the integral</u>:

$$(1.3.7)\qquad c_n(s) = c\int_{\bar{N}} p_{-2\rho-s}(I[\bar{n}])\, d\bar{n} \ , \quad \text{with } c = 1/c_n(0).$$

Define the new integral:

$$(1.3.8)\qquad b_n(s) = \int_{\bar{N}} p_{-s}(I[\bar{n}])\, d\bar{n} \ .$$

Then $c_n(s) = b_n(s + 2\rho)/b_n(2\rho)$, and it is easy to prove an inductive formula for b_n. Note that

$$b_2(s) = B(1/2,\ s - 1/2) = \text{the beta function}.$$

Note also that b_n is a special case of the K-Bessel functions considered earlier. The final <u>formula for the Harish-Chandra c-function</u> is:

$$(1.3.9)\qquad c_n(s) = \frac{\prod_{1 \le i \le j \le n-1} B(\tfrac{1}{2},\ s_i + s_{i+1} + \dots + s_j + \tfrac{1}{2}(j-i+1))}{\prod_{1 \le i \le j \le n-1} B(\tfrac{1}{2},\ \tfrac{1}{2}(j-i+1))}$$

This completes the discussion of Mellin inversion on P_n. It would be useful to add a discussion of the image of the compactly supported functions under the Helgason-Mellin transform; i.e., we need the Paley-Wiener theorem for P_n (cf. Helgason [27]).

<u>EXAMPLE</u>. The Helgason-Mellin transform of $f_X(Y) = |Y|^r \exp\{-\mathrm{Tr}(X^{-1})\}$ for fixed r in $\mathbb{C}$ and X in P_n is: $p_{s^*}(X[k])\Gamma_n(s^*)$, where $s^* = s + (0,\dots, 0, r)$. Moreover f_X is a Gauss-type kernel. It forms a Dirac sequence at Y as r approaches infinity

(cf. Maass [40, §7]), if f_X is normalized to have integral one over P_n. When one thinks of Gauss kernels, it is more natural to multiply f_X by $|X|^{-n}$ and let X approach zero.

For a number theorist, it is natural to try to add up the Mellin transforms of the $f_X(Y)$ over X in $\mathbb{Z}^{n \times n}$ of rank n. Unfortunately, this is a divergent sum in general. So we are led to the next section to do what we want -- matrix Mellin transform the non-singular terms of a theta function.

However, for solving PDEs on P_n, the Helgason-Mellin transform is quite adequate. For example, Helgason has shown in [26] how to solve the wave equation on the symmetric space. But number theory seems to lead us to the Mellin transform over $P_n/GL(n, \mathbb{Z})$ and the next section.

§§2. MELLIN INVERSION ON $P_n/GL(n, \mathbb{Z})$.

§2.1. WHY $P_n/GL(n, \mathbb{Z})$?

The natural generalization to P_n of Riemann's argument in Example 2 of the introduction involves matrix Mellin transforms of Siegel modular forms such as the <u>theta function</u>:

(2.1.1) $$\theta(X,Y) = \sum_{A \in \mathbb{Z}^{n \times m}} \exp\{-\pi \, \mathrm{Tr}\,(X[A]Y)\}, \text{ for } X \text{ in } P_n, \ Y \text{ in } P_m.$$

One example of a matrix Mellin transform of the non-singular terms in theta is <u>Koecher's zeta function</u> (cf. [34]):

(2.1.2) $$Z_{m,n-m}(S|X) = \sum_{A \in \mathbb{Z}^{n \times m} \ \mathrm{rk}\ m/GL(m,\,\mathbb{Z})} |X[A]|^{-s} \quad , \ \mathrm{Re}\ s > n/2,$$

for $1 \leq m \leq n$, X in P_n. When m = 1, this is <u>Epstein's zeta function</u> and when n = m, this a product of Riemann zeta functions (cf. [60]):

$$Z_{n,0}(S|X) = |X|^{-s} \prod_{j=0}^{n-1} \zeta(2s - j) \quad ,$$

which can be identified as the zeta function of the simple algebra of all n × n matrices over $\mathbb{Q}$. The Mellin transform here is not over P_m but instead over

$P_m/GL(m,\mathbb{Z})$, because $\theta(X,Y[U]) = \theta(X,Y)$ for all U in GL(m,Z). Explicitly, the Mellin transform is:

$$(2.1.3)\qquad \int_{P_m/GL(m,\mathbb{Z})} |Y|^s\ \theta_m(X,Y)\ d\mu_n(Y) = 2\ \pi^{-ms-m(m-1)/4} \prod_{j=0}^{m-1} \Gamma(s-j/2)\ Z_{m,n-m}(s|X).$$

Here θ_m denotes the sum over all terms of (2.1.1) such that the rank of A is m and $1 \leq m \leq n$. See Solomon [52] and Bushnell and Reiner [9] for generalizations and applications of such zeta functions to the theory of algebras and combinatorics.

This kind of example motivates the search for an analogue of Hecke's correspondence which would work for Siegel modular forms and Dirichlet series of several variables, which like Koecher's zeta function are Mellin transforms over $P_n/GL(n,\mathbb{Z})$. One needs to add the extra variables to be able to invert the Mellin transform. This inversion was used by Kaori Imai in the case of Siegel modular forms of genus 2 (for $Sp(2,\mathbb{Z})$) to generalize the Hecke correspondence (cf. [29]). Her result says that one has a picture like:

$$\begin{pmatrix}\text{Siegel modular forms}\\ \text{of genus 2 for } Sp(2,Z)\end{pmatrix} \longleftrightarrow \begin{pmatrix}\text{Dirichlet series in 2 variables}\\ \text{with functional equations}\end{pmatrix}.$$

The " $\rightarrow$ " can be found in Maass [40, §16]) -- in fact, for arbitrary genus. Imai proves the " $\leftarrow$ " using the Roelcke-Selberg-Mellin inversion formula on $P_2/GL(2,\mathbb{Z})$.

The main goal of the present section is to present some of the computations necessary for harmonic analysis on $P_n/GL(n,\mathbb{Z})$ from the same point of view that we considered harmonic analysis on P_n in section one. The main result is that the spectral measure on the highest dimensional part of the spectrum is constant. This is clear from (2.3.14) and the asymptotics/functional equations principle. However, there are also lower dimensional parts of the spectrum and we shall not discuss these in complete detail here. The secondary goal of the present section is to examine Fourier expansions of Eisenstein series and this involves the K-Bessel functions of §1.2. These expansions have recently been obtained by Imai in [30] for GL(3).

Harmonic analysis on $P_n/GL(n,\mathbb{Z})$ is developed here independently of harmonic analysis for P_n. This is somewhat unsatisfactory, since it would be nice to connect the two theories using some analogue of the Poisson summation formula (and by this we do not mean the Selberg trace formula). However, even Fourier analysis on $\mathbb{R}$ and $\mathbb{R}/\mathbb{Z}$ are usually developed independently. And usually Poisson summation is deduced from the fact that Fourier series converge to periodic functions. However, the Laplace transform can be used to remedy this situation (cf. Anderson and Stark [1] and Sneddon [51]). It is not at all clear that it is possible to generalize this argument that derives Fourier series from Fourier integrals to derive harmonic analysis on $P_n/GL(n,\mathbb{Z})$ from that on P_n. For many complications arise. A detailed but still concomplete discussion of the Selberg trace formula for discrete subgroups of higher rank Lie groups with fundamental domains of finite volume will soon appear in a book of M.S. Osborne and G. Warner called "Toward the Trace Formula in the Sense of Selberg".

§2.2. THE GEOMETRY OF $P_n/GL(n,\mathbb{Z})$.

In [43, II, pp. 53ff] Minkowski describes a convex cone M_n through the origin that represents $P_n/GL(n,\mathbb{Z})$, up to boundary identifications. The boundary consists of a finite number of hyperplanes which are explicitly described by Minkowski only for $n \le 6$. The definition of <u>Minkowski's fundamental domain</u> is:

$$(2.2.1)\quad M_n = \left\{ Y = (y_{in}) \in P_n \;\middle|\; \begin{array}{l} Y[a] \ge y_{ii},\ \text{if } a \in Z^n \text{ with g.c.d.}(a_i,\dots,a_n)=1 \\ y_{i,i+1} \ge 0,\ \text{for } i=1,\dots,n \end{array} \right\}.$$

Minkowski also computed the <u>Euclidean volume:</u>

$$(2.2.2)\quad \text{vol}\,\{Y \in M_n \mid |Y| \le 1\} = 2(n+1)^{-1} \prod_{j=2}^{n} \Lambda(j/2),\quad \text{where } \Lambda(s)=\pi^{-s}\Gamma(s)\zeta(2s).$$

The noneuclidean volume is infinite. One might wonder whether this geometric result might lead to some arithmetic information about $\zeta(3)$, $\zeta(5)$, etc. So far, the answer seems to be: "NO"!.

For all n define the <u>determinant one surface</u> in the fundamental domain by:

(2.2.3) $$SM_n = \{Y \in M_n \mid |Y| = 1\} .$$

We shall normalize measures on SM_n using (1.1.5).

<u>EXAMPLE</u>. When n=2, the determinant one surface is computed as follows using the Iwasawa decomposition:

$$Y = \begin{pmatrix} y & 0 \\ 0 & 1/y \end{pmatrix} \begin{bmatrix} 1 & x \\ 0 & 1 \end{bmatrix} = Y(x,y) .$$

$$SM_2 = \{Y(x,y) \mid 0 \le x \le 1/2, \quad x^2 + y^2 \ge 1 , \quad y > 0\} .$$

This is the gives half of the familiar fundamental domain for the modular group SL(2,Z), on the upper half plane.

Certain integral formulas are useful for analysis of M_n. These are discussed in [62]. The simplest such formula goes back to Siegel [49,III, pp. 46ff]. This formula says that if $1 \le k < n$, for a sufficiently nice function f on $\mathbb{R}^{n \times k}$:

(2.2.4) $$\mathrm{vol}(SM_n)^{-1} \int_{W \in SM_n} \sum_{N \in \mathbb{Z}^{n \times k}_{rk\ k}} f(w[N])\, dW$$

$$= \prod_{j=n-k+1}^{n} \pi^{j/2}\, \Gamma(j/2)^{-1} \int_{P_k} f(Y)|Y|^{n/2}\, d\mu_k .$$

There are several applications of such formulas. For example, (2.2.4) can be used to give a quick proof of convergence of Koecher's zeta function (2.1.2). It can also be used to obtain a generalization of the Minkowski-Hlawka theorem to the effect that there are matrices Y in SM_n such that

$$\min \{|Y[A]| \mid A \in \mathbb{Z}^{n \times k}_{rk\ k}\} > r\, |Y|^{k/n} , \quad \text{if } r < (n/2\pi e)^k ,$$

and if n is sufficiently large depending on k and k is fixed. See [62] for the details and the more general integral formulas which are needed to show the convergence of Eisenstein series for GL(n,$\mathbb{Z}$).

§2.3. SPECIAL FUNCTIONS ON P_n/GL(n,$\mathbb{Z}$). EISENSTEIN SERIES.

We need to consider the special functions called <u>automorphic forms</u> v for

Γ = some subgroup of $GL(n,Z)$, defined to be functions $v: SP_n \to \mathbb{C}$ such that:

(2.3.1)
1) v is an eigenfunction for all the generalized Laplacians $L(1.1.6)$, so that $Lv = \lambda_L v$ for some non-zero complex number λ_L;
2) $v(Y[A]) = v(Y)$ for all A in Γ and all Y in SP_n;
3) v satisfies some growth condition; e.g., v has at most polynomial growth in the $|Y_j|$ if

$$Y = \begin{pmatrix} Y_j & * \\ * & * \end{pmatrix}, \quad Y_j \in P_j,\ j = 1,\dots, n-1.$$

We shall write $v \in A(\Gamma,\lambda)$ if (2.3.1) holds.

There are many analogies with classical elliptic and Siegel modular forms, as Maass noted in [41], the first paper to consider this new type of automorphic form. Maass has called these automorphic forms "grössencharacters" in [40,§10]. That name could be justified by the fact that Hecke grössencharacters (cf. Hecke [21]) play the same role in harmonic analysis for GL(2) over a number field that forms in $A(GL(n,\mathbb{Z}), \lambda)$ play for harmonic analysis on $P_n/GL(n,\mathbb{Z})$ (cf. Stark [53] and Weil [68]). Motivated by the study of representations of semi-simple real Lie groups, Harish-Chandra gave a very general definition of automorphic form (cf. Borel [7, pp. 199-210]).

In §§1 we saw how to build up the eigenfunctions of the generalized Laplacians on P_n by integrating power functions over $O(n)$ or over $N(m,n-m)$ as in (1.2.9) This led to spherical functions (1.2.23) and K-Bessel functions (1.2.10). The powers s in $p_s(Y)$ give a way of indexing the eigenvalues of a generalized Laplacian L via $Lp_s(Y) = \lambda_L(s)p_s(Y)$. We shall use this sort of indexing when we speak of the dimensionality of the spectrum components. For Mellin inversion on P_n, the spectrum needed was n-dimensional. The inverse transform required integration over n lines in $\mathbb{C}^n$ given by $\mathrm{Re}\ s = -1/2$. We shall see that life is much more complicated for $P_n/GL(n,\mathbb{Z})$, since there are also discrete and lower dimensional spectra appearing in Mellin inversion. However the basic method of constructing GL(n,$\mathbb{Z}$) - invariant eigenfunctions in the highest dimensional part of the spectrum is analogous to the construction of spherical and K-Bessel functions. One must sum power functions over GL(n,$\mathbb{Z}$). But it is not quite that simple.

It is useful to construct the Eigenstein series inductively. Suppose that $\phi \in A(SL(m,\mathbb{Z}),\lambda)$, $Y \in SP_n$, $s \in \mathbb{C}$, $1 \leq m \leq n-1$, Re $s > n/2$, $W^o = |W|^{-1/m}\, W$ for $W \in P_m$. We define the <u>parabolic subgroup</u> P(m,n-m) of GL(n,$\mathbb{Z}$) by:

$$(2.3.2) \qquad P(m,n-m) = \left\{ \begin{pmatrix} A & B \\ 0 & C \end{pmatrix} \,\middle|\, A \in GL(m,\mathbb{Z}),\ C \in GL(n-m,Z),\ B \in \mathbb{Z}^{m\times(n-m)} \right\}.$$

Then the <u>Eisenstein series</u> we want to consider is:

$$(2.3.3)\ E_{m,n-m}(s,\phi|Y) = E(s,\phi|Y) = \sum_{A = (A_1\ *)\, \in\, SL(n,\mathbb{Z})/P(m,n-m)} |Y[A_1]|^s \phi(Y[A_1]^o).$$

When ϕ is an eigenfunction of all the Hecke operators for GL(m,Z), then there is a relation between the Eisenstein series above and the <u>zeta function</u> defined by:

$$(2.3.4) \qquad Z_{m,n-m}(s,\phi|Y) = Z(s,\phi|Y) = \sum_{B \in \mathbb{Z}^{n\times m}\ rk\ m/GL(m,\mathbb{Z})} |Y[B]|^{-s}\ \phi(Y[B]^o),$$

with s, ϕ, Y as in (2.3.3). Note that if $\phi \equiv 1$, this is Koecher's zeta function. The convergence of (2.3.3) and (2.3.4) for Re $s > n/2$ is easily proved using the integral tests of [62].

In order to describe the relation between the Eisenstein series and zeta function above, we must discuss Hecke operators. Let r denote any positive integer. Then the <u>Hecke operator</u> T_r is defined by sending the function $f\colon SP_n \to \mathbb{C}$ to

$$(2.3.5) \qquad T_r f(Y) = \sum_{A \in V_r} f((Y[A])^o)\ , \quad \text{if } Y \in P_n, \quad Y^o = |Y|^{-1/n}\, Y.$$

Here V_r denotes any complete system of representatives for $D_r/GL(n,\mathbb{Z})$, where D_r consists of all $n \times n$ integral matrices of determinant r. If ϕ is an eigenfunction for all the Hecke operators for GL(m,$\mathbb{Z}$), then we have the following <u>relation</u>

between the Eisenstein series and the zeta function:

$$Z(s,\phi|Y) = L_\phi(2s)\ E(s,\phi|Y), \tag{2.3.6}$$

where $T_r\phi = u_\phi(r)\phi$, and $L_\phi(s) = \sum_{r \geq 1} u_\phi(r)\ r^{-s}$, for Re $s > m/4$.

The proof is given in [60].

Maass obtains the analytic continuation of $Z(s,\phi|Y)$ to all s in $\mathbb{C}$ as a meromorphic function with functional equation (cf. [40, §16]). The method is to write $Z(s,\phi|Y)$ times the appropriate gamma factors as a Mellin transform of a theta function. Then one use a trick of Selberg involving differential operators to annihilate the integrals coming from the singular terms of theta. This method makes it rather difficult to study the poles of $Z(s,\phi|Y)$. Another method which retains these poles explicitly is given in [60] for the case $n = 3$. The functional equation relates $Z(s,\phi|Y)$ and $Z(n/2 - s,\phi^*|Y^{-1})$, where $\phi^*(X) = \phi(X^{-1})$.

The asymptotics of Epstein's zeta function $Z_{1,1}(s|Y)$ in (2.1.2) are easily deduced from the definition as a Dirichlet series, in the region of the parameter s, where the series converges. It is possible to obtain more explicit information by examining the Fourier expansion of $Z_{1,1}(s|Y)$ for

$$Y = \begin{pmatrix} y & 0 \\ 0 & 1/y \end{pmatrix} \begin{bmatrix} 1 & x \\ 0 & 1 \end{bmatrix}, \quad \text{as a periodic function of } x:$$

$$\left\{\begin{array}{l} \pi^{-s}\ \Gamma(s)\ Z_{1,1}(s|Y) = y^s\Lambda(s) + y^{1-s}\Lambda(1-s) + 2 \sum_{n \neq 0} e^{2\pi inx}\ c_n(y), \text{ where} \\ c_n(y) = |n|^{s-1/2}\ \sigma_{1-2s}(n) y^{1/2} K_{s-1/2}(2\pi|n|y),\ \sigma_s(n) = \sum_{0 < d|n} d^s\ , \\ \Lambda(s) = \pi^{-s}\ \Gamma(s)\ \zeta(2s). \end{array}\right. \tag{2.3.7}$$

This Fourier expansion is easily deduced from the transformation formula of the theta function, since $Z_{1,1}(s\ Y)$ is a Mellin transform of a theta function (cf. Chowla and Selberg [10]). Because (cf. Lebedev [39])

$$|K_s(y)| \leq (\pi/2y)^{1/2}\ e^{-y}\ , \tag{2.3.8}$$

it follows that Epstein's zeta function has the asymptotics:

(2.3.9) $\pi^{-s}\Gamma(s)\,\zeta(2s)Z_{1,1}(s|Y) \sim y^{s}\,\Lambda(s) + y^{1-s}\,\Lambda(1-s)$, as $y \to \infty$, s fixed.

Kubota obtains a slightly different Fourier expansion in [35]. To see that the two expansions are exactly the same, one needs the identity relating the divisor function and the singular series (cf. p.141 of Hardy's book Ramanujan):

(2.3.10) $$\zeta(2s) \sum_{\substack{c>0,\ d \bmod c \\ (d,c)=1}} c^{-2s} \exp(2\pi imd/c) = \sigma_{1-2s}(m) .$$

The name "singular series" was applied to Fourier coefficients of Eisenstein series for $Sp(n,\mathbb{Z})$ by Siegel and Maass (cf. [40, pp. 300-313] and [49, I, p. 329]).

The Fourier expansion of Epstein's zeta function (2.3.7) has seen multitudes of applications in number theory from its use by Weber (cf. [67, III, p. 526ff]) to derive Kronecker's limit formula:

$$\lim_{s \to 1} \{Z_{1,1}(s|Y) - \tfrac{\pi}{2}(s-1)^{-1}\} = \pi\{\gamma - \log 2\pi - \log(y^{1/2}\,|\eta(x+iy)|^2)\},$$

where $\eta(z)$ is Dedekind's eta function. The proof is an easy exercise in the definition of eta using the infinite product. Then Weber goes on to use the Kronecker limit formula in explicit class field theory for imaginary quadratic fields. Stark used generalizations of (2.3.7) to show that there are exactly 9 imaginary quadratic fields with class number 1 in [54]. The Fourier expansion (2.3.7) can also be used to obtain comparisons of $\zeta(3)$ and $\zeta(4)$, $\zeta(5)$ and $\zeta(6)$, etc. in terms of rapidly converging series of exponentials (cf. [65]). In [16], Grosswald obtains a relation between $\zeta_K(2m)$ and $\zeta_K(2m-1)$ (K = number field) involving a Meijer G-function, which is an inverse Mellin transform:

$$\frac{1}{2\pi i} \int_{\mathrm{Re}\, s = c} \{\Gamma(s)\,\Gamma(s+a/2)\}^n\, x^s\, ds \quad , \quad a = 2m-1, \quad n = [K:\mathbb{Q}] .$$

See the Bateman Manuscript Project [12, I, pp. 206-222] for more facts about Meyer's G-functions. One wonders whether the formulas of Grosswald are the same as those in [65] involving K-Bessel functions. This appears to be true if $K = \mathbb{Q}$, since in that case n = 1, and the above Meyer's G-function is easily seen to be $K_{a/2}(x^2/4)$.

Related Fourier expansions of Eisenstein series have recently been used by Kubota, Patterson and Heath-Brown to study cubic Gauss sums and the old conjecture of Kummer. Goldfeld has also used such expansions to study various questions concerning elliptic curves (cf. Goldfeld's paper in Springer Lecture Notes 751). The method is like the proof of Gauss's conjecture on the average order of class numbers of binary positive quadratic forms which interprets class numbers as Fourier coefficients of Eisenstein series.

In this section we want to consider higher rank analogues of the Fourier expansion (2.3.7). For the computation of the spectral measure in harmonic analysis on M_n only the constant term in the Fourier expansion is needed, but from the point of view of a number theorist, considering all the applications of (2.3.7) just mentioned, it would appear useful to obtain the whole Fourier expansion. This is also useful if one needs to extend the Hecke correspondence between modular forms and Dirichlet series to modular forms for $GL(n,\mathbb{Z})$, in an explicit classical version of the adelic theory of Jacquet, Piatetski-Shapiro and Shalika [32]. Recently Imai has obtained the Fourier expansion of the Eisenstein series for $GL(3,\mathbb{Z})$ (cf. [30]) and we shall discuss these results here.

The general theory of the constant term of these Fourier expansions was treated by Langlands in [37], Harish-Chandra in [19]; and Arthur in [6, I, pp. 253-274]. But these authors do not appear to discuss the other terms of the expansions.

Most authors use the Bruhat decomposition in the derivations of Fourier expansions of Eisenstein series (cf. Kubota [35], Imai and Terras [30]). This leads to complications, illustrated in the case of GL(2) by comparing Kubota's derivation in [35] with that of Bateman and Grosswald [3] or Chowla and Selberg [10].

To find the Fourier expansion of the zeta function $Z_{n,n-1}(s,\phi|Y)$ without the Bruhat decomposition, we need a good substitute; i.e., a good set of representatives for $\mathbb{Z}^{n\times(n-1)}$ rk $(n-1)/GL(n-1,\mathbb{Z})$:

(2.3.11)
$$\begin{cases} \mathbb{Z}^{n\times(n-1)} \text{ rk } (n-1)/GL(n-1,\mathbb{Z}) = S_1 \cup S_2, \text{ where} \\ S_1 = \left\{ \binom{B}{0} \;\middle|\; B \in \mathbb{Z}^{(n-1)\times(n-1)} \text{ rk}(n-1)/GL(n-1,\mathbb{Z}) \right\}, \\ S_2 = \left\{ \binom{B}{{}^tc} \;\middle|\; \begin{array}{l} B=HD,\ H \in GL(n-1,\mathbb{Z})/{}^tP,\ D=\begin{pmatrix} d_1 & & 0 \\ & \ddots & \\ d_{ij} & & d_{n-1} \end{pmatrix} d_2,\dots,d_{n-1}>0 \\ {}^tc=(c_1\ 0\ \dots\ 0), c_1 > 0, \qquad d_{ij} \bmod d_j \end{array} \right\} \\ {}^tP = \text{the lower triangular subgroup of } GL(n-1,\mathbb{Z}). \end{cases}$$

To prove that this set of representatives is correct, write A in $\mathbb{Z}^{n\times(n-1)}$ of rank n-1 as

$$A = \binom{B}{{}^tc} \text{ with } B \text{ in } \mathbb{Z}^{(n-1)\times(n-1)} \text{ and } c \in \mathbb{Z}^{n-1}.$$

If $c = 0$, then A lies in the set of representatives S_1. Otherwise there is a matrix W in $GL(n-1,\mathbb{Z})$ such that

$$AW = \begin{pmatrix} b_1 & B_2 \\ c_1 & 0 \end{pmatrix} \text{ with } c_1 > 0,\ B_2 \in \mathbb{Z}^{(n-1)\times(n-1)},\ b_1 \in \mathbb{Z}^{n-1}.$$

Moreover we can write:

$$\begin{pmatrix} b_1 & B_2 \\ c_1 & 0 \end{pmatrix} = \begin{pmatrix} b_1' & B_2' \\ c_1' & 0 \end{pmatrix} \begin{pmatrix} x & y \\ v & W \end{pmatrix} \text{ if and only if } x = 1 \text{ and } y=0 \in \mathbb{Z}^{n-1}.$$

So we need to take $A = \binom{B}{{}^tc}$ modulo the subgroup of $GL(n-1,\mathbb{Z})$ of matrices of the form

$$\begin{pmatrix} 1 & 0 \\ v & W \end{pmatrix}.$$

Thus W must be in $GL(n-2,\mathbb{Z})$. It must be shown that this puts A in S_2. Elementary divisor theory writes $B \in \mathbb{Z}^{(n-1)\times(n-1)}$ in the form B = BD with H in $GL(n-1,\mathbb{Z})$ and

$$D = \begin{pmatrix} d_1 & 0 \\ d_{12} & D_2 \end{pmatrix},$$

with d_1 in $\mathbb{Z}$ and with a lower triangular, non-singular D_2 in $\mathbb{Z}^{(n-2)\times(n-2)}$. And we can reduce H modulo the lower triangular group ${}^tP$. This completes the proof of (2.3.11) except to check that none of the matrices in $S_1 \cup S_2$ are equivalent modulo $GL(n-1,\mathbb{Z})$, but this is easy.

It is not hard to obtain the asymptotic behavior of $Z(s,\phi|Y)$ for Y with partial Iwasawa decomposition

$$(2.3.12) \qquad Y = \begin{pmatrix} U & 0 \\ 0 & W \end{pmatrix} \begin{bmatrix} I_m & 0 \\ 0 & I_{n-m} \end{bmatrix}, \quad \text{with } U \in P_m,\ W \in P_{n-m},\ Q \in R^{m\times(n-m)}.$$

Write A in $\mathbb{Z}^{n \times m}$ rk m, in the form

$$A = ({}^t{}^B_C)\ , \ B \in Z^{m \times m},\ C \in Z^{m\times(n-m)}\ .$$

Then we have:

$$(2.3.13) \qquad Y\,[{}^t{}^B_C] = U[B + Q^tC] + W[{}^tC]\ .$$

Suppose now that W goes to infinity in the Dirichlet series (2.3.4) for Re $s > n/2$. Then only the terms with $C = 0$ remain and we obtain <u>the asymptotic formula:</u>

$$(2.3.14) \qquad Z(s,\phi|Y) \underset{W \to \infty}{\sim} L_\phi(2s)\ |U|^{-s}\ \phi(U^o), \text{ for x fixed with Re } s > n/2,$$

assuming that ϕ is an eigenfunction of all the Hecke operators (2.3.5) and L_ϕ is the L function corresponding to ϕ as in (2.3.6). Here we are using the notation (2.3.12).

If $n = 3$ and ϕ is a cusp form for $SL(2,\mathbb{Z})$, there are no other members of the constant term in the Fourier expansion of the Eisenstein series $E_{2,1}(s,\phi|Y)$ with respect to the parabolic subgroup P(2,1) and thus (2.3.14) gives the full story of the asymptotics of the Eisenstein series. However when ϕ is itself an Eisenstein series $E_{1,1}$, then there are two more members of the constant term of the Fourier series of $E_{2,1}$ with respect to P(2,1). These facts fit into a general theory of the constant term of the Fourier expansion of the Eisenstein series, as developed by Langlands in [37] (cf. also Harish-Chandra [19] and Arthur [6, pp. 253-274]). The corollary for harmonic analysis on M_3 is that the spectral measure is constant for the continuous spectrum. See also Venkov [73] and the book of Warner and Osborne, to appear.

Suppose that $\phi_r \in A(SL(2,Z), r(r-1))$ is a <u>cusp form</u>. Then we can write the Fourier expansion of $\phi_r(U)$ for

$$U = \begin{pmatrix} y & 0 \\ 0 & 1/y \end{pmatrix} \begin{bmatrix} 1 & 0 \\ x & 1 \end{bmatrix} , \quad y > 0,\ x \in R$$

in the form:

$$\phi_r(U) = \sum_{n \neq 0} \exp(2\pi inx)\, \alpha_n y^{1-r} k_{1,1}(r \mid I_2, \pi ny) , \tag{2.3.15}$$

where $k_{1,1}$ is defined by (1.2.10) and is essentially the usual K-Bessel function.

Suppose that Y is given by (2.3.12) with m = 2, n = 3. Then the Eisenstein series $E_{2,1}(s,\phi_r|Y)$ with cusp form ϕ_r has <u>Fourier expansion</u> given by:

$$Z(s,\phi_r|Y) = L_{\phi_r}(2s)\phi_r(U^0)|U|^{-s} + \sum_{A,c,d_1,d_2,k} \alpha_k c^{2-2s-r} d_2^{\,r-2s} \exp(2\pi i\, {}^t QAm) \times k_{2,1}\left((s-\tfrac{r}{2},r) \,\middle|\, \begin{pmatrix} U[{}^tA^{-1}] & 0 \\ 0 & W \end{pmatrix}, \pi m\right) . \tag{2.3.16}$$

where ${}^tm = c(d_1,k/d_2)$ and the sum is over the $A \in SL(2,\mathbb{Z})/P(1,1), c > 0, d_1 \in \mathbb{Z}$, $0 < d_2/k, k \neq 0$. The parabolic subgroup P(1,1) is defined in (2.3.2) and the Bessel function $k_{2,1}$ is defined in (1.2.10). The proof of (2.3.16) is given in Imai and Terras [30]. It makes use of an inductive formula for $k_{2,1}$ as an integral involving $k_{1,1}$ -- a formula discovered by Imai.

Not every automorphic form for $GL(n,\mathbb{Z})$ is an Eisenstein series. A <u>cusp form</u> v for $GL(n,\mathbb{Z})$ is defined to be an automorphic form such that all m with $1 \leq m \leq n-1$, the constant term in the Fourier expansion of v with respect to P(m,n-m) vanishes. This definition is analogous to that of Siegel cusp forms or parabolic forms as Andrianov calls them in [2]. So far as I know, no explicit examples of cusp forms for $GL(n,\mathbb{Z})$ are known, even for n=2. However Maass gives examples for congruence subgroups of $SL(2,\mathbb{Z})$ in [41] -- examples derived from the existence of Hecke L-functions of real quadratic fields. Adelic analogues of this example appear in [32, pp. 255-256]. Gelfand and Piatetski-Shapiro have shown the spectrum of the generalized Laplacians is discrete on the space of cusp forms (cf. Harish-Chandra [19, Thm.2, 2, Ch.I] or Godement [7, pp. 225-234]).

REFERENCES

1. R. Anderson and H. Stark, Oscillation theorems, Proc. this Conference.

2. A.N. Andrianov, Dirichlet series with Euler product in the theory of Siegel modular forms of genus 2, Proc. Steklov Inst. Math. 112 (1971).

3. P.T. Bateman and E. Grosswald, On Epstein's zeta function, Acta Arith. 9 (1964), 365-373.

4. T. Bengtson, Matrix K-Bessel Functions, Ph.D. Thesis, UCSD.

5. T.S. Bhanu-Murti, Plancherel's measure for the factor space SL(n,R)/SO(n), Soviet Math. Dokl. 1 (1960), 860-862.

6. A. Borel and W. Casselman, Automorphic forms, representations and L-functions, Proc. Symp, Pure Math. 33 AMS, Providence, 1976.

7. A Borel and G. Mostow, Algebraic groups and discontinuous subgroups, Proc. Symp. Pure Math. 9, AMS, Providence, 1966.

8. M. Born and K. Huang, Dynamical theory of crystal lattices, Oxford U. Press, London, 1956.

9. C.J. Bushnell and I. Reiner, Zeta functions of arithmetic orders and Solomon's conjectures, preprint.

10. S. Chowla and A. Selberg, On Epstein's zeta function, J. Reine Angew. Math. 227 (1967), 86-110.

11. R. Courant and D. Hilbert, Methods of mathematical physics, Wiley-Intersceince, N.Y., 1961, Vol. I.

12. A. Erdélyi, ed., Higher transcendental functions, McGraw-Hill, N.Y., 1953, Vol. I.

13. R.H. Farrell, Techniques of multivariate calculus, Lecture Notes in Math. 520, Springer, N.Y., 1976.

14. D. Goldfeld and C. Viola, Mean values of L-functions associated to elliptic, Fermat and other curves at the centre of the critical strip, J. Number Theory 11 (1979), 305-320.

15. E. Grosswald, Die Werte der Riemannschen Zetafunktion an ungeraden Argumentstellen, Nachr. Akad. Wiss. Göttingen, Math. Phys. Kl II (1970), 9-13.

16. _______, Relations between the values at integral arguments of Dirichlet series that satisfy functional equations, Proc. Symp. Pure Math. 24, AMS Providence, 1973, 111-122.

17. _______, On some generalizations of theorems by Landau and Polya, Israel J. Math. 3 (1965), 211-220.

18. _______, Topics from the theory of numbers, Macmillan, N.Y., 1966.

19. Harish-Chandra, Automorphic forms on a semisimple Lie group, Lecture Notes in Math. 62, Springer, N.Y., 1968.

20. _______, Spherical functions on a semisimple Lie group I, II, Amer. J. Math. 80 (1958), 241-310, 553-613.

21. E. Hecke, Mathematische Werke, Vandenhoeck und Ruprecht, Göttingen, 1970.

22. _______, Dirichlet series, modular functions, and quadratic forms (IAS Lectures), Edwards Bros., Ann Arbor, 1938.

23. D. Hejhal, The Selberg trace formula for PSL(2,R), I, II, Lecture Notes in Math., 548, ?, Springer, N.Y., 1976,?

24. _______, Some observations concerning eigenvalues of the Laplacian and Dirichlet L-series, Durham symposium on Analytic Number Theory, 1979.

25. S. Helgason, Differential geometry and symmetric spaces, Academic, N.Y., 1962.

26. _______, Lie groups and symmetric spaces, in Battelle Rencontres, edited by DeWitt and Wheeler, Benjamin, N.Y., 1968.

27. _______, Functions on symmetric spaces, Proc. Symp. Pure Math. 26, AMS, Providence, 1973, 101-146.

28. C. Herz, Bessel functions of matrix argument, Ann. of Math. (2) 61 (1955), 474-523.

29. K. Imai, Generalization of Hecke's correspondence to Siegel modular forms, Amer. J. Math. 102 (1980), 903-936.

30. _______, and Terras, Fourier expansions of Eisenstein series for GL(3), preprint.

31. K. Imai, On a Mellin transform of modular forms for GL(3), preprint.

32. H. Jacquet, I.I. Piatetski-Shapiro and J. Shalika, Automorphic forms on GL(3), I, II, Annals of Math. 109 (1979), 169-212, 213-258.

33. A.J. James, Special functions of matrix and single argument in statistics, in Theory and applications of special functions, edited by R. Askey, Academic, N.Y., 1975, 497-520.

34. M. Koecher, Über Dirichlet-Reihen mit Funktionalgleichung, J. Reine Angew. Math. 192 (1953), 1-23.

35. T. Kubota, Elementary theory of Eisenstein series, Wiley, N.Y., 1973.

36. S. Lang, $SL_2(R)$, Addison-Wesley, Reading, Mass, 1975.

37. R.P. Langlands, On the functional equations satisfied by Eisenstein series, Lecture Notes in Math. 544, Springer, N.Y., 1976.

38. A.F. Lavrik, On functional equations of Dirichlet functions, Math. U.S.S.R. Izvestija 31 (1967), 421-432.

39. N.N. Lebedev, Special functions and their applications, Dover, N.Y., 1972.

40. H. Maass, Siegel's modular forms and Dirichlet series, Lecture Notes in Math. 216, Springer, N.Y., 1971.

41. _______, Über eine neue Art von nichtanalytischen automorphen Funktionen und die Bestimmung Dirichletscher Reihen durch Funktionalgleichung, Math. Ann. 121 (1949), 141-183.

42. M.L. Mehta, Random matrices and the statistical theory of energy levels, Academic, N.Y., 1967.

43. H. Minkowski, Gesammelte Abhandlungen, Chelsea, N.Y., 1967.

44. H.L. Montgomery, The pair correlation of zeros of the zeta function, Proc. Symp. Pure Math. 24, AMS, Providence, 1973, 181-193.

45. W. Roelcke, Über die Wellengleichung bei Grenzkreisgruppen erster Art, Sitzber. Akad. Heidelberg, Math.-naturwiss. Kl. (1953/55), 4 Abh.

46. P. Sarnak, Applications of the Selberg trace formula to some problems in number theory and Riemann surfaces, Thesis, Stanford, 1980.

47. A. Selberg, Harmonic analysis and discontinuous groups in weakly symmetric Riemannian spaces with applications to Dirichlet series, J. Ind. Math. Soc. 20 (1956), 47-87.

48. _______, Bemerkninger om et Multipelt Integral, Norsk Mat. Tidsskr. 26 (1944), 71-78.

49. C.L. Siegel, Gesammelte Abh., Springer, N.Y., 1966, Vols. I, II.

50. _______, Lectures on quadratic forms, Tata Inst., Bombay, 1956.

51. I.N. Sneddon, The use of integral transforms, McGraw-Hill, N.Y., 1972.

52. L. Solomon, Partially ordered sets with colors, in Proc. Symp. Pure Math. 34, AMS Providence, R.I., 1979, 309-329.

53. H. Stark, Algebraic number theory course, MIT & UCSD.

54. _______, On the problem of unique factorization in complex quadratic fields, Proc. Symp. Pure Math. 7, AMS, Providence, 1969, 41-56.

55. G. Strang, Linear algebra and its applicatins, Academic, N.Y., 1976.

56. T. Tamagawa, On Selberg's trace formula, J. Fac. Sci. U. Tokyo, I, 7 (1960), 363-386.

57. J. Tate, The general reciprocity law, Proc. Symp. Pure Math. 28, AMS, Providence, 1976, 311-322.

58. A. Terras, The minima of quadratic forms and the behavior of Epstein and Dedekind zeta functions, J. Number Theory 12 (1980), 258-272.

59. _______, Applications of special functions for the general linear group to number theory, Sém. Delange-Pisot-Poitou , 1976/77, n°23.

60. _______, On automorphic forms for the general linear group, to appear in Rocky Mt. J. of Math.

61. _______, Harmonic analysis on symmetric spaces and applications, in preparation.

62. _______, Integral formulas and integral tests for series of positive matrices Pacific J. Math. 89 (1980), 471-490.

63. _______, Fourier coefficients of Eisenstein series of one complex variable for the special linear group, T.A.M.S. 205 (1975), 97-114.

64. _______, Noneuclidean harmonic analysis, preprint.

65. _______, A relation between $\zeta_K(s)$ and $\zeta_K(s-1)$ for any algebraic number field K, in Algebraic Number Fields, edited by A. Fröhlich, Academic, N.Y., 1977, 475-484.

66. R. Terras, Incomplete gamma functions in several dimensions, preprint.

67. H. Weber, Algebra, Chelsea, N.Y., 1908, Vol. III.

68. A. Weil, Dirichlet series and automorphic forms, Lecture Notes in Math. 189, Springer, N.Y., 1971.

69. _______, Discontinuous subgroups of classical groups, U. of Chicago, 1958.

70. _______, Über die Bestimmung Dirichletscher Reihen durch Funktionalgleichung, Math. Ann. 168 (1967), 149-156.

71. E.P. Wigner, On a generalization of Euler's angles, in Group Theory and its Applications, edited by E.M. Loebl, Academic, N.Y., 1968, 119-129.

72. J. Wishart, The generalized product moment distribution in samples from a normal multivariate population, Biometrika 20A (1928), 32-43.

73. A.B. Venkov, On the trace formula for SL(3,Z), J. Sov. Math. 12 (1979), 384-424.

74. M.F. Vignéras, L'équation fonctionnelle de la fonction zêta de Selberg du groupe modulaire PSL(2,Z), Astérisque 61 (1979), 235-249.

C-012

Math. Dept.

U.C.S.D.

La Jolla, CA. 92093